Non-Linear Parametric Optimization

by B. Bank, J. Guddat, D. Klatte, B. Kummer, and K. Tammer

With 5 figures

Akademie-Verlag · Berlin
1982

Verfasser:
Doz. Dr. Bernd Bank
Prof. Dr. Jürgen Guddat
Dr. Diethard Klatte
Doz. Dr. Bernd Kummer
Humboldt-Universität Berlin, Sektion Mathematik
Doz. Dr. Klaus Tammer
Technische Hochschule Leipzig

Erschienen im Akademie-Verlag, DDR-1086 Berlin,
Leipziger Str. 3–4
Lektor: Dipl.-Math. Gesine Reiher

Lizenznummer: 202 · 100/405/82
Gesamtherstellung: VEB Druckerei „Thomas Müntzer“,
5820 Bad Langensalza
Bestellnummer: 7629877 (6636) · LSV 1085
Printed in GDR
DDR 45,– M
ISSN 0076-5430

Dedicated to our esteemed teacher and friend

Prof. František Nožička

whose many years of activity at the Humboldt-Universität made this book possible

Preface

This volume is intended for readers who, whether they be mathematicians, workers in other fields or students, are familiar with the basic concepts and methods of mathematical optimization. The subject matter is concerned with optimization problems in which some or all of the individual data involved are seen as variable quantities. Such problems, which here are referred to as "parametric optimization problems", play an important role in both the theory and solution procedures of mathematical optimization as well as in numerous practical fields (inexact models, postoptimal decisions and vector optimization to name a few).

The continuity of parameter-dependent constraint sets, solution sets and extreme values of the corresponding optimization problems is studied, which for the main part leads to an analysis of concrete point-to-set mappings. We also consider properties of those parameter sets for which the corresponding characteristic mappings are "well-behaved" (for example continuous) in the case of special problem classes (convex quadratic problems, linear complementarity problems, non-linear optimization problems with parameter-dependent objective functions, mixed-integer linear problems). These investigations constitute a theoretical basis for solution procedures given for certain important classes of parametric problems and which procedures all involve the simplex technique.

As compared to the theory of linear parametric optimization, which today may be seen as essentially complete, no unified or comprehensive exposition of the foundations of non-linear parametric optimization has hitherto been published. The present book is an attempt to fill this gap, it contains results of research carried out by the authors in the last seven years in collaboration with other members and students of the mathematical optimization department of the Sektion Mathematik of the Humboldt-Universität Berlin. Known results are taken into account and included in the text (see the General Introduction), we however are conscious of the fact, and this was indeed not our objective, that not all questions connected with this area of research have been covered.

We express our sincere thanks to all colleagues and former students who directly or indirectly contributed to the making of this book, including above all Prof. František Nožička.

Special thanks are due to Dr. Monika Noack for compiling the comprehensive bibliography, to Dr. Kim Rose for his assistance in preparing the final version of the manuscript, to Dr. Ralf Gollmer, Dr. Pavel Kleinmann, and Ralph Lehmann for assisting in proof-reading as well as to our meticulous typist Andrea Behrend. We further gratefully acknowledge the support of Gesine Reiher and Dr. Reinhard Höppner extended in the name of the publishers Akademie-Verlag, and finally the care taken by the printers in setting-up the book in print.

We especially remark that comments and criticisms are always welcome; these should be directed to the authors at the following address: 1086 Berlin, Humboldt-Universität, Sektion Mathematik, Postfach 1297.

Berlin, May 1982 The Authors

Contents

1. General Introduction

Problems of optimization arise in many scientific, technological and economic fields. The models employed, like any mathematical model, only describe facets of the real-life situation and then only at some level of approximation. In the following considerations we assume that the functional relations between the various quantities of interest have been sufficiently well established. The quality of an optimization model then depends strongly on the accuracy of the initial data supplied. Applications in economics are particularly affected by the objective difficulties surrounding the collection of this data. We thus find that, from the point of view of the mathematics of the problem three basic questions become apparent.

Firstly, one must determine for which models (here we confine ourselves to optimization problems) the intuitively reasonable assumption that the accuracy of the solutions obtained increases with the degree of approximation of the initial data, is in fact justified.

Secondly, one must ascertain whether an exact solution of a fixed problem may be viewed as an approximation to the solutions of the problems which correspond to small perturbations of the basic data (demand, prices, resources, technical parameters etc.), this is certainly of interest if the calculation of a solution entails much time and expense.

Finally, assuming that at least one of the two previous questions may be answered positively, estimates of the degree to which variations of the initial data influence the behaviour of the solutions would appear to be desirable, in other words we are looking for a quantitative description of the behaviour of the solutions and an error analysis.

In order to provide a basis for the study of these questions we consider the parametric optimization problem

$$(\mathrm{P}_\lambda) \qquad \min \{f(x, \lambda) \mid x \in M(\lambda)\}\,, \qquad \lambda \in \Lambda\,,$$

where $M(\lambda) \subset X$, X and Λ are metric spaces and f is a function $f\colon X \times \Lambda \to \mathbb{R} \cup \{+\infty, -\infty\}$. One may think of the parameter vector λ as representing the initial data of an optimization problem. In our attempt to achieve a unified and lucid exposition we only consider metric spaces, however in particular cases possibilities of a further generalization of the results obtained are indicated.

The first study in the field of parametric optimization appears to be that of A. S. Manne in the year 1953, since then at least 400 individual papers have appeared (cf. the Bibliography) and the field has greatly expanded, particularly in the last fifteen years.

Two lines of development have become apparent, the first of which follows directly from the basic questions outlined above: certain stability properties of the *constraint set mapping*

$$M : \lambda \to M(\lambda) ,$$

the *extreme value function*

$$\varphi : \lambda \to \varphi(\lambda) =_{\text{Df}} \inf_{x \in M(\lambda)} f(x, \lambda) ,$$

and the *optimal set mapping*

$$\psi : \lambda \to \psi(\lambda) =_{\text{Df}} \{x \in M(\lambda) \mid f(x, \lambda) = \varphi(\lambda)\}$$

are examined for general classes of optimization problems (P_λ). The second consists of looking for solution procedures for certain classes of parametric optimization problems or in elaborating the theoretical foundations of such solution procedures. In the following we give a short description of the problems involved in either case.

By stability of the mappings M, ψ, and φ we mean in particular certain continuity attributes of these mappings; in this restricted sense we speak of "qualitative stability". The concepts of semicontinuity according to HAUSDORFF or BERGE are applied to the point-to-set mappings M and ψ in the same manner as in the literature.[1]) Returning to the first of the aforementioned principal questions we note that the upper semicontinuity of the mapping ψ and the continuity of the function φ are seen as indicating that the problem (P_λ) is "well-behaved". On the other hand the lower semicontinuity of ψ at the point considered is of decisive importance in answering the second question. In this respect the study of the relations between the continuity properties of the three characteristic mappings is of particular interest.

The following example is intended to show that the first two basic questions raised above are indeed not trivial.

Example 1.1. Consider the problem

$$\min \{\lambda^2 x^2 - 2\lambda(1 - \lambda)\, x \mid x \geqq 0\}$$

where x and λ are non-negative real numbers. The unique solution is $\hat{x}(\lambda) = (1 - \lambda)/\lambda$ and $\varphi(\lambda) = -(1 - \lambda)^2$ if $\lambda \in (0, 1)$. However if $\lambda \geqq 1$ the solution is $\psi(\lambda) = \{0\}$ and $\varphi(\lambda) = 0$, whereas if $\lambda = 0$ we find $\psi(\lambda) = \{x \mid x \geqq 0\}$. Thus neither the extreme value function nor the mapping ψ is continuous at $\lambda = 0$. As far as φ is concerned we even have the paradox that $|\varphi(\lambda) - \varphi(0)|$ increases if $\lambda > 0$ decreases for small positive values of λ and is zero otherwise (namely for $\lambda \geqq 1$). The mapping ψ is not lower semicontinuous since, for example, the limit of the distance between the solution $x = 100$ for $\lambda = 0$ and the solution sets $\psi(\lambda)$ for $\lambda > 0$ is obviously not zero.

Important contributions to the theory of qualitative stability of parametric optimization problems are due to C. BERGE 1959 [2], G. DEBREU 1959 [1], G. DANTZIG, J. FOLKMAN, and N. SHAPIRO 1967 [1], J. P. EVANS and F. J. GOULD 1970 [1], R. R. MEYER 1970 [1], H. J. GREENBERG and W. P. PIERSKALLA 1972 [1], 1975 [2], E. G. GOL'ŠTEĬN 1971 [3], W. KRABS 1972 [1], 1973, [2], [3], W. W. HOGAN 1973 [3], S. M. ROBINSON 1975 [4], 1976 [6], 1979 [10], S. DOLECKI 1977 [2], 1978 [3], [4], B. BROSOWSKI 1976 [1], 1980 [4], and other authors. We also refer the reader to B. KUMMER

[1]) cf. Section 2.2.

1977 [1], [2], [3], 1978 [4], D. KLATTE 1977, 1979 [1], [2], [3], and B. BANK 1978 [1], 1979 [2].

Although they lie outside the scope of this volume further studies should be included in what we called the first line of development of parametric optimization. These include differentiability conditions for the extreme value function which are related to what is usually termed "differential stability" or "marginal value" in the literature. We refer the interested reader to H. D. MILLS 1956 [1], A. C. WILLIAMS 1963 [1], E. G. GOL'ŠTEĬN 1971 [3], E. S. LEVITIN 1974, 1976 [3], [6], V. F. DEM'JANOV and A. B. PEVNIĬ 1972—74 [1], [2], [3], W. W. HOGAN 1973 [1], J. GAUVIN and J. W. TOLLE 1977 [1], and F. LEMPIO and H. MAURER 1979 [1]. Further studies are concerned with quantitative stability considerations relating to the Lipschitzian continuity or the rate of continuity of the mappings ψrand φ or the Kuhn-Tucker set mapping. By Kuhn-Tucker mapping we understand the mapping which assigns to each parameter λ the set of Kuhn-Tucker solutions of the corresponding optimization problem. Besides the classic paper of A. J. HOFFMANN 1952 [1] we refer to work by D. W. WALKUP and R. J.-B. WETS 1969 [2], P. BRUNS 1972 [1], J. W. DANIEL 1973 [2], S. M. ROBINSON 1973, 1974 1975, 1976, 1977, 1979, 1980 [2], [3], [4], [5], [8], [9], [11], W. HAGER 1976 [1], E. S. LEVITIN 1976 [5], and S. DOLECKI 1977, 1978 [2], [3], [4]. We also note a paper by P. KLEINMANN 1978 [1].

An important motivation for the study of qualitative and quantitative stability properties of parametric optimization problems comes from other areas of mathematics. Thus for instance the continuity of φ and ψ plays an important role in studies of the convergence of solution procedures in non-linear optimization, some references in which general solution procedures are given are W. I. ZANGWILL 1969 [1], R. R. MEYER 1970 [1], G. G. L. MEYER 1977, 1979 [2], [3], J. PIRZL 1971 [1], [2], P. HUARD 1975 [3]; the convergence of these procedures is demonstrated by using continuity properties of point-to-set mappings. The rate of convergence of imbedding procedures may be determined by using the results of quantitative stability studies (for instance S. M. ROBINSON 1974, 1979 [3], [9], D. ELLENBERG 1979 [1], and C. RICHTER 1980 [1]).

The continuity properties of ψ and φ are discussed in a number of papers and employed a) in the analysis of optimal control problems by discretization and regularization (A. N. TIHONOV 1965, 1966 [1], [3], J. CULLUM 1969, 1971 [1], [2], A. WIERZBICKI 1970 [2], B. M. BUDAK and E. M. BERKOVIČ 1971 [1], [2], J. W. DANIEL 1971 [1], W. KRABS 1972, 1973 [1], [2], H. ESSER 1972 [1], and A. DONTCHEV 1975 [2]), b) in approximation theory (P. J. LAURENT 1972 [1], W. KRABS 1973 [5], 1975 [6], J. REINDLMEYER 1975 [1], B. BROSOWSKI, F. DEUTSCH, and G. NÜRNBERGER 1979 [1]) and c) in incorrectly formulated optimization problems (cf. A. N. TIHONOV 1966 [2], A. N. TIHONOV and V. JA. ARSENIN 1974 [1], V. G. KARMANOV 1975 [1], I. I. EREMIN and N. N. ASTAF'JEV 1976 [1], and V. V. FEDOROV 1979 [2]).

Sensitivity analysis, i.e. the analysis of the influence of model errors on certain characteristic quantities of the model plays an important part in mathematical modelling (optimization problems and control problems) and in economics and technology (cf. G. DEBREU 1959 [1], V. L. MAKAROV and A. M. RUBINOV 1973 [1], W. HILDENBRAND 1974 [1], and A. WIERZBICKI 1977 [5]).

The relation to stochastic optimization may be found in, for example R. J.-B. WETS 1966 [2], B. BEREANU 1976 [5], P. KALL and W. OETTLI 1975 [1].

The interrelations of properties of the extreme value function and duality statements of non-linear optimization have been worked out by R. T. ROCKAFELLAR 1967, 1971 [1], [4], I. EKELAND and R. TEMAM 1978 [1], F. G. GOL'ŠTEĬN 1971 [3], and J. L. JOLY and P. J. LAURENT 1971 [1]. It is well-known that duality theory is of great importance with respect to the development and study of solution procedures in non-linear optimization. Generalized Lagrangian and penalty functions for non-convex problems are examined in numerous studies (for instance J. D. ROODE 1969 [1], M. BELLMORE, H. J. GREENBERG, and J. J. JARVIS 1970 [1], F. J. GOULD 1969 [1], H. J. GREENBERG 1973 [1], R. T. ROCKAFELLAR 1974 [7], A. P. WIERZBICKI and S. KURCYUSZ 1977 [1]) involving a generalization of the approach used for convex functions (M. J. D. POWELL 1969 [1], M. R. HESTENES 1975 [1], R. T. ROCKAFELLAR 1973 [5], [6], 1976 [7], E. G. GOL'ŠTEĬN, N. V. TRET'YAKOV 1979 [1], D. P. BERTSEKAS 1975 [1], and B. W. KORT and D. P. BERTSEKAS 1976 [1]). Here properties of the extreme value function of the parametric optimization problem associated with the initial problem also play a role. In this context we also refer to the paper of C. GROSSMANN and G. SCHÖNINGER 1977 [1] on sentitivity and applicability of penalty and barrier methods.

Qualitative and quantitative continuity investigations relating to the optimal set mapping may in a certain sense be included in the study of parametric generalized equations of the form

$$0 \in \Gamma(z, \lambda), \qquad z \in Z, \tag{1.1}$$

where $\lambda \in \Lambda$ denotes the parameter; fundamental and relatively strong results have been given by HOANG TUY 1974, 1977 [1], [2], S. M. ROBINSON 1976, 1978, 1979, 1980 [5], [8], [9], [11] and a number of other authors, these results however lie outside the scope of this book, we only note that the system (1.1) as a rule reflects the Kuhn-Tucker condition for a (convex) optimization problem:

$$z = (x, y) \in \mathbb{E}_n \times \mathbb{E}_m, \qquad Z = \mathbb{E}_n \times \mathbb{E}_m^+,$$

$$\Gamma(z, \lambda) =_{\mathrm{Df}} \left\{\nabla_x f(x, \lambda) + \sum_{i=1}^{m} y_i \nabla_x g_i(x, \lambda)\right\} \times \{g(x) + \Delta(y)\} \tag{1.2}$$

where the functions f, g_i, $i = 1, \dots, m$, determine the problem

$$\min \{f(x, \lambda) \mid x \in \mathbb{E}_n, g_i(x) \leqq 0, i = 1, \dots, m\}$$

and the set $\Delta(y)$ is given by

$$\Delta(y) =_{\mathrm{Df}} \{\eta \in \mathbb{E}_m^+ \mid \eta_i = 0 \text{ if } y_i > 0\}.$$

One easily confirms that $z = (x, y)$ is a solution of (1.1) if and only if (x, y) is a Kuhn-Tucker point. Propositions and hypotheses relating to the generalized equation (1.1) thus apply to the solution set of the primal as well as the associated dual problem. Further implications are possible using this general approach if stronger assumptions than those employed here are made, in particular if compactness of the primal as well as the dual problem is assumed and the linearization is "well-behaved" at the Kuhn-Tucker points of the limiting problem. Quantitative studies and marginal values are also of interest in this respect. Work on perturbed Kuhn-Tucker points by V. FIACCO and G. P. MCCORMICK 1968 [1], S. M. ROBINSON 1974 [3], R. L. ARMACOST and A. V. FIACCO 1975 [2], [3], A. V. FIACCO 1976 [3], E. S. LEVITIN 1974, 1975 [3], [4], and others may also be included in this context, however in this book these results only play a role in relation to the second line of development indicated above and which is concerned for the main part with special optimization problems in spaces of finite dimensions. These problems are closely connected with the development of solution procedures. In numerous studies (here we only mention the monographs of W. DINKELBACH 1969 [1] and T. GAL 1973 [6]) the suitability of known algorithms, in particular the simplex method, for determining a solution of the problem (P_λ) for each λ contained in the so-called *solubility set* $\mathfrak{A} = \{\lambda \in \Lambda \mid \psi(\lambda) \neq \emptyset\}$ is examined; in general the parametric problem (P_λ) represents an infinite class of optimization problems, the objective can therefore only be achieved if closed solutions may be found for certain subclasses containing an infinite number of problems. In this respect F. NOŽIČKA's concept of a local stability set (cf. F. NOŽIČKA 1972 [1], F. NOŽIČKA et al. 1974 [2]) introduced for linear parametric optimization appears to be fruitful. This method-independent approach enables one to uniquely divide the solubility set into a finite number of local stability sets. The concept may be also profitably applied to more general classes of problems (convex quadratic optimization problems, J. GUDDAT 1974, 1976 [1], [3], D. KLATTE 1977 [1], linear complementarity problems, K. TAMMER 1979 [9], mixed-integer linear optimization problems, B. BANK 1978, 1979 [1], [2], linear hyperbolic programming, E.-CH. TAMMER 1973, 1977 [1], [3]) as well as to linear optimization problems with parameter-dependent constraint matrices (L. GRYGAROVÁ 1975 [3], D. KLATTE 1977, 1979 [1], [2], [4], R. GOLLMER 1980 [1]).

Corresponding solution procedures have been worked out for most of the classes of problems listed above and with which one may calculate a finite partitioning of the solubility set using local stability sets. We refer in particular to the book of C. VAN DE PANNE 1975 [3] in which methods may be found for dealing with linear complementarity problems having parameter-dependent right-hand sides of the constraints. These methods are based on the simplex technique and are presumably just as effective as similar procedures for linear optimization problems with parameter-dependent objective functions and right-hand sides of the constraints. For problems of greater generality than the linear parametric complementarity problems already mentioned we do not appear to be able to achieve the objective named above, at least not in any effective fashion (cf. F. GUERRA 1980 [1]). In this respect procedures for approximating a selection function are of special importance and we refer in particular to work by E. S. LEVITIN 1974 [2], A. V. FIACCO 1976 [3], R. L. ARMACOST and A. V. FIACCO 1975, 1976 [2], [3], M. KOJIMA 1979 [2], and M. LIPS 1979 [1].

An important part is played in many applications by solution procedures for parametric optimization problems, here the influence of perturbations of the initial data of an optimization problem may be quantified. Many further applications may be found in a volume edited by K. LOMMATZSCH 1979 [2]. Solution procedures for certain parameter-dependent auxiliary problems are also needed in the imbedding methods of P. WOLFE 1959 [1], S. HOUTHAKKER 1960 [1], C. E. LEMKE 1965 [1], K. RITTER 1965 [3], and K. TAMMER 1976 [3] as well as in the partitioning procedures for the solution of fixed optimization problems (cf. B. BANK, R. MANDEL, and K. TAMMER 1979 [1]). We would also like to draw special attention to the possibilities of applying procedures for solving parametric optimization problems to the study of vector optimization problems (cf. W. DINKELBACH 1969 [1], J. FOCKE 1973 [1], M. ZELENY 1974 [1] and J. GUDDAT 1979 [5]) and of decision problems in stochastic optimization (cf. K. TAMMER 1978, 1979 [7], [8], [9]). These applications are particularly suited to computing techniques in the case of linear and special quadratic problems, which also applies to higher dimensions.

The concept of a local stability set mentioned earlier is strongly related to the qualitative stability studies which characterize the first line of development noted above.

One finds that the upper semicontinuity of the optimal set mapping ψ and the continuity of the extreme value function φ hold in quite general classes of parametric optimization problems (P_λ). On the other hand the lower semicontinuity of the mapping ψ is in general not given globally even in linear optimization problems with parameter-dependent objective functions. The question arises of whether certain subsets of the solubility set may be distinguished in such a fashion that properties over and above the upper semicontinuity of ψ and the continuity of φ are given (for example lower semicontinuity of ψ). One thus arrives in natural fashion at the concept of a local stability set for various classes of parametric optimization problems. On the other hand the lower semicontinuity of the mapping ψ is closely connected to the question of whether a continuous selection function of the optimal set mapping exists, and this question is crucial for the possibility of constructing suitable solution procedures. One may conclude that the two lines of development outlined above are closely related.

In Chapter 2 we present basic concepts and fundamental general properties of point-to-set mappings. We are not so much concerned with the greatest possible generality of the theorems as with transparent proofs and providing theorems needed in the following chapters (MICHAEL's selection theorems for example are not proved in Section 2.3 for arbitrary metric spaces as inverse image sets but rather for a certain subclass which however suffices for our applications).

In Chapter 3 we examine semicontinuity properties of special point-to-set mappings M of the form

$$M(\lambda) = \{x \in X \mid g_i(x, \lambda) \leqq 0, i \in I\} \cap \Gamma(\lambda) .$$

As in the following chapters we try to avoid dispensable compactness conditions and attempt to show that the conditions employed are in fact necessary.

Chapter 4 is concerned with continuity properties of the extreme value function φ and the optimal set mapping ψ for more or less general classes of problems (P_λ). In particular we consider non-convex problems with constraint sets contained in a compact metric space, furthermore convex problems for which no Slater condition is satisfied at the parameter point of interest, and finally mixed-integer quadratic problems with parameter-dependent linear parts of the objective function and the constraints.

In Chapter 5 the characteristic parameter sets (convexity set, feasible parameter set, solubility set, local stability set) for various classes of parametric optimization problems in finite-dimensional spaces (i.e. parametric optimization problems with fixed constraint sets, parameter-dependent convex quadratic problems, parametric linear complementarity problems and mixed-integer linear parametric optimization problems) are studied.

In Chapter 6 we discuss various approaches for a numerical analysis of parametric optimization problems. Two algorithms for determining a partitioning for parameter-dependent linear complementarity problems are presented, one for the multi-parameter and the other for the single-parameter case. For more general classes a local approximation of a *selection function* of the optimal set mapping is discussed. The relation to the concepts of FIACCO and LEVITIN is shown, some of our results are more far-reaching than those of the latter.

In the Appendix various concepts and results are compiled which are used in the text but which themselves are not associated with the field of parametric optimization.

Chapters 4, 5 and 6 are each provided with a separate introduction which also contains further references. The references of the Bibliography are denoted by I, II, and III, where I means that parametric optimization in the sense considered here is concerned. II refers to related literature and/or applications, and III identifies basic literature. In I we attempted to list as many individual studies in the field of optimization as possible, without evaluating the works. A classification of the works that we consider to be essential is given, in addition to the General Introduction, still in the introductions to the Chapters 4, 5, and 6 as well as at appropriate places in the text, of course. We deliberately did not aim at being comprehensive in II and III. Let us remark that this Bibliography represents the level, as we knew it, of November 1980.

2. Basic Concepts and Definitions

2.1. Notation and Basic Concepts

The majority of the concepts employed in this book derive from convex analysis and functional analysis, here for the main part we follow R. T. ROCKAFELLAR [2], J. STOER, CH. WITZGALL [1], and E. ZEIDLER [1] respectively.

Sets, functions, and quantifiers. All sets are denoted by capital letters. *Inclusion* of sets is symbolized by $A \subset B$ which may also mean that A is equal to B. The *power set*, i.e. the set of all subsets of a set A, is denoted by 2^A. If f is a *function* defined on A with values in B we write

$$f\colon A \to B\,.$$

In particular B may be a power set 2^C; in this case f is also called a *point-to-set mapping*. $A \times B$ is the set of all ordered pairs (a, b), where $a \in A$ and $b \in B$. Certain sets have standard symbols, e.g.

$\mathbb{E}_n$ — *n-dimensional Euclidean space,*
$\mathbb{R}$ — *set of all real numbers,*
$\mathbb{R}^+$ — *set of all non-negative real numbers,*
$\mathbb{Z}$ — *set of all integers .*

Further examples may be found in the List of Symbols at the end of this section.

Logical propositions are abbreviated by stenographic symbols for logical quantifiers and operations as follows:

$\exists$ — *there exists,*
$\forall$ — *for all,*
$a \Rightarrow b$ — *a implies b,*
$a \Leftrightarrow b$ — *a if and only if b*

where a and b are propositions.

Sets are defined in curly brackets, for example

$$M = \{x \in \mathbb{R} \mid x \geqq 0,\, x = 2k,\, k \in \mathbb{Z}\}$$

signifies that M is the set of all non-negative even integers. The *number of elements* or *cardinality* of a set M is denoted by $|M|$.

Metric spaces. A *metric space* (X, d_X) is defined by a set X and a function $d_X\colon X \times X \to \mathbb{R}^+$ with the properties:

$$d_X(x, y) = 0 \Leftrightarrow x = y ,$$
$$d_X(x, y) = d_X(y, x) ,$$
$$d_X(x, z) \leqq d_X(x, y) + d_X(y, z) .$$

The numerical value $d_X(x, y)$ is the *distance* between x and y, the function $d = d_X$ is the *metric* of the space. If the metric is understood we also say simply "metric space X". Every subset M of a metric space (X, d) is itself a metric space with the metric function d restricted to $M \times M$.

A subset $M \subset X$ is

(i) *bounded*, if $\sup \{d(x, y) \mid x \in M, y \in M\} < \infty$;

(ii) *open*, if there exists for each point $x \in M$ an $\varepsilon > 0$ such that the set $U_\varepsilon\{x\}$ of all points $y \in X$ having $d(x, y) < \varepsilon$ is contained in M;

(iii) *closed*, if M is the complement of an open set, i.e. if N is open and $M = X \setminus N$. In a metric space X a subset M is closed if and only if for each $x \in X$ and each infinite sequence $\{x^n\} \subset M$ the proposition

$$\lim_{n \to \infty} d(x, x^n) = 0 \Rightarrow x \in M$$

is true.

We will frequently make use of the *distance between a point x and a set $M \subset X$*, which is defined by $d(x, M) = \inf_{y \in M} d(x, y)$ and which by convention is given the value $+\infty$ if $M = \emptyset$. Contrary to common usage we denote by *accumulation point* of a set M any point x having $d(x, M) = 0$. The sets

$$\text{cl } M = \{x \in X \mid d(x, M) = 0\} ,$$
$$\text{bd } M = \text{cl } M \cap \text{cl } (X \setminus M)$$

are called the *closure* and the *boundary* of the set M respectively. The closure cl M is the intersection of all closed subsets N with $M \subset N$.

Further, a point $x \in M$ is an *interior* point of M if an open set $\Omega \subset X$ exists and $x \in \Omega \subset M$ holds. The set of all interior points is denoted by int M, and any open set Ω containing x is a *neighbourhood* of x.

A subset N of the set M is said to be *dense* in M if $M \subset \text{cl } N$. If a countable dense subset $N \subset M$ exists then M is *separable*. A sufficient condition for M to be separable is the existence of a countable set $N \subset X$ with $M \subset \text{cl } N$.

A subset M of the space X is called *compact*, if every non-empty countable subset $N \subset M$ possesses an accumulation point x in M. Compact sets are closed and separable. A subset M of a metric space X is compact if and only if the covering theorem of Heine-Borel is satisfied: If $\{\Omega_i \mid i \in I\}$ is an arbitrary family of open subsets satisfying $M \subset \bigcup_{i \in I} \Omega_i$ then there exists a finite subset $J \subset I$ such that $M \subset \bigcup_{i \in J} \Omega_i$ holds. In particular a metric space itself may be separable or compact. The main reason we use compactness here is to demonstrate that certain sequences $\{x^t\}_{=1,2,\ldots}$, $x^t \in X$, converge. Instead of $\lim_{t \to \infty} d(x, x^t) = 0$ we also write $x^t \to x$ omitting $t \to \infty$ for convenience if the limit is not to be understood otherwise. In order to avoid double indices we will take the expression *without loss of generality** (marked with an asterisk) to mean that an infinite sequence $\{x^{t_k}\}_{k=1,2,\ldots}$ with a certain property may be chosen from a given sequence $\{x^t\}$, for example: Since $x^t \in M (\forall t)$ and since M is compact we have without loss of generality *

$$x^t \to x \in M .$$

A metric space X for which the Cauchy condition of convergence is fulfilled (i.e., a sequence $\{x^t\}$ converges to an element $x \in X$ if for each $\varepsilon > 0$ there exists a $t(\varepsilon)$ such that $d(x^t, x^s) < \varepsilon$ for all $s, t \geqq t(\varepsilon)$) is called *complete*.

Let X be a metric space and $f: X \to \mathbb{R} \cup \{-\infty, +\infty\}$ be a (improper) functional. f is then *upper* (*lower*) *semicontinuous* at a point $x^0 \in X$ if for each sequence $\{x^t\} \subset X$, $x^t \to x^0$ the inequality

$$\overline{\lim}\, f(x^t) \leqq f(x^0) \qquad (\underline{\lim}\, f(x^t) \geqq f(x^0))$$

holds. $\overline{\lim}$ and $\underline{\lim}$ denote the *upper* and *lower limit* respectively, i.e.

$$\overline{\lim}\, f(x^t) = \inf \{\alpha \mid f(x^t) \leqq \alpha \text{ for almost all } t\}\ ,$$

$$\underline{\lim}\, f(x^t) = \sup \{\alpha \mid f(x^t) \geqq \alpha \text{ for almost all } t\}\ .$$

As we are dealing with arbitrary sequences $x^t \to x^0$ we also find: A function f is upper (lower) semicontinuous at a point x^0 if $x^t \to x^0$ implies

$$\underline{\lim}\, f(x^t) \leqq f(x^0) \qquad (\overline{\lim}\, f(x^t) \geqq f(x^0))\ .$$

We say that a function f is upper (lower) semicontinuous *on* $M \subset X$ if f is upper (lower) semicontinuous at all points $x \in M$. If f is both upper and lower semicontinuous we say that the improper functional is *continuous*. In the following we abbreviate the terms upper and lower semicontinuous by *u.s.c.* and *l.s.c.* respectively.

Let X and Y be metric spaces. A mapping $f: X \to Y$ is said to be *continuous* at a point $x^0 \in X$ if the following holds:

$$(\{x^t\} \subset X \text{ and } x^t \to x^0) \Rightarrow f(x^t) \to f(x^0)\ .$$

If f is continuous at all points $x \in M$ then f is said to be continuous *on* M.

Normed spaces. A real linear space X (we refer the reader to the relevant literature for the definition of a linear space) is called a *normed space* if there exists a function $||\cdot||$ on X called the *norm* with values in $\mathbb{R}^+$ satisfying the conditions

$$||x|| = 0 \Leftrightarrow x = 0 \quad (\text{zero element of } X)\ ,$$

$$||x + y|| \leqq ||x|| + ||y||\ ,$$

$$||\alpha x|| = |\alpha| \cdot ||x||$$

where $x, y \in X$ and $\alpha \in \mathbb{R}$.

A metric in X may be defined by $d(x, y) = ||x - y||$ and X is thus a metric space. A complete normed space is called a *Banach space*.

If X is a normed space then X^* denotes its associated *dual space*, which consists of all continuous linear functions $f: X \to \mathbb{R}$ defined on X with the norm

$$||f||^* = \sup_{||x||=1} |f(x)|$$

and the relations

$$(f + g)(x) = f(x) + g(x)\ , \qquad (\alpha f)(x) = \alpha \cdot f(x)\ .$$

A subset M of the normed space X is called

(i) a *subspace*, if for arbitrary $x, y \in M$ and $\alpha \in \mathbb{R}$ also $x + y$ and αx are contained in M,

(ii) a *cone*, if for arbitrary $x \in M$ and $\alpha \in \mathbb{R}^+$ also M contains αx, and

(iii) *convex*, if for arbitrary $x, y \in M$ and $\lambda \in (0, 1)$ also M contains $\lambda x + (1 - \lambda) \cdot y$.

If M and N are subsets of X, then

$$M + N =_{\mathrm{Df}} \{m + n \mid m \in M, n \in N\}$$

denotes their *Minkowski sum*. A subset M for which a point $x \in X$ exists such that $\{x\} + M$ is a subspace, is called an *affine subspace*. *Affine cones* are defined in similar fashion. Then M is a non-empty affine cone if and only if an $x^0 \in M$ exists such that $x \in M$ and $\alpha \in \mathbb{R}^+$ imply that $x^0 + \alpha(x - x^0) \in M$. In this case x^0 is called a *vertex point*, the set of all vertex points is the *vertex set* of M. The vertex set is an affine subspace.

The following sets are associated with a given subset $M \subset X$:

aff M — intersection of all affine subspaces containing M,
con M — intersection of all cones containing M,
conv M — intersection of all convex sets containing M,
ri M — set of all points $x \in M$ for which there exists an open set $\Omega \subset X$ such that $x \in \Omega \cap \text{aff}\, M \subset M$

and are called the *affine*, *conic* and *convex hulls* and *relative interior* respectively of M. Besides the trivial case $M = \emptyset$ the set ri M may also be empty if M is a convex set. A collection of $n + 1$ points x^0 to x^n is *affinely independent* (or equivalently the *points* are affinely independent) if $\sum_{i=1}^{n} t_i x^i = 0$, $t_i \in \mathbb{R}$, $\sum_{i=0}^{n} t_i = 0$, imply $t_0 = t_1 = \dots = t_n = 0$, otherwise it is *affinely dependent*. The points x^0 to x^n are affinely independent if and only if the system of equations

$$\sum_{i=0}^{n} \alpha_i(x^i - x^0) = 0\,, \qquad \alpha_i \in \mathbb{R}\,, \qquad i = 1, \dots, n\,,$$

has only the trivial solution $\alpha_1 = \dots = \alpha_n = 0$ (linear independence of the *vectors* $(x^i - x^0)$). An affine subspace S is *n-dimensional* if it contains $n + 1$ affinely independent points and if any collection of $n + 2$ points is affinely dependent. We use the formal expression dim $S = n$ to indicate this situation. A convex subset $M \subset X$ is n-dimensional (dim $M = n$) if its affine hull is n-dimensional. In this case we have ri $M \neq \emptyset$. In particular ri $\{x\} = \{x\}$.

To every subset M of X there corresponds a cone in the dual space X^*:

$$M^p = \{f \in X^* \mid f(x) \leqq 0 \ \ \forall x \in M\}\,.$$

M^p is the *polar cone* of the set M and is convex and closed in X^*.

Furthermore, to any non-empty convex subset M of the normed space X there corresponds the cone

$$O^+M =_{\text{Df}} \{u \in X \mid \exists x \in M : x + \lambda u \in M \ \ \forall \lambda \geqq 0\}\,.$$

This convex cone is the *recession cone* of the set M. If M is convex and closed then O^+M is also closed and one has

$$O^+M = \{u \in X \mid x + \lambda u \in M \ \ \forall x \in M \ \ \forall \lambda \geqq 0\}\,.$$

If a non-empty convex set $M \subset X$ is bounded then obviously $O^+M = \{0\}$; if M is closed and a subset of a finite dimensional space the reverse is true, i.e. $O^+M = \{0\}$ implies that the convex set M is bounded.

Convex functions. Let M be a convex subset of a normed space X and

$$f\colon M \to \mathbb{R} \cup \{-\infty, +\infty\}$$

be a function whose *effective domain* $\operatorname{dom} f =_{\mathrm{Df}} \{x \in M \mid f(x) \neq +\infty\}$ is convex; f is then a *convex function* if for arbitrary $x, y \in \operatorname{dom} f$, $\lambda \in (0, 1)$ the inequality

$$f(\lambda x + (1 - \lambda) y) \leqq \lambda f(x) + (1 - \lambda) f(y) \tag{2.1.1}$$

holds. We set the right-hand side equal to $-\infty$ if $f(x)$ or $f(y)$ takes this value. If f is finite on *dom* f and the inequality (2.1.1) is strictly satisfied für $x \neq y$ then f is *strictly* convex. If instead of (2.1.1) we impose the weaker condition

$$f(\lambda x + (1 - \lambda) y) \leqq \max \{f(x), f(y)\} \tag{2.1.2}$$

we define the class of *quasiconvex functions*. If (2.1.2) is strictly satisfied for $f(x) \neq f(y)$ then f is *strictly quasiconvex*. Obviously $f\colon M \to \mathbb{R} \cup \{-\infty, +\infty\}$ is quasiconvex if and only if each set of the form $N_f(c) =_{\mathrm{Df}} \{x \in M \mid f(x) \leqq c\}$, where $c \in \mathbb{R} \cup \{-\infty, +\infty\}$, is convex.

Euclidean Spaces. The *Euclidean space* $\mathbb{E}_n$ may be identified with the set of all real n-tuples

$$x = \begin{pmatrix} x_1 \\ \vdots \\ x_n \end{pmatrix}$$

if a norm is defined by

$$\|x\| = \sqrt{\sum_{i=1}^{n} x_i^2}$$

and a scalar product by

$$x^{\mathsf{T}} y = \sum_{i=1}^{n} x_i y_i \,.$$

The symbol T stands for "transpose" as is the case generally for matrices. By setting

$$x \leqq y \Leftrightarrow x_i \leqq y_i \,, \qquad i = 1, \dots, n \,,$$

we establish a *partial ordering* $\leqq$ in $\mathbb{E}_n$.

For convenience we denote the $(n + m)$-tuple derived from a pair of vectors $x \in \mathbb{E}_n$ and $y \in \mathbb{E}_m$ by

$$(x, y) = (x_1, \dots, x_n, y_1, \dots, y_m)^{\mathsf{T}} \,.$$

An (m, n)-matrix may be viewed as being an element of the space $\mathbb{E}_{n \cdot m}$, and it therefore makes sense to assign a norm $\|A\|$ to the matrix A and to speak of the continuity of A as a function which establishes a correspondence between elements λ of a metric space Λ and (m, n)-matrices $A(\lambda)$ (m, n are fixed).

An (n, n)-matrix C is said to be *positive definite* if for arbitrary $x \in \mathbb{E}_n$, $x \neq 0$, one has

$$x^{\mathsf{T}} C x > 0 \,.$$

If $x^{\mathsf{T}} C x \geqq 0$ for all $x \in \mathbb{E}_n$ then C is *positive semi-definite*. Further classes of matrices which are of interest in relation to complementarity problems are discussed in the Appendix.

Optimization problems. By

(P) $\quad \min \{f(x) \mid x \in X, \; g_i(x) \leqq 0, i \in I\}$

we symbolize the problem of finding in the subset of all points $x \in X$ satisfying the conditions $g_i(x) \leqq 0$, $i \in I$, a point at which the function $f = f(x)$ attains its lowest value. Occasionally "inf" is substituted for "min" in order to indicate that the existence of a solution of the problem (P) is not assumed. The set

$$M = \{x \in X \mid g_i(x) \leqq 0, i \in I\}$$

of all feasible points is called the *feasible region*, the function $f\colon X \to \mathbb{R}$ is the *objective function*.

If I is a finite index set then we say that the point x satisfies the *Slater condition* or x is a *Slater point* if $x \in X$ and $g_i(x) < 0$ hold for all $i \in I$. Further constraint qualifications may be found in the Appendix (A5).

It goes without saying that we use basic results of the theory of optimization problems with "fixed" data in this discussion of parameter dependent optimization, and refer to the literature in this context, we would however like to note two important facts:

(i) If X is a normed space and if all the functions $f, g_i\colon X \to \mathbb{R}$ are quasiconvex then M and the *solution set* ψ of the problem (P) are convex.

(ii) If $X = \mathbb{E}_n$, if all the functions $f, g_i\colon X \to \mathbb{R}$ are convex and differentiable, and further if I is finite and there exists a Slater point x^0, then the following is true: A point $x \in \mathbb{E}_n$ is a solution of the problem (P) if and only if there exists a point $u \in \mathbb{E}_{|I|}$ such that

$$\begin{aligned}
&\nabla f(x) + \sum_{i \in I} u_i \nabla g_i(x) = 0\,,\\
&g_i(x) \leqq 0\,,\\
&u \geqq 0\,,\\
&\sum_{i \in I} u_i g_i(x) = 0
\end{aligned}$$

hold (local Kuhn-Tucker conditions). By $\nabla f(x)$ we denote the gradient of the function f at the point x.

Finally, if f is a real function of the two variables x and λ then $\nabla_x f(\bar{x}, \bar{\lambda})$ is the gradient at the point $\bar{x}$ of the function $f(\cdot, \bar{\lambda})$ defined as depending only on x.

List of Symbols

$\subset$	— inclusion of sets including equality,
$a \Rightarrow b$	— a implies b,
$a \Leftrightarrow b$	— a if and only if b,
$\forall$	— for all,
$\exists$	— there exists,
$=_{\mathrm{Df}}$	— definition,
$\mathbb{R}$	— set of all real numbers,
$\mathbb{R}^+$	— set of all non-negative real numbers,
$\mathbb{Z}$	— set of all integers,
$\mathbb{E}_n$	— n-dimensional Euclidean space,
$\mathbb{H}^{m,n}$	— set of all (m, n)-matrices provided with the Euclidean norm,
A^{T}	— transpose of the matrix A,
$\operatorname{int} A$	— (topological) interior of the set A,
$x \in \operatorname{int} A$	— interior point,
$\operatorname{ri} A$	— relative interior of the set A,
$x \in \operatorname{ri} A$	— relatively interior point,

bd A	— boundary of the set A,
cl A	— closure of the set A,
aff A	— affine hull of the set A,
con A	— conic hull of the set A,
conv A	— convex hull of the set A,
A^p	— polar cone of the set A,
X^*	— dual space corresponding to the space X,
Λ	— parameter set (metric space),
$x^\mathsf{T}y$	— scalar product,
φ	— extreme value function,
ψ	— optimal set mapping,
M	— constraint set mapping,
2^X	— power set of the set X,
$\overline{\lim}$	— upper limit (limes superior),
$\underline{\lim}$	— lower limit (limes inferior),
$x^t \to x$	— convergence of x^t to x for $t \to \infty$,
*	— after "without loss of generality" signifies that a proposition is true subject to the choice of an infinite subsequence,
$\mathfrak{A}$	— solubility set,
$\mathfrak{B}$	— feasible parameter set (feasible region is non-empty),
$d(x, A) = \inf\limits_{y \in A} d(x, y)$	— distance between a point $x \in X$ (X is a metric space with metric d) and a set $A \subset X$,
$U_\varepsilon A$	— ε-neighbourhood of the set A,
$\dfrac{\partial f(x^0)}{\partial x_i}$	— partial derivative of a function $f: E_n \to R$ by x_i, the function is differentiable at x^0 and the derivative is taken at x^0,
$\nabla_x f(x^0, \lambda^0)$	— gradient of a function $f: E_n \times E_k \to R$ relative to x taken at the point (x^0, λ^0),
$J(x^0, \lambda^0) = \begin{pmatrix} \dfrac{\partial f_1(x^0, \lambda^0)}{\partial x_1} & \cdots & \dfrac{\partial f_1(x^0, \lambda^0)}{\partial x_n} \\ \cdots & \cdots & \cdots \\ \dfrac{\partial f_m(x^0, \lambda^0)}{\partial x_1} & \cdots & \dfrac{\partial f_m(x^0, \lambda^0)}{\partial x_n} \end{pmatrix}$	— Jacobian matrix of a function $f: E_n \times E_k \to E_m$ relative to x and at the point (x^0, λ^0),
$\nabla^2 f_x(x^0, \lambda^0) = \begin{pmatrix} \dfrac{\partial^2 f(x^0, \lambda^0)}{\partial x_1\, \partial x_1} & \cdots & \dfrac{\partial^2 f(x^0, \lambda^0)}{\partial x_1\, \partial x_n} \\ \cdots & \cdots & \cdots \\ \dfrac{\partial^2 f(x^0, \lambda^0)}{\partial x_n\, \partial x_1} & \cdots & \dfrac{\partial^2 f(x^0, \lambda^0)}{\partial x_n\, \partial x_n} \end{pmatrix}$	— Hessian matrix relative to x at the point (x^0, λ^0),
$\nabla f(x^0)$	— gradient of a function $f: E_n \to R$ at x^0, the function is differentiable at x^0,
$f\|_A$	— function f is restricted to a set A,
$\|A\|$	— cardinality of the set A,
u.s.c.	— upper semicontinuous,
l.s.c.	— lower semicontinuous.

2.2. Point-to-Set Mappings: Definitions and Elementary Properties

This section is concerned with continuity properties of point-to-set, or in different terminology, multivalued mappings. The reason we devote space to this field despite the fact that a large number of studies have already been published in this context, is twofold: firstly, we collect a number of results which are of central importance in the following chapters, and secondly, definitions and notation vary considerably in the literature thus necessitating an explicit statement of the terminology used here. The results are not presented in the most general form in order that they may be applied directly to the stability problems in which we are interested. For a similar reason we only consider from the multitude of theorems on multivalued mappings those which are relevant to the present work, thus differentiability conditions do not appear in this selection of well-known results as they are not needed in what follows.

Let (X, d_X) and (Λ, d_Λ) be metric spaces. By a *point-to-set mapping* of Λ into X we understand a (set-valued) function Γ which assigns a (possibly empty) subset $\Gamma\lambda$ of X to each element $\lambda \in \mathbb{R}$. Γ is thus a mapping of Λ into the power set 2^X of X, thus we may write

$$\Gamma: \Lambda \to 2^X .$$

If A is a subset of X and $\varepsilon > 0$ we denote the *ε-neighbourhood* of the set A by

$$U_\varepsilon A =_{\mathrm{Df}} \{x \in X \mid \inf_{y \in A} d_X(x, y) < \varepsilon\}$$

where

$$d_X(x, A) = \inf_{y \in A} d_X(x, y) \quad \text{(see pp. 19 or 24, too).}$$

The infimum of the empty set is defined in usual manner as being equal to $+\infty$. To avoid misunderstandings we denote an ε-neighbourhood of a subset B of Λ by $V_\varepsilon B$.

Continuity definitions. A point-to-set mapping $\Gamma: \Lambda \to 2^X$ is

(i) *closed* at a point λ^0 if for each pair of sequences $\{\lambda^t\} \subset \Lambda$ and $\{x^t\} \subset X$, $t = 1, 2, \dots$, with the properties

$$\lambda^t \to \lambda^0 , \qquad x^t \in \Gamma\lambda^t , \qquad x^t \to x^0$$

it follows that $x^0 \in \Gamma\lambda^0$;

(ii) *upper semicontinuous* (according to Berge, or simply: B) at a point λ^0, if for each open set Ω containing $\Gamma\lambda^0$ there exists a $\delta = \delta(\Omega) > 0$ such that

$$\Gamma\lambda \subset \Omega \qquad \forall \lambda \in V_\delta\{\lambda^0\}$$

holds;

(iii) *lower semicontinuous* (according to Berge, or simply: B) at a point λ^0, if for each open set Ω satisfying $\Omega \cap \Gamma\lambda^0 \neq \emptyset$ there exists a $\delta = \delta(\Omega) > 0$ such that

$$\Gamma\lambda \cap \Omega \neq \emptyset \qquad \forall \lambda \in V_\delta\{\lambda^0\}$$

holds;

(iv) *upper semicontinuous* (according to Hausdorff, or simply: H) at a point λ^0, if for each $\varepsilon > 0$ there exists a $\delta > 0$ such that

$$\Gamma\lambda \subset U_\varepsilon\Gamma\lambda^0 \qquad \forall \lambda \in V_\delta\{\lambda^0\}$$

holds;

(v) *lower semicontinuous* (according to HAUSDORFF, or simply: H) at a point λ^0, if for each $\varepsilon > 0$ there exists a $\delta > 0$ such that

$$\Gamma\lambda^0 \subset U_\varepsilon \Gamma\lambda \qquad \forall \lambda \in V_\delta\{\lambda^0\}$$

holds; and

(vi) *strongly lower semicontinuous* at a point λ^0, if for each $x \in \Gamma\lambda^0$ there exist an $\varepsilon > 0$ and a $\delta > 0$ such that

$$U_\varepsilon\{x\} \subset \Gamma\lambda \qquad \forall \lambda \in V_\delta\{\lambda^0\}$$

holds.

For convenience we employ the following abbreviations throughout the text:

u.s.c.-B for upper semicontinuous (B),

l.s.c.-B for lower semicontinuous (B)

as well as "u.s.c.-H", "l.s.c.-H" and "strongly l.s.c." whose meanings are then obvious.

Further, a point-to-set mapping Γ is

(vii) *continuous* at λ^0 if it is u.s.c.-H and l.s.c.-B at λ^0;

(viii) *B-continuous* at λ^0 if it is u.s.c.-B and l.s.c.-B at λ^0; and

(ix) *H-continuous* at λ^0 if it is u.s.c.-H and l.s.c.-H at λ^0.

We would like to point out that, with the exception of "closed" the notation varies in the literature. Here we follow S. DOLECKI for the main part with respect to semicontinuity, he however does not use "strongly lower semicontinuous". The term "open" frequently appears in the literature in the same sense as our "l.s.c.-B" (cf. for instance R. R. MEYER [1], W. HOGAN [3]).

The following relations are immediate consequences of the above definitions:

u.s.c.-B $\Rightarrow$ u.s.c.-H ,

l.s.c.-H $\Rightarrow$ l.s.c.-B ,

strongly l.s.c. $\Rightarrow$ l.s.c.-B .

Further implications can only be made if additional hypotheses are introduced as in the remainder of this section.

Elementary properties of point-to-set mappings. As already indicated we are interested in continuity properties of a mapping $\Gamma: \Lambda \to 2^X$ at a point $\lambda^0 \in \Lambda$.

Lemma 2.2.1. *If the mapping Γ is u.s.c.-H at λ^0 and if the set $\Gamma\lambda^0$ is closed then the mapping Γ is closed at λ^0.*

Proof. If the point x^0 were not contained in the closed set $\Gamma\lambda^0$ (note the definition of the closedness of Γ) then there would exist an $\varepsilon > 0$ such that $x^0 \notin U_\varepsilon \Gamma\lambda^0$; since $x^t \to x^0$ it would follow that $x^t \notin U_\alpha \Gamma\lambda^0$ for $\alpha = \frac{1}{2}\varepsilon$ and almost all t, which implies that $\Gamma\lambda^t \setminus U_\alpha \Gamma\lambda^0 \neq \emptyset$ for almost all t and which, because of $\lambda^t \to \lambda^0$ thus contradicts the first hypothesis (Γ is u.s.c.-H at λ^0). □

Lemma 2.2.2. *Let the set $\Gamma\lambda_0$ be closed. Γ is u.s.c.-B at λ^0 if and only if the following is true for all sequences $\{\lambda^t\} \subset \Lambda$, $\lambda^t \to \lambda^0$,* and *$\{x^t\} \subset X$, $x^t \in \Gamma\lambda^t \setminus \Gamma\lambda^0$: The sequence $\{x^t\}$ has an accumulation point contained in $\Gamma\lambda^0$.*

Proof. Assume that Γ is u.s.c.-B at λ^0. If no accumulation point of the sequence $\{x^t\}$ exists then the set A of all points x^t is closed and thus (for $\Omega =_{\mathrm{Df}} X \setminus A$)

$$x^t \in \Gamma\lambda^t \setminus \Omega \ \forall t\,, \qquad \lambda^t \to \lambda^0$$

which contradicts the assumption.

If on the other hand $\{x^t\}$ possesses an accumulation point x^0 this point is contained in $\Gamma\lambda^0$ because Γ is closed (Lemma 2.2.1). Further, assume that Γ is not u.s.c.-B at λ^0. There then exists an open set Ω containing $\Gamma\lambda^0$ and there also exist sequences $\{\lambda^t\}$, $\lambda^t \to \lambda^0$, and $\{x^t\}$, $x^t \in \Gamma\lambda^t \setminus \Omega$. As every accumulation point x^0 of the sequence $\{x^t\}$ is contained in the closed set $X \setminus \Omega$ this contradicts $x^0 \in \Gamma\lambda^0$. □

In the case of closed image sets $\Gamma\lambda$ the previous lemma clearly indicates the difference between a mapping which is closed at λ^0 and one which is u.s.c.-B: whereas closedness means that

$$\lambda^t \to \lambda^0\,, \qquad x^t \in \Gamma\lambda^t \setminus \Gamma\lambda^0,$$

implies the relation $x^0 \in \Gamma\lambda^0$ for any accumulation point x^0 if such a point exists, upper semicontinuity in the sense of Berge in addition presupposes the existence of an accumulation point.

Lemma 2.2.3

(1) *Γ is u.s.c.-B at λ^0 if Γ is closed at λ^0 and X is compact;*
(2) *Γ is u.s.c.-B at λ^0 if Γ is u.s.c.-H at λ^0 and $\Gamma\lambda^0$ is compact;*
(3) *Γ is l.s.c.-H at λ^0 if Γ is l.s.c.-B at λ^0 and* cl *$\Gamma\lambda^0$ is compact.*

Proof. (1) If the mapping is closed at λ^0 the set $\Gamma\lambda^0$ must be closed from which by Lemma 2.2.2 the first part of the lemma follows.

(2) Apply Lemma 2.2.2 and consider sequences $\lambda^t \to \lambda^0$, $x^t \in \Gamma\lambda^t \setminus \Gamma\lambda^0$. As Γ is u.s.c.-H there exist $y^t \in \Gamma\lambda^0$ such that $d(x^t, y^t) \to 0$. The sequence $\{y^t\}$ possesses by hypothesis an accumulation point x^0 in $\Gamma\lambda^0$. Obviously x^0 is also an accumulation point of $\{x^t\}$.

(3) Assume $\varepsilon > 0$, since cl $\Gamma\lambda^0$ is compact one can then find a finite number of points x^k such that

$$x^k \in \Gamma\lambda^0 \quad \text{and} \quad \sup_{x \in \Gamma\lambda^0} \min_k d_X(x, x^k) < \tfrac{1}{2}\varepsilon$$

hold. If according to the above definition of lower semicontinuity (B) one chooses $\Omega_k = U_{\varepsilon/2}\{x^k\}$ then one finds that a $\delta_k > 0$ exists such that one has

$$\Gamma\lambda \cap \Omega_k \neq \emptyset \qquad \forall \lambda \in V_{\delta_k}\{\lambda^0\}\,.$$

If $\delta = \min\limits_k \delta_k$ and $\lambda \in V_\delta\{\lambda^0\}$ it then follows that

$$\forall x \in \Gamma\lambda^0 \ \exists x^k \in \Gamma\lambda^0\colon d_X(x, x^k) < \frac{\varepsilon}{2} \quad \text{and} \quad x^k \in U_{\varepsilon/2}\Gamma\lambda$$

and thus

$$\Gamma\lambda^0 \subset U_\varepsilon \Gamma\lambda \qquad \forall \lambda \in V_\delta\{\lambda^0\}\,. \ \square$$

Lemma 2.2.4. *Let I be an arbitrary index set and let $\Gamma_i\colon \Lambda \to 2^X$, $i \in I$, be mappings which are closed at λ^0. The mapping defined by*

$$(\cap\, \Gamma_i)\,\lambda = \bigcap_{i \in I} \Gamma_i\lambda$$

is then closed at λ^0.

Proof. The lemma is an immediate consequence of the definitions. □

Conclusion 2.2.4.1. *If Γ is closed at λ^0 and K ($\subset X$) is a compact set then the mapping*

$$\Gamma'\lambda =_{\mathrm{Df}} \Gamma\lambda \cap K\,, \qquad \lambda \in \Lambda\,,$$

is u.s.c.-B at λ^0.

Proof. Apply Lemmas 2.2.3 and 2.2.4. □

Lemma 2.2.5. *Let the mappings $\Gamma_1\colon \Lambda \to 2^X$ and $\Gamma_2\colon \Lambda \to 2^X$ be l.s.c.-B at λ^0 and strongly l.s.c. at λ^0 respectively. The mappings*

$$(\Gamma_1 \cap \Gamma_2)\,\lambda =_{\mathrm{Df}} \Gamma_1\lambda \cap \Gamma_2\lambda$$

and

$$\mathrm{cl}(\Gamma_1 \cap \Gamma_2)\,\lambda =_{\mathrm{Df}} \mathrm{cl}(\Gamma_1\lambda \cap \Gamma_2\lambda)$$

are then l.s.c.-B at λ^0.

Proof. We verify the proposition for $\Gamma = \mathrm{cl}(\Gamma_1 \cap \Gamma_2)$ (the case $\Gamma_1 \cap \Gamma_2$ is easier to prove).

Let Ω be an open set satisfying $\Omega \cap \Gamma\lambda^0 \neq \emptyset$ and let x be a point contained in this intersection. By definition of Γ there exists for each $\varepsilon > 0$ a point $y^\varepsilon \in \Gamma_1\lambda^0 \cap \Gamma_2\lambda^0$ with $d(x, y^\varepsilon) < \varepsilon$. For sufficiently small ε one has $y^\varepsilon \in \Omega$. As Γ_2 is strongly l.s.c. at λ^0 there exist an open neighbourhood Ω^ε of y^ε contained entirely in Ω and a $\delta_2 > 0$ such that $\Omega^\varepsilon \subset \Gamma_2\lambda\ \forall \lambda \in V_{\delta_2}\{\lambda^0\}$. Since Γ_1 is l.s.c.-B at λ^0 one has $\Gamma_1\lambda \cap \Omega^\varepsilon \neq \emptyset\ \forall\lambda \in V_{\delta_1}\{\lambda^0\}$ for some $\delta_1 > 0$. If we choose $\delta = \min\{\delta_1, \delta_2\}$ it follows that

$$\emptyset \neq \Gamma_1\lambda \cap \Gamma_2\lambda \cap \Omega^\varepsilon \subset (\Gamma_1 \cap \Gamma_2)\,\lambda \cap \Omega \qquad \forall\lambda \in V_\delta\{\lambda^0\}$$

holds. □

A simple consequence of the last lemma is given by

Corollary 2.2.5.1. *Let Γ_1, Γ_2, Γ_3 be mappings of Λ into 2^X with the properties*

(1) *Γ_1 is l.s.c.-B at λ^0,*
(2) *Γ_2 is strongly l.s.c. at λ^0,*
(3) *$\Gamma_2\lambda \subset \Gamma_3\lambda \qquad \forall\lambda \in \Lambda$.*

$\Gamma_1 \cap \Gamma_3$ is then l.s.c.-B at λ^0 if $\Gamma_1\lambda^0 \cap \Gamma_3\lambda^0 \subset \mathrm{cl}(\Gamma_1\lambda^0 \cap \Gamma_2\lambda^0)$ holds.

Proof. It is sufficient to note that $\mathrm{cl}(\lambda_1 \cap \Gamma_2)$ is l.s.c.-B at λ^0 and that we obviously have for all λ

$$\mathrm{cl}(\Gamma_1 \cap \Gamma_2)\,\lambda \subset \mathrm{cl}(\Gamma_1 \cap \Gamma_3)\,\lambda\,. \ \square$$

2.3. Continuous Selection Functions for Lower Semicontinuous Point-to-Set Mappings

The question of whether or not a *continuous selection function*, i.e. a continuous mapping

$$f\colon \Lambda \to X \quad \text{where} \quad f(\lambda) \in \Gamma\lambda \qquad \forall\lambda \in \Lambda$$

exists for a given mapping

$$\Gamma\colon \Lambda \to 2^X$$

is basically topological in nature, and has been examined in the very interesting and comprehensive studies of E. MICHAEL [1], [2], [3] which lead to a characterization of paracompact sets and lower semicontinuous (B) mappings.

In contrast to the above author we are here only concerned with sufficient conditions for the existence of continuous selection functions and assume that Λ is a compact metric space. We are thus able to slightly simplify the proofs of Theorems 2.3.2 and 2.3.1 in comparison with the corresponding theorems of E. MICHAEL which are formulated in stronger form. The three selection theorems we present here are of sufficient generality for a large number of applications since they may be extended in elementary fashion to domains which can be represented as the union of a countable family of compact sets (in a metric space).

Continuous selection functions for the optimal set mapping ψ appear to be of particular interest in relation to parametric optimization problems; they assign a unique solution $x(\lambda)$ of the problem (P_λ) to each parameter λ and as a consequence of their continuity they, in a certain sense, exhibit stable behaviour if the parameters are subjected to small perturbations.

The class of all problems for which such a function exists is however relatively small, and it is essentially determined by the conditions that ψ is l.s.c.-B and all optimal sets are non-empty and convex. If on the other hand one considers ε-optimal solutions (i.e. $x(\lambda)$ is feasible and attains the optimal value of the objective function except for some positive ε) then the strong condition of lower semicontinuity of ψ may to some extent be avoided and be replaced by the lower semicontinuity of the constraint set mapping.

Throughout this section we impose the following conditions:

(A1) Λ is a compact metric space,

(A2) X is a Banach space,

(A3) $\Gamma: \Lambda \to 2^X$ is l.s.c.-B, and

(A4) the image sets $\Gamma\lambda$ are non-empty and convex for all $\lambda \in \Lambda$.

One should note that

(A1) is of technical nature and may be weakened (see the remarks following Theorem 2.3.3),

(A2) cannot be weakened in any of the following theorems,

(A3) means in an equivalent form that for each pair $\lambda^0 \in \Lambda$, $x^0 \in \operatorname{cl} \Gamma\lambda^0$ there exists a function $f_{\lambda^0,x^0}: \Lambda \to X$ which is continuous at λ^0 and which satisfies

$$f_{\lambda^0,x^0}(\lambda) \in \Gamma\lambda \qquad \forall \lambda \in \Lambda\,, \qquad \lambda \neq \lambda^0$$

and $f_{\lambda^0,x^0}(\lambda^0) = x^0$; Γ must therefore be l.s.c.-B if for each such pair (λ^0, x^0) there exists a selection function which is continuous on Λ and which assigns the point x^0 to λ^0.

As far as

(A4) is concerned one may show that if convexity is dispensed with, continuous mappings Γ without continuous selection functions exist, as the following example demonstrates: Λ is the unit ball in $\mathbb{E}_2$ and

$$\Gamma\lambda =_{\mathrm{Df}} \{x \in \Lambda \| \; \|x - \lambda\| \geqq \tfrac{1}{2}\}\,.$$

Γ is thus continuous but a continuous selection function does not exist as this would possess a fixed point (by BROUWER's fixed point theorem).

Finally, we employ the following abbreviations and definitions:

$C(\Lambda, X)$ — Banach space of all *continuous mappings f of Λ into X* with $|f| = \max\limits_{\Lambda} \|f(\lambda)\|$,

$I(M)$ — *supporting interior* of a convex closed set M in a Banach space. A point $x \in M$ is contained in $I(M)$ if and only if it is not contained in any supporting set S of M. A non-empty, convex and closed subset S of M is called a supporting set if $S \neq M$ and if each open line segment in M whose intersection with S is non-empty lies entirely in S (V. KLEE's definition [1]).

Theorem 2.3.1.[1]) *Let* (A1) *to* (A4) *hold. There then exists a function $f \in C(\Lambda, X)$ such that*

$$f(\lambda) \in \operatorname{cl} \Gamma\lambda \qquad \forall \lambda \in \Lambda . \tag{2.3.1}$$

Proof. This proof is divided into two parts, the first of which demonstrates that for each $\varepsilon > 0$ there exists an $f^\varepsilon \in C(\Lambda, X)$ with $d(f^\varepsilon(\lambda), \Gamma\lambda) < \varepsilon$ for all $\lambda \in \Lambda$. In the second part a sequence of functions $\{f^k\}$ is constructed which converges in $C(\Lambda, X)$ to a function f satisfying (2.3.1).

Part I. For fixed $\varepsilon > 0$ and arbitrary $f \in C(\Lambda, X)$ we define

$$\Lambda(f) = \{\lambda \in \Lambda \mid d(f(\lambda), \Gamma\lambda) < \varepsilon\} .$$

Since Γ is l.s.c.-B these sets are open in Λ. Because of (A1) and $\Lambda = \bigcup\limits_f \Lambda(f)$ there exist a finite number of functions f_n in $C(\Lambda, X)$ such that

$$\Lambda = \bigcup_n \Lambda(f_n) . \tag{2.3.2}$$

We define

$$\alpha_n(\lambda) = \frac{d(\lambda, A_n)}{\sum\limits_t d(\lambda, A_t)}, \qquad \lambda \in \Lambda ,$$

where A_n are the closed complement sets $\Lambda \setminus \Lambda(f_n)$. The denominator of the fraction is by (2.3.2) non-zero and the functionals α_n are thus continuous, which also applies to the function f defined by

$$f(\lambda) = \sum_n \alpha_n(\lambda) f_n(\lambda) .$$

For each fixed λ we have from this definition

$$f(\lambda) \in \operatorname{conv} \{f_n(\lambda) \mid d(f_n(\lambda), \Gamma\lambda) < \varepsilon\} .$$

The desired inequality $d(f(\lambda), \Gamma\lambda) < \varepsilon$ follows from the convexity of $\Gamma\lambda$.

Part II. Now set $\varepsilon_k = 2^{-k}$, $k = 1, 2, \ldots$, and $\Gamma_0 = \Gamma$. Starting with $k = 1$ we construct a sequence $\{f^k\}$ in $C(\Lambda, X)$ and a sequence $\{\Gamma_k\}$ of mappings of Λ into 2^X according to the prescription:

$$f^k \in C(\Lambda, X) \quad \text{and} \quad d(f^k(\lambda), \Gamma_{k-1}\lambda) < \varepsilon_k \quad \forall \lambda , \tag{2.3.3}$$

$$\Gamma_k\lambda = \Gamma_{k-1}\lambda \cap U_{\varepsilon_k}\{f^k(\lambda)\} \quad \forall \lambda . \tag{2.3.4}$$

The existence of f^k follows from the first part of the proof if Γ_{k-1} satisfies (A3) and (A4). This is trivial for Γ_0 and follows for Γ_k from the continuity of f^k and (2.3.3) (cf.

[1]) This theorem, as well as the following, may be found in a stronger version in E. MICHAEL [1] where in addition Lemma 2.3.1 is also proved.

Lemma 2.2.5). One obtains from (2.3.3) and (2.3.4) for arbitrary natural numbers k and m:

$$|f^k - f^{k+m}| < \varepsilon_k + \dots + \varepsilon_{k+m} < \varepsilon_{k-1} \,.$$

The functions f^k thus constitute a Cauchy sequence, the limit

$$f = \lim_{k\to\infty} f^k$$

exists in the complete space $C(\Lambda, X)$ and (2.3.1) follows from $d\big(f^k(\lambda), \Gamma\lambda\big) \leqq \varepsilon_k$. □

One should note that the conditions (A1) to (A4) are not sufficient to guarantee the existence of a continuous function having the genuine selection property $f(\lambda) \in \Gamma\lambda \ \forall \lambda$. E. Michael illustrates this by an example in [1] where all the sets $\Gamma\lambda$ are open and X is a non-separable Banach space[1]). The possibility that suggests itself is to look for a genuine selection function — if such a function exists — amongst certain interior points of the convex set F of all $f \in C(\Lambda, X)$ satisfying (2.3.1). In fact however this set does not even necessarily possess an algebraically interior point in respect to its affine linear hull as is shown by the following example:

$$\Lambda = [0, 1] \,, \qquad X = \mathbb{R} \,,$$

$$\Gamma\lambda = \begin{cases} [0, \lambda] & \text{if} \quad \lambda \notin \left\{1, \dfrac{1}{2}, \dots, \dfrac{1}{k}, \dots\right\}, \\ \left\{\dfrac{1}{k}\right\} & \text{if} \quad \lambda = \dfrac{1}{k} \,. \end{cases}$$

In this example one may show that for each continuous selection function f for Γ there exists a second continuous selection g (dependent on f) such that the function $h_\varepsilon =_{\mathrm{Df}} f + \varepsilon(f - g)$ does not satisfy the condition (2.3.1) for any positive ε. We thus have $h_\varepsilon \notin F \ \ \forall \varepsilon > 0$. g may be constructed by choosing a point λ^k in the open interval $\big(1/(k+1), (1/k)\big)$ such that $f(\lambda^k) > 1/(k+1)$ and then defining g piece-wise by continuous functions g^k defined in the k-th interval and satisfying

$$0 \leqq g^k(\lambda) \leqq \lambda \,, \qquad g^k(\lambda^k) = 0 \,,$$

$$g^k\left(\frac{1}{k+1}\right) = \frac{1}{k+1} \,, \qquad g^k\left(\frac{1}{k}\right) = \frac{1}{k} \,.$$

Detailed investigations are therefore necessary in order to establish the existence of genuine continuous selection functions if certain sets $\Gamma\lambda$ are not closed. A useful instrument in this problem for the case that X is a separable Banach space is provided by the following lemma.

Lemma 2.3.1. *If $A = \{a_1, a_2, \dots, a_k, \dots\}$ is a countable dense subset of a convex bounded set M in a Banach space then the point*

$$x = \sum_{k=1}^{\infty} 2^{-k} a_k$$

is contained in the supporting interior $I(\mathrm{cl}\, M)$.

In addition $I(\mathrm{cl}\, M)$ is non-empty for every separable, non-empty and convex set M in a Banach space.

---

[1]) X consists of all real functions $x = x(t)$ defined on [0, 1] which are not equal to zero in at most a countable subset of the domain and for which $||x|| = \sum_t |x(t)| < \infty$. Further, $\Lambda = [0, 1]$ and

$$\Gamma\lambda = \{x \mid x(\lambda) > 0\} \,.$$

Proof. The second part of the lemma follows from the first, since for each open ball B, $B \cap M \neq \emptyset$, one may easily demonstrate that $I(\mathrm{cl}(B \cap M)) \subset I(\mathrm{cl}\, M)$.

Writing x in the form

$$x = 2^{-k} a_k + (1 - 2^{-k})\, b_k$$

where

$$b_k = \sum_{\substack{m=1\\ m \neq k}}^{\infty} \frac{2^{-m}}{1 - 2^{-k}} a_m$$

we find that b_k is contained in cl M as it is an infinite convex combination of the points a_m, $m \neq k$. Thus if S is the supporting set of cl M then $a_k \in S$ for all k follows from $x \in S$. The set cl M = cl A would then also be a subset of S which contradicts $S \neq$ cl M. □

Theorem 2.3.2. *If X is a separable Banach space and if* (A1), (A3), *and* (A4) *are fulfilled then a function $f \in C(\Lambda, X)$ exists with the property*

$$f(\lambda) \in I(\mathrm{cl}\, \Gamma\lambda) \qquad \forall \lambda \in \Lambda\,. \tag{2.3.5}$$

Proof. Let f_0 be a function whose existence is established by Theorem 2.3.1 and which has images in cl $\Gamma\lambda$ and assume $\varepsilon_0 > 0$. The mapping

$$\Gamma_0\lambda =_{\mathrm{Df}} \Gamma\lambda \cap U_{\varepsilon_0}\{f_0(\lambda)\}$$

then also satisfies (A3) and (A4) and because of the continuity of f_0 on the compact set Λ there exists a ball B in X containing all the sets $\Gamma_0\lambda$, $\lambda \in \Lambda$. Finally, we also have $I(\mathrm{cl}\, \Gamma_0\lambda) \subset I(\mathrm{cl}\, \Gamma\lambda)$ so that it suffices to find a function $f \in C(\Lambda, X)$ that satisfies (2.3.5) where Γ_0 is substituted for Γ.

We will now construct a countable set of continuous selection functions f_k, $k = 1, 2, \dots$, for cl Γ_0 such that each set

$$A(\lambda) =_{\mathrm{Df}} \{f_k(\lambda) \mid k = 1, 2, \dots\}$$

is dense in cl $\Gamma_0\lambda$. By Lemma 2.3.1

$$f = \sum_{k=1}^{\infty} 2^{-k} f_k$$

is then the desired function.

If the countable set $X_0 = \{x_1, x_2, \dots, x_m, \dots\}$ is dense in X then the sets defined by

$$\Lambda_{m,l} = \{\lambda \in \Lambda \mid \mathrm{cl}\, \Gamma_0\lambda \cap U_{1/l}\{x_m\} \neq \emptyset\}$$

are open since Γ_0 is l.s.c.-B (l is a natural number). Each of these sets may be represented as the union of a countable family of closed sets:

$$\Lambda_{m,l} = \bigcup_{n=1}^{\infty} C_{m,l,n}\,.$$

We now introduce the mappings $\Gamma_{m,l,n}$:

$$\Gamma_{m,l,n}\lambda = \begin{cases} \mathrm{cl}\, \Gamma_0\lambda & \text{if } \lambda \in \Lambda \setminus C_{m,l,n}\,, \\ \mathrm{cl}\, \Gamma_0\lambda \cap U_{1/l}\{x_m\} & \text{if } \lambda \in C_{m,l,n}\,. \end{cases}$$

Since $C_{m,l,n}$ is closed, the restriction $\Gamma_{m,l,n}|_{C_{m,l,n}}$ is l.s.c.-B and the image sets $\Gamma_{m,l,n}\lambda$ are contained in cl $\Gamma_0\lambda$ for $\lambda \in C_{m,l,n}$, we find that $\Gamma_{m,l,n}$ is also a mapping satisfying (A3) and A(4). By Theorem 2.3.1 there then exist functions $f_{m,l,n} \in C(\Lambda, X)$ satisfying $f_{m,l,n}(\lambda) \in \mathrm{cl}\, \Gamma_{m,l,n}\lambda$ for all $\lambda \in \Lambda$.

If we now keep $\lambda \in \Lambda$ and $x \in \mathrm{cl}\,\Gamma_0\lambda$ fixed there exist for each $\varepsilon > 0$ an l such th a $1/l < \varepsilon$ and an m such that $x \in U_{1/l}\{x_m\}$. Hence $\lambda \in \Lambda_{m,l}$ and there further exists an n such that $\lambda \in C_{m,l,n}$. The choice of $f_{m,l,n}$ thus guarantees that

$$||x - f_{m,l,n}(\lambda)|| \leqq \frac{2}{l} < 2\varepsilon$$

holds. It follows that the countable set

$$\{f_{m,l,n}(\lambda) \mid m, l, n \in \{1, 2, \ldots\}\}$$

is in fact dense in $\mathrm{cl}\,\Gamma_0\lambda$; it is also clear that a countable set of selection functions f_k having the above property exists, thus completing the proof. □

From the previous theorem we find

Corollary 2.3.2.1. *Under the conditions of Theorem* 2.3.2 *there exists a function* $f \in C(\Lambda, X)$ *such that*

$$f(\lambda) \in \Gamma\lambda \qquad \forall \lambda \in \Lambda$$

if the following is true for each image set $\Gamma\lambda$: $\Gamma\lambda$ *is closed or contains an interior point or is finite dimensional.*

Proof. It is easily seen that in each of the three cases $I(\mathrm{cl}\,\Gamma\lambda) \subset \Gamma\lambda$ holds, since $I(\mathrm{cl}\,\Gamma\lambda) = \mathrm{int}\,\Gamma\lambda \subset \Gamma\lambda$ if $\mathrm{int}\,\Gamma\lambda \neq \emptyset$ and $I(\mathrm{cl}\,\Gamma\lambda) = \mathrm{ri}\,\Gamma\lambda \subset \Gamma\lambda$ if $\dim \Gamma\lambda < \infty$. □

The following stronger version of Theorem 2.3.2 in addition uses

Lemma 2.3.2.[1] *If the conditions* (A1), (A2), *and* (A3) *are satisfied, then the set*

$$\Lambda_0 =_{\mathrm{Df}} \{\lambda \in \Lambda \mid \Gamma\lambda \textit{ is separable}\}$$

is closed and the set

$$X_0 =_{\mathrm{Df}} \bigcup_{\lambda \in \Lambda_0} \mathrm{cl}\,\Gamma\lambda$$

is separable in X.

Proof. (Λ_0 is closed.) Assume $\lambda^t \in \Lambda_0$, $t = 1, 2, \ldots$, and $\lambda^0 = \lim \lambda^t$. That λ^0 is an element of Λ_0 may then be demonstrated as follows: Let A_t be countable dense subsets of $\Gamma\lambda^t$, $t \geqq 1$, $\varepsilon > 0$ and let $x \in \Gamma\lambda^0$. As Γ is l.s.c.-B we have for large t

$$\Gamma\lambda^t \cap U_\varepsilon\{x\} \neq \emptyset$$

and thus $A_t \cap U_\varepsilon\{x\} \neq \emptyset$.

Defining the countable set $A = \bigcup A_t$ we obtain $x \in U_\varepsilon A$, i.e. there exists in X a countable set A such that $\Gamma\lambda^0 \subset \mathrm{cl}\,A$. As is well-known there then also exists a countable dense subset of $\Gamma\lambda^0$.

(X_0 is separable in X.) Let $A(\lambda)$ be a countable dense subset of $\mathrm{cl}\,\Gamma\lambda$, $\lambda \in \Lambda^0$, and let Λ_{00} be countable and dense in Λ_0 (Λ_0 is compact and thus separable). $B =_{\mathrm{Df}} \bigcup_{\lambda \in \Lambda_{00}} A(\lambda)$ is then countable. For arbitrary $\lambda \in \Lambda_0$, $x \in \mathrm{cl}\,\Gamma\lambda$, $\varepsilon > 0$ it then follows from (A3) that a neighbourhood $V(\lambda)$ of λ exists in Λ such that

$$\Gamma\mu \cup U_\varepsilon\{x\} \neq \emptyset \qquad \forall \mu \in V(\lambda)\,.$$

One notes that $V(\lambda)$ contains a point μ of Λ_{00}. Since $A(\mu)$ is dense in $\Gamma\mu$ we then have

$$A(\mu) \cap U_\varepsilon\{x\} \neq \emptyset$$

[1] The Lemma still holds and the proof is similar if the following conditions are imposed: Λ is a separable metric space and X is a metric space.

and thus

$$x \in U_\varepsilon B .$$

Because x was chosen arbitrarily from X_0 we obtain the assertion $X_0 = \mathrm{cl}\, B$. □

Theorem 2.3.3. *Under the conditions* (A1) *to* (A4) *there exists a function* $f \in C(\Lambda, X)$ *with the properties*

$$f(\lambda) \in \mathrm{cl}\, \Gamma\lambda \qquad \forall \lambda \in \Lambda , \tag{2.3.1}$$

$$f(\lambda) \in I(\mathrm{cl}\, \Gamma\lambda) \qquad \forall \lambda \in \Lambda^0 \tag{2.3.2}$$

where Λ_0 *is the set of all* $\lambda \in \Lambda$ *with separable* $\Gamma\lambda$.

Proof. Since the set $X_0 = \bigcup_{\lambda \in \Lambda_0} \mathrm{cl}\, \Gamma\lambda$ is separable there exists a separable closed subspace V of X containing X_0. Because Λ_0 is closed and as a subset of Λ also compact, Theorem 2.3.2 may be applied to the mapping Γ restricted to Λ_0: there thus exists a function $f_0 \in C(\Lambda_0, X)$ such that $f_0(\lambda) \in I(\mathrm{cl}\, \Gamma\lambda)$ $\forall \lambda \in \Lambda_0$. Because of $f_0(\lambda) \in \mathrm{cl}\, \Gamma\lambda$ and the continuity of f_0 the mapping defined by

$$\Gamma_0\lambda = \begin{cases} \{f_0(\lambda)\} & \text{if} \quad \lambda \in \Lambda_0 , \\ \Gamma\lambda & \text{if} \quad \lambda \in \Lambda \setminus \Lambda_0 \end{cases}$$

is also l.s.c.-B.

Theorem 2.3.1 therefore implies the existence of a function $f \in C(\Lambda, X)$ which satisfies (2.3.1) with respect to Γ_0 and which thus obviously exhibits the above properties. □

The compactness conditions imposed on the inverse image spaces in the theorems of this section may be weakened. It is sufficient to simply assume that Λ is a subset of a metric space and may be represented as the union

$$\Lambda = \bigcup_{k=1}^{\infty} \Lambda_k , \quad k \text{ is a natural number} ,$$

of compact sets Λ_k. The sequence of sets $\{\Lambda_k\}$ may be taken to be non-decreasing (otherwise consider sets $\Lambda_k' = \bigcup_{n \leq k} \Lambda_k$ with this property). If f_k is a function fulfilling any one of the above theorems with respect to the mapping Γ restricted to Λ_k then one may define

$$\Gamma_k\lambda = \begin{cases} \{f_k(\lambda)\} & \text{if} \quad \lambda \in \Lambda_k , \\ \Gamma\lambda & \text{if} \quad \lambda \in \Lambda_{k+1} \setminus \Lambda_k , \end{cases}$$

apply the theorem to Γ_k, and obtain a function f_{k+1} having the corresponding selection property with respect to $\Gamma|_{\Lambda_{k+1}}$. In addition f_{k+1} and f_k are identical on Λ_k, thus the sequence $\{f_k\}$ may be used to define a function

$$f(\lambda) = f_k(\lambda) \quad \text{if} \quad \lambda \in \Lambda_k$$

which then exhibits the corresponding property.

If in particular Λ is a subset of E_n then there exists a representation as the union of a countable family of compact sets if there is simply a countable family of open and closed sets from which Λ may be obtained by a countable sequence of union and intersection operations.

3. Point-to-Set Mappings with Given Structure

3.1. Continuity of Parametric Inequalities with Compactness or Constraint Qualifications

Various questions arising in the theory and application of parametric optimization motivate a study of the continuity of solution sets of systems of parametric inequalities. The present section is devoted to such point-to-set mappings with (partially) given structure.

Let us consider a point-to-set mapping $M \cap \Gamma \colon \Lambda \to 2^X$ as $(M \cap \Gamma)\,\lambda =_{\mathrm{Df}} M(\lambda) \cap \Gamma\lambda$ and which satisfies the following general conditions:

Λ and X are metric spaces; (3.1.1)

$$M(\lambda) =_{\mathrm{Df}} \{x \in X \mid g_i(x, \lambda) \leqq 0 \ \forall i \in J\}\,, \qquad \lambda \in \Lambda\,, \tag{3.1.2}$$

where g_i is, for each $i \in J$, a real-valued function defined on $X \times \Lambda$;

J is an index set and may be infinite if not otherwise specified; (3.1.3)

Γ is a given point-to-set mapping of Λ into 2^X. (3.1.4)

While one may show that $M \cap \Gamma$ is a closed mapping under relatively weak conditions, those needed to establish the upper or lower semicontinuity of $M \cap \Gamma$ are more restrictive. In order to demonstrate that $M \cap \Gamma$ is u.s.c.-B (or u.s.c.-H) at a point λ^0 one must in general assume that a compact set $K \subset X$ exists such that for all parameters in a certain neighbourhood $V\{\lambda^0\}$ the sets $M(\lambda)$ are contained in K. Hypotheses of essentially different nature are necessary if one wishes to prove that $M \cap \Gamma$ is l.s.c.-B at $\lambda^0 \in \Lambda$; one finds that the structure assigned to the mappings, in particular the description of the sets $M(\lambda)$, is of decisive importance in this respect since constraint qualification of type similar to the Slater condition must be imposed on the set $M(\lambda^0)$.

In formulating the theorems of this section the authors have attempted to use continuity conditions which are as weak as possible, and furthermore only to consider point-to-set mappings of the form $M \cap \Gamma$, i.e., the solution sets of the system of parametric inequalities (3.1.2) are restricted to the sets $\Gamma\lambda$. This particular methodological approach enables one to find a broad field of applications for the results obtained, thus the continuity properties of the optimal set mappings and of the solution function of a parametric optimization problem which are studied in Chapter 4 follow directly from the theorems of this section.

The following is based on work by C. Berge [2], G. B. Dantzig, J. Folkman, and N. Shapiro [1], J. P. Evans and F. J. Gould [1], H. J. Greenberg and W. P. Pierskalla [2], R. R. Meyer [1], W. W. Hogan [3], and others.

Theorem 3.1.1. *If the functions g_i, $i \in J$, are l.s.c. on $X \times \{\lambda^0\}$ where $\lambda^0 \in \Lambda$ then $M \cap \Gamma$ is closed at λ^0 provided Γ is closed at λ^0.*

Proof. By Lemma 2.2.4 is suffices to show that M is closed at λ^0. If $\{\lambda^k\} \subset \Lambda$ and $\{x^k\} \subset X$ are arbitrary sequences with $\lambda^k \to \lambda^0$, $x^k \to x^0$ and $g_i(x^k, \lambda^k) \leqq 0$ it follows that

$$g_i(x^0, \lambda^0) \leqq \underline{\lim}\, g_i(x^k, \lambda^k) \leqq 0$$

since the g_i are by hypothesis l.s.c. on $X \times \{\lambda^0\}$. Thus $x^0 \in M(\lambda^0)$, q.e.d. □

Theorem 3.1.2. *Let g_i, $i \in J$, be l.s.c. functions on $X \times \{\lambda^0\}$.*
Then
(1) *$M \cap \Gamma$ is u.s.c.-B (and thus u.s.c.-H) at λ^0 if both Γ is closed at λ^0 and there exists a compact set $K \subset X$ and a $\delta > 0$ such that*

$$(M \cap \Gamma)\,\lambda \subset K \qquad \forall \lambda \in V_\delta\{\lambda^0\};$$

(2) *$M \cap \Gamma$ is u.s.c.-B at λ^0 if both Γ is u.s.c.-H at λ^0 and the set $\Gamma\lambda^0$ is compact and nonempty.*

Proof. The first proposition follows immediately from Theorem 3.1.1 and Lemma 2.2.3. To prove (2) assume that $M \cap \Gamma$ is not u.s.c.-B at λ^0. There then exists an open set Ω containing $(M \cap \Gamma)\,\lambda^0$ and a sequence $\{\lambda^t\} \subset \Lambda$, $\lambda^t \to \lambda^0$ with

$$(M \cap \Gamma)\,\lambda^t \setminus \Omega \neq \emptyset \quad \forall t\,.$$

Let x^t be, for all t, elements of the above difference set. As Γ is u.s.c.-H at λ^0 there exist $y^t \in \Gamma\lambda^0$ with $d(x^t, y^t) \to 0$, $t \to \infty$. Furthermore, there exists in the compact set $\Gamma\lambda^0$ an accumulation point x^0 of all y^t which at the same time must be an accumulation point of the sequence $\{x^t\}$. As M is closed at λ^0 (Theorem 3.1.1) x^0 is contained in $M(\lambda^0) \cap \Gamma\lambda^0$ and thus also in Ω. It follows that the open set Ω contains an infinite number of the points x^t, which contradicts $x^t \notin \Omega$. □

The significance of the compactness condition hypothesized in proposition (1) of the previous theorem is shown by Example 3.1.3, which illustrates the fact that it does not suffice to merely assume that the set $(M \cap \Gamma)\,\lambda^0$ is compact.

Lemma 3.1.1. *Let $X = \mathbb{E}_n$, $\lambda^0 \in \Lambda$, and $x_0 \in M(\lambda^0) \cap S$ where $S \subset \mathbb{E}_n$ is a convex compact set. Further, let g_i for each $i \in J$ have the form*

$$g_i(x, \lambda) = f_i(x, \lambda) - \tau_i(\lambda)$$

where, for all $i \in J$, the functions f_i and τ_i have the following properties:

$$\begin{aligned}
&f_i\colon \mathbb{E}_n \times \Lambda \to \mathbb{R} \quad \text{is l.s.c. on } \mathbb{E}_n \times \{\lambda^0\}\,,\\
&\tau_i\colon \Lambda \to \mathbb{R} \quad \text{is u.s.c. at } \lambda^0\,,\\
&f_i(x^0, \cdot)\colon \Lambda \to \mathbb{R} \quad \text{is u.s.c. at } \lambda^0 \text{ and}\\
&f_i(\cdot, \lambda)\colon \mathbb{E}_n \to \mathbb{R} \quad \text{is quasiconvex on } \mathbb{E}_n \qquad \forall \lambda \in \Lambda\,.
\end{aligned}$$

Finally, let an $\varepsilon > 0$, a sequence $\{\lambda^k\} \subset \Lambda$ with $\lambda^k \to \lambda^0$, and a sequence $\{x^k\} \subset \mathbb{E}_n$ with $x^k \in M(\lambda^k)$ be given such that

$$x^k \notin \operatorname{cl} U_\varepsilon S \quad \forall k$$

holds. For each k the intersection of the line segment conv $\{x^0, x^k\}$ *with the boundary of the set $U_\varepsilon S$ then contains only one point, and the sequence $\{z^k\}$ of these points has an* *accumulation point z^* with the property*

$$z^* \in M(\lambda^0) \cap \text{bd}\, U_\varepsilon S .$$

Proof. Because of the assumption $x^k \notin \text{cl}\, U_\varepsilon S$, the convexity of $U_\varepsilon S$ and $U_\varepsilon\{x^0\} \subset U_\varepsilon S$ the intersection of each line segment conv $\{x^0, x^k\}$ with the boundary of the set $U_\varepsilon S$ contains only one element and there exist numbers α_k, $0 < \alpha_k < 1$, such that

$$z^k = \alpha_k x^k + (1 - \alpha_k)\, x^0 \in \text{bd}\, U_\varepsilon S \quad \forall k \tag{3.1.5}$$

holds. Since bd $U_\varepsilon S$ is compact there exists a z^* such that without loss of generality *

$$z^k \to z^* , \qquad k \to \infty , \quad \text{and} \quad z^* \in \text{bd}\, U_\varepsilon S . \tag{3.1.6}$$

Let $i \in J$ be arbitrary. The functions $f_i(\cdot, \lambda^k)$ are quasiconvex on $\mathbb{E}_n$ and it thus follows from (3.1.5) and the condition $x^k \in M(\lambda^k)$ that

$$f_i(z^k, \lambda^k) \leqq \max \{f_i(x^k, \lambda^k), f_i(x^0, \lambda^k)\} \leqq \max \{\tau_i(\lambda^k), f_i(x^0, \lambda^k)\} .$$

The continuity conditions for f_i and τ_i and (3.1.6) then give

$$\begin{aligned} f_i(z^*, \lambda^0) &\leqq \underline{\lim}\, f_i(z^k, \lambda^k) \leqq \max \{\overline{\lim}\, \tau_i(\lambda^k), \overline{\lim}\, f_i(x^0, \lambda^k)\} \\ &\leqq \max \{\tau_i(\lambda^0), f_i(x^0, \lambda^0)\} \leqq \tau_i(\lambda^0) . \end{aligned} \tag{3.1.7}$$

As (3.1.7) is valid for all $i \in J$ and taking (3.1.6) into account $z^* \in M(\lambda^0) \cap \text{bd}\, U_\varepsilon S$ follows. □

In the next theorem we consider a point-to-set mapping $M \cap \Gamma$ which fulfills the conditions (3.1.1) to (3.1.4), in the case of a system of convex parametric inequalities in $\mathbb{E}_n$.

Theorem 3.1.3. *If $\lambda^0 \in \Lambda$ and $X = \mathbb{E}_n$ the mapping $M \cap \Gamma$ is u.s.c.-B at λ^0 under the following additional hypotheses:*

(1) *Γ is closed at λ^0 and l.s.c.-B at λ^0, the sets $\Gamma\lambda$ are convex for all $\lambda \in \Lambda$,*

(2) *$g_i(x, \lambda) = f_i(x, \lambda) - \tau_i(\lambda)$ for all $i \in J$ and all $\lambda \in \Lambda$ where the functions $f_i: \mathbb{E}_n \times \Lambda \to \mathbb{R}$ are l.s.c. on $\mathbb{E}_n \times \{\lambda^0\}$ and $\tau_i: \Lambda \to \mathbb{R}$ are u.s.c. at λ^0 for all $i \in J$,*

(3) *there exists a point $x^0 \in (M \cap \Gamma)\, \lambda^0$ such that $f_i(x^0, \cdot)$ is u.s.c. at λ^0 for each $i \in J$,*

(4) *the functions $f_i(\cdot, \lambda)$ are quasiconvex for all $i \in J$ and all $\lambda \in \Lambda$, and*

(5) *$(M \cap \Gamma)\, \lambda^0$ is a bounded set.*

Proof. We shall demonstrate that a real number $\delta > 0$ exists such that

$$(M \cap \Gamma)\, \lambda \subset K =_{\text{Df}} \text{cl}\, U_1((M \cap \Gamma)\, \lambda^0) \qquad \forall \lambda \in V_\delta\{\lambda^0\} \tag{3.1.8}$$

holds; it then follows from Theorem 3.1.2 that $M \cap \Gamma$ is u.s.c.-B at λ^0.

Assume that (3.1.8) is not valid. There then exist, for each $k = 1, 2, \ldots$ a $\lambda^k \in V_{1/k}\{\lambda^0\}$ and an $x^k \in \mathbb{E}_n$ with the properties

$$x^k \in (M \cap \Gamma)\, \lambda^k \quad \text{and} \quad x^k \notin K \quad \forall k .$$

If we now set $S = (M \cap \Gamma)\, \lambda^0$ the conditions of Lemma 3.1.1. are fulfilled for λ^0, x^0, $\{\lambda^k\}$ and $\{x^k\}$ because of hypotheses (2), (3), (4), and (5). There thus exist a $z^* \in \mathbb{E}_n$ and numbers $\alpha^k \in (0, 1)$ such that

$$z^* \in M(\lambda^0) \cap \text{bd}\, K , \tag{3.1.9}$$

$$z^k = \alpha_k x^k + (1 - \alpha_k)\, x^0 \in \text{bd}\, K , \tag{3.1.10}$$

and without loss of generality *

$$z^k \to z^* , \qquad k \to \infty , \tag{3.1.11}$$

hold. Since Γ is l.s.c.-B at λ^0 there exists a sequence $\{y^k\} \subset E_n$ with $y^k \to x^0$ and $y^k \in \Gamma\lambda^k$ for all k. Using the coefficients α_k $\forall k$ of (3.1.10) we define

$$\bar{y}^k = \alpha_k x^k + (1 - \alpha_k)\, y^k\,.$$

From the convexity of $\Gamma\lambda^k$ it follows that $\bar{y}^k \in \Gamma\lambda^k$ and by (3.1.11) and $\bar{y}^k = z^k + (1 - \alpha^k)(y^k - x^0)$ finally $\bar{y}^k \to z^*$. As Γ is closed at λ^0 one finds

$$z^* \in \Gamma\lambda^0\,. \tag{3.1.12}$$

(3.1.9.) and (3.1.12) then imply $z^* \in (M \cap \Gamma)\,\lambda^0 \cap \operatorname{bd} K$. By definition of K however every point of the set $(M \cap \Gamma)\,\lambda^0$ must be an interior point of K and we have thus arrived at a contradiction: the above assumption is false, (3.1.8) is valid, and the proof is completed. □

The convexity condition imposed on the sets $\Gamma\lambda$ $\forall\lambda \in \Lambda$ in Theorem 3.1.3 may be weakened if upper semicontinuity (H) at λ^0 is substituted for the closedness of Γ at λ^0.

Corollary 3.1.3.1. *The mapping $M \cap \Gamma$ is u.s.c.-B at λ^0 under the conditions* (2), (3), (4), *and* (5) *of Theorem* 3.1.3 *if $X = E_n$, λ^0 is a point such that $\Gamma\lambda^0$ is a closed and convex subset of E_n and Γ is u.s.c.-H and l.s.c.-B at λ^0.*

Proof. We define $\Gamma'\lambda = \operatorname{conv} \Gamma\lambda$ $\forall\lambda \in \Lambda$. One easily confirms that Γ' is u.s.c.-H at λ^0. By Lemma 2.2.1 Γ' is closed at λ^0 since the set $\Gamma'\lambda^0 = \Gamma\lambda^0$ is closed. It follows that $M \cap \Gamma'$ satisfies the conditions (1) to (5) if $(M \cap \Gamma')\,\lambda^0 = (M \cap \Gamma)\,\lambda^0$ is taken into account. $M \cap \Gamma'$ is thus u.s.c.-B at λ^0. Since we have $(M \cap \Gamma)\,\lambda \subset (M \cap \Gamma')\,\lambda$ for all $\lambda \in \Lambda$ it then follows that $M \cap \Gamma$ is also u.s.c.-B at λ^0, q.e.d. □

An immediate consequence of Theorems 3.1.3, 3.1.2, and Lemma 2.2.3 is given by

Corollary 3.1.3.2. *If a point-to-set mapping $\Gamma\colon \Lambda \to 2^{E_n}$ is closed and l.s.c.-B at $\lambda^0 \in \Lambda$ and if the sets $\Gamma\lambda$ are convex for all $\lambda \in \Lambda$ then Γ is H-continuous at λ^0 provided that the set $\Gamma\lambda^0$ is bounded and non-empty.*

The following examples show that each of the conditions of Theorem 3.1.3 is essential, i.e. if any one of them is not satisfied the theorem is in general no longer valid independently of whether the remaining conditions are satisfied or not.

Example 3.1.1. Let B be the space of all bounded infinite sequences $x = (x_1, x_2, \ldots)$ of real numbers with the norm $\|x\| = \sup\{|x_k| \mid k = 1, 2, \ldots\}$ and let W be the union of the sequence $\{1/k\}_{k=1,2,\ldots}$ with the number 0. Consider the sets $X = B$, $\Lambda = W$, $\Gamma 0 = \{x \in B \mid x_k \leqq 1\ \forall k\}$, $\Gamma(1/k) = \operatorname{cl}\operatorname{conv}(\Gamma 0 \cup \{z^k\})$ where $z^k = (z_1^k, z_2^k, \ldots)$ is defined by

$$z_i^k = \begin{cases} 1 & \text{if } i \neq k\,, \quad i \leqq 2k\,, \\ 1 + 2^{-i} & \text{if } i = k\,, \\ 0 & \text{otherwise}\,. \end{cases}$$

Assume further $g(x, \lambda) \equiv g(x) =_{\mathrm{Df}} -\sum_{k=1}^{\infty} 2^{-k} x_k$ and

$$M(\lambda) \equiv M =_{\mathrm{Df}} \{x \in B \mid g(x) \leqq -1\}\,.$$

We then have $M \cap \Gamma 0 = \{e\}$ where $e_i = 1$, $i = 1, 2, \ldots$, and $z^k \in M \cap \Gamma(1/k)$ satisfies

$$\|z^k - e\| = 1 \quad \forall k\,.$$

The conditions (1) to (5) of Theorem 3.1.3 are satisfied for $\lambda^0 = 0$; $M \cap \Gamma$ is however not u.s.c.-B at λ^0. The reason is that $X = B$ is not a finite dimensional space as was assumed in the theorem.

Example 3.1.2. Assume $X = \Lambda = \mathbb{R}$,

$$\Gamma\lambda \equiv G =_{\mathrm{Df}} \{x \in \mathbb{R} \mid -1 \leqq x \leqq +1\} \qquad \forall \lambda \in \lambda ,$$

$$g(x, \lambda) = x + \tau(\lambda) \quad \text{where} \quad \tau(\lambda) = \begin{cases} 0 & \text{if } \lambda = 0 , \\ -1 & \text{if } \lambda \neq 0 , \end{cases}$$

and $M(\lambda) = \{x \in \mathbb{R} \mid g(x, \lambda) \leqq 0\}$. We obviously then have $(M \cap \Gamma)\, 0 = [-1, 0]$ and $(M \cap \Gamma)\, \lambda = G\ \forall \lambda \neq 0$, i.e. $M \cap \Gamma$ is not closed and not u.s.c.-B at $\lambda^0 = 0$. In this example g is not l.s.c. on $\mathbb{R} \times \{0\}$.

Example 3.1.3. Assume $X = \Lambda = \mathbb{R}$, $\lambda^0 = 0$, $M(\lambda) = \{x \in \mathbb{R} \mid g(x, \lambda) \leqq 0\}$, and

$$\varrho(\lambda) = \begin{cases} 0 & \text{if } \lambda = 0 , \\ 1/\lambda & \text{if } \lambda \neq 0 ; \end{cases}$$

the following two special cases a) and b) lead to $(M \cap \Gamma)\, 0 = \{0\}$ and $(M \cap \Gamma)\, \lambda = \{1/\lambda\}$ $(\lambda \neq 0)$ and thus to the conclusion that $M \cap \Gamma$ is not u.s.c.-B at λ^0.

a) If $\Gamma\lambda = \{\varrho(\lambda)\}\ \forall \lambda \in \mathbb{R}$ then Γ is not l.s.c.-B at λ^0. Assuming $g(x, \lambda) \equiv 0$ we find that $M \cap \Gamma \equiv \Gamma$ is not l.s.c.-B at λ^0.

b) If $\Gamma\lambda \equiv \mathbb{R}$ and $g(x, \lambda) = \big(x - \varrho(\lambda)\big)^2\ \forall x\ \forall \lambda$ all conditions of Theorem 3.1.3 are fulfilled with the exception of (3).

Similar simple examples may be constructed to show that Theorem 3.1.3 is in general not valid if the compactness condition (5) or the convexity conditions of (1) and (4) are omitted. In the next Theorem X is again assumed to be an arbitrary metric space.

Theorem 3.1.4. *Let $J = \{1, \dots, m\}$ and $\lambda^0 \in \Lambda$. There further exist for each $i \in J$ a function $\delta_i \colon \Lambda \to \mathbb{R}$ which is continuous at λ^0, such that*

$$g_i(x, \lambda^0) - g_i(x, \lambda) \leqq \delta_i(\lambda) \quad \forall x \in X \quad \text{and} \quad \delta_i(\lambda^0) = 0$$

hold. M is then u.s.c.-H at λ^0 if the mapping

$$\Phi(y) =_{\mathrm{Df}} \{x \in X \mid g_i(x, \lambda^0) \leqq y_i\ \forall i \in J\} , \qquad y \in \mathbb{E}_m ,$$

is u.s.c.-H at $y = 0$. Furthermore, $M(\lambda) \subset \Phi\big(\delta(\lambda)\big)$ where $\delta(\lambda) = \big(\delta_1(\lambda), \dots, \delta_m(\lambda)\big)\ \forall \lambda \in \Lambda$ and $M(\lambda^0) = \Phi(0)$.

Proof. The hypotheses of the theorem immediately give $M(\lambda) \subset \Phi\big(\delta(\lambda)\big)$ and $M(\lambda^0) = \Phi(0)$. Because of the upper semicontinuity (H) of Φ and the continuity of δ_i we then find

$$M(\lambda^k) \subset \Phi\big(\delta(\lambda^k)\big) \subset U_\varepsilon \Phi\big(\delta(\lambda^0)\big) = U_\varepsilon \Phi(0) = U_\varepsilon M(\lambda^0)$$

for each $\varepsilon > 0$ and each sequence $\{\lambda^k\} \subset \Lambda$ with $\lambda^k \to \lambda^0$ and for sufficiently large k, q.e.d. □

Note. The theorem is similarly valid if "u.s.c.-H" is replaced by "u.s.c.-B".

Section 3.4 contains conditions for the continuity of point-to-set mappings defined by inequalities which have parameter-dependent right-hand sides. Using the condition established in Theorem 3.1.4 one may extend these results to a broader class of systems of parametric inequalities.

The fact that the intersection of a mapping which is l.s.c.-B at λ^0, with a mapping that is strongly l.s.c. is itself then l.s.c.-B at λ^0 (Lemma 2.2.5), is of not inconsiderable importance in ascertaining under what conditions a mapping $M \cap \Gamma$, which fulfills the conditions (3.1.1) to (3.1.4), is l.s.c.-B at a certain parameter point λ^0. We denote the mapping which assigns to each $\lambda \in \Lambda$ the set of Slater points of $M(\lambda)$ by M^0:

$$M^0(\lambda) =_{\mathrm{Df}} \{x \in X \mid g_i(x, \lambda) < 0 \ \forall i \in J\}, \qquad \lambda \in \Lambda. \tag{3.1.13}$$

This mapping is of particular importance in determining whether $M \cap \Gamma$ is l.s.c.-B.

Theorem 3.1.5. *Let M and Γ be mappings which satisfy* (3.1.1) *to* (3.1.4). *In addition let J be a finite index set, Γ be l.s.c.-B at λ^0, and let*

(1) *the functions g_i, $i \in J$, be u.s.c. on $M^0(\lambda^0) \times \{\lambda^0\}$ and*
(2) $(M \cap \Gamma)\,\lambda^0 \subset \mathrm{cl}(M^0 \cap \Gamma)\,\lambda^0$.

The mapping $M \cap \Gamma$ is then l.s.c.-B at λ^0.

Proof. The mapping M^0 is strongly l.s.c. since if $x^0 \in M^0(\lambda^0)$ the existence of positive ε_i and δ_i such that

$$g_i(x, \lambda) < 0 \qquad \forall (x, \lambda) \in U_{\varepsilon_i}\{x^0\} \times V_{\delta_i}\{\lambda^0\}$$

holds, follows from (1). If $\varepsilon = \min \varepsilon_i$, $\delta = \min \delta_i$ we then have

$$U_\varepsilon\{x^0\} \subset M^0(\lambda) \qquad \forall \lambda \in V_\delta\{\lambda^0\}.$$

Employing Corollary 2.2.5.1, where we set

$$\Gamma_1 = \Gamma, \qquad \Gamma_2 = M^0, \qquad \Gamma_3 = M$$

we find that the proof follows from (2). □

Restricting the mapping M to $M \cap \Gamma$ is particularly useful if certain systems of parametric inequalities for which the set of Slater points is empty, are considered; if those inequalities which are actually fulfilled as equalities at $\lambda^0 \in \Lambda$ are collected under the mapping Γ one may, without additional hypotheses, then show that Γ is l.s.c.-B for a certain subclass of convex functions. For example it is sufficiently well known that the solution sets $\Gamma\lambda$ of linear systems of equations in E_n: $Ax = \lambda$, $\lambda \in \Lambda = E_m$, are H-continuously dependent on λ (on the feasible parameter set $\mathfrak{B} = \{\lambda \mid \Gamma\lambda \neq \emptyset\}$).

These questions are examined in the following sections of this chapter.

Hypothesis (2) of Theorem 3.1.5 is very restrictive, and we now give an example to indicate that, in general, it cannot be dispensed with.

Example 3.1.4. If $\Lambda = X = R$, $\Gamma\lambda \equiv \{0\} \cup \{1\}$, and $M(\lambda) = \{x \in R \mid x \leqq \lambda\}$, $\lambda \in \Lambda$, then we have $(M \cap \Gamma)\,1 = \{0, 1\}$, $(M^0 \cap \Gamma)\,1 = \{0\}$, and $(M \cap \Gamma)\,\lambda = \{0\}$ for $0 \leqq \lambda < 1$, $M \cap \Gamma$ is thus not l.s.c.-B at $\lambda^0 = 1$. All conditions of Theorem 3.1.5 are satisfied (for $\lambda^0 = 1$) except (2).

The next example is intended to illustrate the fact that the inclusion $(M \cap \Gamma)\,\lambda^0 \subset \mathrm{cl}(M^0 \cap \Gamma)\,\lambda^0$ (in the case $M^0(\lambda^0) \cap \Gamma\lambda^0 \neq \emptyset$) is a sufficient but not a necessary condition for the lower semicontinuity of $M \cap \Gamma$.

Example 3.1.5. Assume $\Lambda = R$, $X = E_2$, $\Gamma\lambda \equiv E_2 \ \forall \lambda$

$$g_1(x, y) = \begin{cases} y - x^2 & \text{if} \quad x \leqq 0, \\ y & \text{otherwise}, \end{cases}$$

$$g_2(x, y) = \begin{cases} -y - x^2 & \text{if} \quad x \leqq 0, \\ -y & \text{otherwise}, \end{cases}$$

and

$$M(\lambda) = \{(x, y) \in E_2 \mid g_1(x, y) \leqq \lambda^2,\ g_2(x, y) \leqq -\lambda^2\}.$$

Obviously $M^0(\lambda) = \{(x, y) \mid x < 0, -x^2 + \lambda^2 < y < x^2 + \lambda^2\}$ and

$$M(\lambda) = \{(x, y) \mid x \leqq 0, -x^2 + \lambda^2 \leqq y \leqq x^2 + \lambda^2\} \cup \{(x, \lambda^2) \mid x \geqq 0\} .$$

$M \cap \Gamma$ is thus l.s.c.-B at $\lambda^0 = 0$ despite the fact that condition (3) of the fundamental theorem is not satisfied: $\{(x, 0) \mid x > 0\} \cap \mathrm{cl}(M^0 \cap \Gamma)\, 0 = \emptyset$.

From the two preceding examples one would expect that certain convexity conditions imposed on the set $(M \cap \Gamma)\,\lambda^0$ would establish the property $(M \cap \Gamma)\,\lambda^0 \subset \mathrm{cl}(M^0 \cap \Gamma)\,\lambda^0$ (if $(M^0 \cap \Gamma)\,\lambda^0 \neq \emptyset$), and this is in fact the case.

We assume that X is a real linear space and X' a non-empty convex subset of X. A function $f\colon X \to \mathbb{R}$ is called *strictly quasiconvex* on X' if for any two points $x \in X'$ and $y \in X'$ and for every α with $0 < \alpha < 1$ the inequality $f(\alpha x + (1 - \alpha)\, y) < f(y)$ follows from $f(x) < f(y)$.

An immediate consequence of the definition is that every function which is convex on X' is also strictly quasiconvex on that subset. If a function f is l.s.c. and strictly quasiconvex on X' then it is also quasiconvex, i.e. every level set

$$\{x \in X' \mid f(x) \leqq t\}\,, \qquad t \in \mathbb{R}\,,$$

is convex (cf. H. J. Greenberg and W. P. Pierskalla [1])

An important property of functions which are strictly quasiconvex on X' is given as follows:

$$\begin{aligned}&\text{If } x, y \in X',\ x \neq y, \text{ are points for which } f(x) = f(y)\\ &\text{then there exists at most one } z \in \mathrm{ri\,conv}\,\{x, y\} \text{ with } f(z) > f(x)\,.\end{aligned} \tag{3.1.14}$$

If there were two numbers α, β satisfying $0 < \alpha < \beta < 1$ and corresponding points $z^\alpha = \alpha x + (1 - \alpha)\, y$, $z^\beta = \beta x + (1 - \beta)\, y$ satisfying the inequalities $f(z^\alpha) > f(x)$ and $f(z^\beta) > f(x)$ and if without loss of generality $f(z^\alpha) \leqq f(z^\beta)$ then z^β would be contained in ri conv $\{z^\alpha, x\}$ and thus because of the strict quasiconvexity we would have $f(z^\beta) < f(z^\alpha)$, which contradicts the previous inequality.

Theorem 3.1.6. *Let X be a real linear normed space, Γ be l.s.c.-B at $\lambda^0 \in \Lambda$, $(M^0 \cap \Gamma)\,\lambda^0$ be non-empty and $\Gamma\lambda^0$ be a convex set. Further, let, for all $i \in J = \{1, \dots, m\}$, the functions g_i be u.s.c. on $M^0(\lambda^0) \times \{\lambda^0\}$ and the functions $g_i(\cdot, \lambda^0)$ be strictly quasiconvex on $\Gamma\lambda^0$.*

The mapping $M \cap \Gamma$ is then l.s.c.-B at λ^0.

Proof. We apply Theorem 3.1.5, according to which it is sufficient to show that $(M \cap \Gamma)\,\lambda^0 \subset \mathrm{cl}(M^0 \cap \Gamma)\,\lambda^0$. By hypothesis there exists a point $\bar{x} \in (M^0 \cap \Gamma)\,\lambda^0$. We assume that $x \in (M \cap \Gamma)\,\lambda^0$ is an arbitrary but fixed point and x^α are points given by $x^\alpha = \alpha\bar{x} + (1 - \alpha)\, x$, $0 < \alpha < 1$. Because of (3.1.14) there then exists an $\alpha_i > 0$ such that

$$g_i(x^\alpha, \lambda^0) \leqq g_i(\bar{x}, \lambda^0) < 0 \quad \text{for all} \quad \alpha \in (0, \alpha_i)$$

if $g_i(x, \lambda^0) = g_i(\bar{x}, \lambda^0)$. If on the other hand $g_i(x, \lambda^0) \neq g_i(\bar{x}, \lambda^0)$ we have by the definition of a strictly quasiconvex function

$$g_i(x^\alpha, \lambda^0) < \max\,\{g_i(x, \lambda^0), g_i(\bar{x}, \lambda^0)\} \leqq 0$$

for all $\alpha \in (0, 1)$ and we then set $\alpha_i = 1$. Thus

$$g_i(x^\alpha, \lambda^0) < 0 \qquad \forall i \in J \qquad \forall \alpha \in (0, \min \alpha_i)\,.$$

Taking the convexity of $\Gamma\lambda^0$ into account we then have $x^\alpha \in M^0(\lambda^0) \cap \Gamma\lambda^0$ for all $\alpha \in (0, \min \alpha_i)$ and the theorem follows in the limit $\alpha \to 0$. □

We end this section with supplementary necessary and sufficient conditions for the case that every local minimum of a function f with respect to a non-empty set $X' \subset X$ is also a global minimum. It is known from the literature[1]) that the function f has this property if and only if its level set mapping $N_{f,X'}: \mathfrak{G} \to 2^X$, where

$$N_{f,X'}(\alpha) =_{\mathrm{Df}} \{x \in X' \mid f(x) \leqq \alpha\}, \qquad \alpha \in \mathfrak{G}, \tag{3.1.15}$$

and

$$\mathfrak{G} =_{\mathrm{Df}} \{\alpha \in \mathbb{R} \mid N_{f,X'}(\alpha) \neq \emptyset\}, \tag{3.1.16}$$

is l.s.c.-B on $\mathfrak{G}$. This may be demonstrated in the following simple manner:

Theorem 3.1.7. *Let X and X^1 be a metric space and a non-empty subset of that space respectively. The following statements are equivalent for a real-valued function $f: X \to \mathbb{R}$:*

(1) *every local minimum of f with respect to X' is a global minimum with respect to X';*

(2) *the level set mapping $N_{f,X'}$ (3.1.15) is l.s.c.-B on $\mathfrak{G}$;*

(3) *if α is a real number such that the set*

$$N^0_{f,X'}(\alpha) =_{\mathrm{Df}} \{x \in X' \mid f(x) < \alpha\}$$

is non-empty then $N_{f,X'}(\alpha) \subset \mathrm{cl}\, N^0_{f,X'}(\alpha)$ holds.

Proof. (1) $\Rightarrow$ (3). Assume $\alpha \in \mathfrak{G}$ and that there exists an $x^0 \in X'$ such that $f(x^0) < \alpha$. If $x \in X'$ is a point satisfying $f(x) < \alpha$ we immediately have

$$x \in N^0_{f,X'}(\alpha) \subset \mathrm{cl}\, N^0_{f,X'}(\alpha) .$$

If on the other hand $x \in X'$ is a point such that $f(x) = \alpha$ there then exists for each $\varepsilon > 0$ a point $x_\varepsilon \in X' \cap U_\varepsilon\{x\}$ with $f(x_\varepsilon) < f(x)$: if this were not true then one could find a real number $a > 0$ such that $f(y) \geqq f(x)$ for all $y \in X' \cap U_a\{x\}$, which would mean that x is a local minimum point contrary to $f(x) = \alpha > f(x^0)$ and condition (1). Because of $f(x_\varepsilon) < 0$ we thus have $x \in \mathrm{cl}\, N^0_{f,X'}(\alpha)$.

(3) $\Rightarrow$ (1). Assume that (1) is not true. Then there exist points $x^0 \in X'$ and $x' \in X'$ as well as an $\varepsilon > 0$ such that $f(x') > f(x^0)$ and

$$f(x) \geqq f(x') \qquad \forall x \in U_\varepsilon\{x'\} \cap X' . \tag{3.1.17}$$

Hence $N^0_{f,X'}(f(x')) = \{x \in X' \mid f(x) < f(x')\}$ is non-empty. By hypothesis (3) we then have $N_{f,X'}(f(x')) \subset \mathrm{cl}\, N^0_{f,X'}(f(x))$. Because of $x' \in N_{f,X'}(f(x'))$ there exists a sequence $\{x^k\} \subset X'$ with $x^k \to x'$ and $f(x^k) < f(x')$ which contradicts (3.1.17) and thus verifies (1).

(2) $\Rightarrow$ (3). Let $\alpha \in \mathfrak{G}$ and let there exist an $x^0 \in X'$ with $f(x^0) < \alpha$. There then exists a $\beta' > 0$ such that $f(x^0) < \alpha - \beta$, and thus $(\alpha - \beta) \in \mathfrak{G}$ holds for all β satisfying $0 \leqq \beta < \beta'$. As $N_{f,X'}$ is l.s.c.-B at α one can find for each $x \in N_{f,X'}(\alpha)$ a sequence $\{x^k\} \subset X'$ with the properties $x^k \to x^0$ and $f(x^k) \leqq \alpha - 1/k < \alpha$ (for sufficiently large k). $N_{f,X'}(\alpha) \subset \mathrm{cl}\, N^0_{f,X'}(\alpha)$ follows.

(3) $\Rightarrow$ (2). Let $\alpha_0 \in \mathfrak{G}$. If $f(x) \geqq \alpha_0$ for all $x \in X'$ we obviously have $N_{f,X'}(\alpha) = \emptyset$ for all $\alpha < \alpha_0$ and $N_{f,X'}(\alpha) \supset N_{f,X'}(\alpha_0)$ for all $\alpha \geqq \alpha_0$ from which it follows that $N_{f,X'}$ is l.s.c.-B at α_0. If on the other hand $N^0_{f,X'}(\alpha^0)$ is non-empty there then exists for each $x^0 \in N^0_{f,X'}(\alpha^0)$ a $\delta = \delta(\alpha^0, x^0)$ such that $f(x^0) < \alpha$ holds for all α with $|\alpha - \alpha_0| < \delta$. Consequently $N^0_{f,X'}$ is l.s.c.-B at α^0. Thus it follows from assumption (3) and Corollary 2.2.5.1 that $N_{f,X'}$ is l.s.c.-B at α_0. □

In conclusion it should be noted that every function which is strictly quasiconvex on X' acquires the equivalent properties (1) to (3) of Theorem 3.1.7 if X and X' are assumed to be a linear normed space and a non-empty subset of that space respectively.

---

[1]) cf. e.g. I. Zang and M. Avriel [1]

3.2. Solution Sets of Special Parametric Systems of Inequalities Satisfying Convexity Conditions

In Section 3.1 conditions were established for the continuity (i.e. continuity in some sense) of the solution sets $M(\lambda)$ of systems of parametric convex inequalities with regard to the parameter λ. In the case of such systems in a Euclidean space E_n these conditions may be given in a more concrete form, as was apparent in Theorems 3.1.3 and 3.1.6. Whereas the mapping is closed at a parameter point $\lambda^0 \in \Lambda$ and, if $M(\lambda^0)$ is compact also u.s.c.-B under quite general conditions, the lower semicontinuity of M at λ^0 in general only follows if $M(\lambda^0)$ satisfies a constraint qualification of Slater type. The main aim of this section is to find criteria for M which do not include such constraint qualifications; the study of parametric systems of inequalities defined by convex weakly analytic functions is of particular importance in achieving this objective.

In this section we consider the point-to-set mapping $M: \Lambda \to 2^{E_n}$

$$M(\lambda) =_{\mathrm{Df}} \{x \in E_n \mid g_i(x, \lambda) \leqq 0, i = 1, \dots, m\}, \qquad \lambda \in \Lambda, \tag{3.2.1}$$

subjected to the general conditions

$$\Lambda \text{ is a metric space}, \tag{3.2.2}$$

$$g_i: E_n \times \Lambda \to R \quad \text{is continuous on } E_n \times \Lambda \text{ for each} \quad i \in \{1, \dots, m\}, \tag{3.2.3}$$

$$g_i(\cdot, \lambda): E_n \to R \quad \text{is convex on } E_n \text{ for each } i \in \{1, \dots, m\} \text{ and all } \lambda \in \Lambda. \tag{3.2.4}$$

We would like to remark here that the convexity condition (3.2.4) is not exploited to the full in all theorems of this section. Bearing Theorems 3.1.3 and 3.1.6 in mind it often suffices to assume that the $g_i(\cdot, \lambda)$ are quasiconvex or strictly quasiconvex (on E_n). We have decided against such a more general approach in favour of a unified exposition.

We now introduce some concepts and notations. Let X be a non-empty convex subset of E_n. The set

$$O^+X =_{\mathrm{Df}} \{u \in E_n \mid x + \alpha u \in X \ \forall x \in X \ \forall \alpha \in R: \alpha \geqq 0\} \tag{3.2.5}$$

is called the *recession cone* of X. Likewise the set

$$L^+X =_{\mathrm{Df}} \{u \in E_n \mid x + \alpha u \in X \ \ \forall x \in X \ \ \forall \alpha \in R\} \tag{3.2.6}$$

is the *vertex set* of the recession cone of X. L^+X is obviously a subspace of E_n. Assuming $I \subset \{1, \dots, m\}$, the point-to-set mapping $M^I: \Lambda \to 2^{E_n}$ is defined by

$$M^I(\lambda) =_{\mathrm{Df}} \{x \in E_n \mid g_i(x, \lambda) \leqq 0, i \in I\}, \qquad \lambda \in \Lambda. \tag{3.2.7}$$

If $I = \emptyset$ we then set $M^\emptyset(\lambda) = E_n$. The *characteristic index set* of a convex subset $X_\lambda \subset M(\lambda)$, $\lambda \in \Lambda$, is given by

$$\operatorname{ch} X_\lambda =_{\mathrm{Df}} \{i \in \{1, \dots, m\} \mid g_i(x, \lambda) = 0 \qquad \forall x \in X_\lambda\}. \tag{3.2.8}$$

Lemma 3.2.1. *Let $\lambda \in \Lambda$ be fixed and $M(\lambda)$ be a mapping which fulfills the conditions* (3.2.1) *to* (3.2.4). *If X_λ is a non-empty convex subset of $M(\lambda)$ then*

$$\operatorname{ri} X_\lambda \subset \{x \in E_n \mid g_i(x, \lambda) = 0 \ \forall i \in \operatorname{ch} X_\lambda, \ g_j(x, \lambda) < 0 \ \forall j \notin \overline{\operatorname{ch} X_\lambda}\}$$

holds.

Proof. Assume that the inclusion is false. Then by definition (3.2.8) there exist a $z \in \operatorname{ri} X_\lambda$ and a $k \notin \overline{\operatorname{ch} X_\lambda}$ such that $g_k(z, \lambda) = 0$. Again by (3.2.8) one may find a $y \in X_\lambda$ with $g_k(y, \lambda) < 0$ and, because of $z \in \operatorname{ri} X_\lambda$, there exists an $\alpha > 0$ such that

$z + \alpha(z - y) \in X_\lambda$. This however leads to

$$0 = g_k(z, \lambda) \leqq \frac{1}{1+\alpha} g_k(z + \alpha(z - y), \lambda) + \frac{\alpha}{1+\alpha} g_k(y, \lambda) < 0$$

and thus to a contradiction. □

The following theorem specializes the results of Section 3.1 to the case of a system of parametric inequalities given by (3.2.1) to (3.2.4).

Theorem 3.2.1. *Let a point-to-set mapping M satisfy the conditions* (3.2.1) *to* (3.2.4) *and let $\lambda^0 \in \Lambda$. Then*

(1) *M^I is closed at λ^0 for each $I \subset \{1, \dots, m\}$;*

(2) *if $M^I(\lambda^0)$ is a bounded and non-empty set and $\emptyset \neq I \subset \{1, \dots, m\}$ it follows that M^I is u.s.c.-B at λ^0;*

(3) *if $M(\lambda^0) \neq \emptyset$ and $I = \text{ch } M(\lambda^0)$ it follows that M is l.s.c.-B at λ^0 if the same holds for M^I, in particular M^I is l.s.c.-B at λ^0 if $I = \emptyset$;*

(4) *if $M(\lambda^0) \neq \emptyset$, $I = \text{ch } M(\lambda^0)$ and M^I is l.s.c.-B at λ^0 it follows that M is H-continuous at λ^0 provided that $M(\lambda^0)$ is a bounded set; and finally,*

(5) *if $M(\lambda^0)$ is non-empty and there exists an $i \in \text{ch } M(\lambda^0)$ such that $g_i(\cdot, \lambda^0)$ is strictly convex then it follows that M is H-continuous at λ^0.*

Proof. (1) and (2) follow immediately from Theorems 3.1.1 and 3.1.3 respectively.

(3) follows from Theorem 3.1.6 if one sets $\Gamma\lambda =_{\text{Df}} M^I(\lambda)$, $\lambda \in \Lambda$, and notes that, because of (3.2.1) to (3.2.4) and $M(\lambda^0) \neq \emptyset$, one may without difficulty obtain from Lemma 3.2.1 the relation

$$\begin{aligned} \text{ri } M(\lambda^0) &= \{x \in E_n \mid g_i(x, \lambda) = 0, i \in I, g_i(x, \lambda) < 0, j \in \{1, \dots, m\} \backslash I\} \\ &= \{x \in E_n \mid g_i(x, \lambda) \leqq 0, i \in I, g_i(x, \lambda) < 0, j \in \{1, \dots, m\} \backslash I\}. \quad (3.2.9) \end{aligned}$$

(4) follows immediately from (3) and Corollary 3.1.3.2.

(5) may be demonstrated by noting that the set $\{x \in E_n \mid g_i(x, \lambda^0) = 0\}$ contains only one element since $g_i(\cdot, \lambda^0)$ is strictly convex and $i \in \text{ch } M(\lambda^0)$. By Theorem 3.1.3 M is then u.s.c.-H at λ^0 and since $M(\lambda^0)$ contains only one element M is then also .s.c.-H at λ^0. □

We now turn our attention to the question of when the mapping M^I, $I = \text{ch } M(\lambda^0)$, $\lambda^0 \in \Lambda$, is l.s.c.-B at λ^0. The convexity condition (3.2.4) implies

$$M^I(\lambda^0) = \{x \in E_n \mid g_i(x, \lambda^0) = 0, i \in I\}, \quad I = \text{ch } M(\lambda^0), \quad \lambda^0 \in \Lambda, \quad (3.2.10)$$

thus no Slater point exists and Theorem 3.1.6 cannot be evoked. In order to prove the following theorem we need various basic results from convex analysis and which are presented here without proof (we refer the reader to R. T. ROCKAFELLAR [2], § 8).

Let $f: E_n \to R$ be a convex function on E_n. The set

$$N_f(\alpha) =_{\text{Df}} \{x \in E_n \mid f(x) \leqq \alpha\}, \quad \alpha \in R, \quad (3.2.11)$$

is called the *α-level set* of f. If $\alpha \in R$, $\beta \in R$ are any two numbers satisfying $N_f(\alpha) \neq \emptyset$ and $N_f(\beta) \neq \emptyset$ then the sets $O^+N_f(\alpha)$ and $O^+N_f(\beta)$ are identical and it follows that $L^+N_f(\alpha) = L^+N_f(\beta)$. In particular we have

$$N_f(\alpha) \neq \emptyset \Rightarrow L^+N_f(\alpha) = L^+N_f(f(x)) \quad \forall x \in E_n. \quad (3.2.12)$$

If the functions $f_i: E_n \to R$, $i = 1, \dots, r$, are convex and if $\bigcap_{i=1}^{r} N_{f_i}(\alpha) \neq \emptyset$, $\alpha \in R$, then

$$L^+\left(\bigcap_{i=1}^{r} N_{f_i}(\alpha)\right) = \bigcap_{i=1}^{r} L^+N_{f_i}(\alpha). \quad (3.2.13)$$

Theorem 3.2.2. *Let $\lambda^0 \in \Lambda$ and $I = \text{ch } M(\lambda^0)$ satisfy the following conditions:*

(1) *$M^I(\lambda)$ is non-empty for all $\lambda \in \Lambda$;*[1])

(2) *$M^I(\lambda^0)$ is an affine subspace of $\mathbb{E}_n$, and*

(3) *$\dim L^+M^I(\lambda) = \dim L^+M^I(\lambda^0)$ for all $\lambda \in \Lambda$.*

The mappings M^I and M are then l.s.c.-B at λ^0.

Proof. Bearing proposition (3) of Theorem 3.2.1 in mind we need only demonstrate that M^I is l.s.c.-B at λ^0. Assume that $\{\lambda^k\} \subset V_\delta\{\lambda^0\}$ is an arbitrary sequence with $\lambda^k \to \lambda^0$. One must then show that for each point $x^0 \in M^I(\lambda^0)$ a sequence $\{x^k\} \subset \mathbb{E}_n$ exists such that $x^k \to x^0$ and $x^k \in M^I(\lambda^k)$ hold (for sufficiently large k). Assume that $x^0 \in M^I(\lambda^0)$ is arbitrary. Because of hypothesis (1) and the general conditions (3.2.3) and (3.2.4) the sets $M^I(\lambda^k)$ are non-empty, convex and closed for all k. There thus exists a unique sequence $\{x^k\} \subset \mathbb{E}_n$ such that

$$||x^k - x^0|| = \min \{||y^k - x^0|| \mid y^k \in M^I(\lambda^k)\}, \qquad x^k \in M^I(\lambda^k) \quad \forall k, \tag{3.2.14}$$

holds. It is thus only necessary to show that $x^k \to x^0$. If this is not the case then there exists an $\varepsilon > 0$ such that, without loss of generality *, we have

$$||x^k - x^0|| \geqq 2\varepsilon \quad \forall k. \tag{3.2.15}$$

Because of (3.2.3) and (3.2.4) Lemma 3.1.1 may be applied to the sequences $\{\lambda^k\}$ and $\{x^k\}$ and to the points λ^0 and x^0 by substituting $\{x^0\}$ for S in the lemma, which then implies that sequences $\{\tilde{x}^k\} \subset \mathbb{E}_n$ and $\{\alpha_k\} \subset \mathbb{R}$ exist such that

$$\tilde{x}^k = \alpha_k x^k + (1 - \alpha_k)\, x^0 \quad \text{and} \quad 0 < \alpha_k < 1 \quad \forall k \tag{3.2.16}$$

hold together with $||\tilde{x}^k - x^0|| = \varepsilon$, and hence there exists without loss of generality * an $x^* \in \mathbb{E}_n$ such that

$$\tilde{x}^k \to x^* \in M^I(\lambda^0) \cap \text{bd } U_\varepsilon\{x^0\}. \tag{3.2.17}$$

If $\dim L^+M^I(\lambda^0) = 0$ (because of condition (2) this infers $M^I\lambda^0 = \{x^0\}$) one obtains a contradiction. Therefore assume $d = \dim L^+M^I(\lambda^0) \geqq 1$. According to condition (3) there exists for each k an orthonormal system of d linearly independent vectors, i.e. vectors satisfying

$$u^{jk} \in L^+M^I(\lambda^k) \quad \text{where} \quad ||u^{jk}|| = 1 \quad \text{and}$$
$$u^{jk\mathsf{T}}u^{lk} = 0 \qquad \forall j, \quad l \in \{1, \dots, d\}. \tag{3.2.18}$$

Since $\{u \in \mathbb{E}_n \mid ||u|| = 1\}$ is compact there also exists for each $j \in \{1, \dots, d\}$ and without loss of generality * a $u^{j*} \in \mathbb{E}_n$ such that

$$u^{jk} \to u^{j*}, \qquad ||u^{j*}|| = 1 \quad \text{and} \quad u^{j*\mathsf{T}}u^{l*} = 0 \qquad \forall, l, j \in \{1, \dots, d\} \tag{3.2.19}$$

hold. Taking (3.2.3) and (3.2.4) in account we find from (3.2.19), (3.2.16), (3.2.17), the relation $x^k \in M^I(\lambda^k)$, and $u^{jk} \in L^+M^I(\lambda^k)$ $\forall k$, $\forall j \in \{1, \dots, d\}$ that

$$\begin{aligned} &g_i(x^* + \alpha u^{j*}, \lambda^0) = \lim g_i(\tilde{x}^k + \alpha u^{jk}, \lambda^k) \\ &\leqq \max \{0, \lim g_i(x^0 + \alpha u^{jk}, \lambda^k)\} \\ &\forall i \in I \qquad \forall j \in \{1, \dots, d\} \qquad \forall \alpha \in \mathbb{R} \end{aligned} \tag{3.2.20}$$

holds. The properties (3.2.12) and (3.2.13) give

$$L^+M^I(\lambda^k) = \bigcap_{i \in I} L^+N_{g_i(\cdot, \lambda_k)}(g_i(x^0, \lambda^k))$$

---

[1]) It would suffice to assume that (1) and (3) hold for a certain neighbourhood $V\{\lambda^0\}$, i.e. to restrict the conditions to $\Lambda = V\{\lambda^0\}$.

which with $u^{jk} \in L^+M^I(\lambda^k)$ implies

$$g_i(x^0 + \alpha u^{jk}, \lambda^k) \leqq g_i(x^0, \lambda^k) \qquad \forall i \in I \qquad \forall j \in \{1, \dots, d\} \qquad \forall \alpha \in \mathbb{R}\,.$$

It follows from (3.2.20) and $\lim g_i(x^0, \lambda^k) = g_i(x^0, \lambda^0) = 0$ that

$$g_i(x^* + \alpha u^{j*}, \lambda^0) \leqq 0 \qquad \forall\, i \in I \qquad \forall j \in \{1, \dots, d\} \qquad \forall \alpha \in \mathbb{R}\,.$$

Finally, (3.2.17) yields

$$u^{j*} \in L^+M^I(\lambda^0) \qquad \forall j \in \{1, \dots, d\}\,. \tag{3.2.21}$$

Because of (3.2.19) and condition (3) the vectors u^{j*}, $j \in \{1, \dots, d\}$, constitute a basis of $L^+M^I(\lambda^0)$. Condition (2) gives $M^I(\lambda^0) = x^0 + L^+M^I(\lambda^0)$, there thus exist numbers $\beta_1, \dots, \beta_d \in \mathbb{R}$ such that

$$x^* = x^0 + \sum_{j=1}^{d} \beta_j u^{j*} \tag{3.2.22}$$

is satisfied. We set

$$y^k = x^k - \sum_{j=1}^{d} \beta_j u^{jk} \quad \forall k$$

from which $y^k \in M^I(\lambda^k)\ \forall k$ follows by definition of $L^+M^I(\lambda^k)$. We then immediately have

$$\left\|\sum_{j=1}^{d} \beta_j u^{jk}\right\| = \left\|\sum_{j=1}^{d} \beta_j u^{j*}\right\| = \|x^* - x^0\| \quad \forall k$$

and thus

$$\|x^k - y^k\|^2 = \left\|\sum_{j=1}^{d} \beta_j u^{jk}\right\|^2 = \varepsilon^2 \quad \forall k\,, \tag{3.2.23}$$

hence there exists a k' such that

$$\left(\sum_{j=1}^{d} \beta_j u^{jk}\right)^{\mathrm{T}} \left(\sum_{j=1}^{d} \beta_j u^{j*}\right) > \frac{\varepsilon^2}{2} \qquad \forall k \geqq k' \tag{3.2.24}$$

holds. Employing the relations (3.2.17), (3.2.22), and (3.2.24) we can find a $k'' \geqq k'$ such that

$$\left(\sum_{j=1}^{d} \beta_j u^{jk}\right)^{\mathrm{T}} (\tilde{x}^k - x^0) > \frac{\varepsilon^2}{2} \qquad \forall k \geqq k''\,.$$

It follows from this and (3.2.16) that

$$\left(\sum_{j=1}^{d} \beta_j u^{jk}\right)^{\mathrm{T}} (x^k - x^0) = \frac{1}{\alpha_k}\left(\sum_{j=1}^{d} \beta_j u^{jk}\right)^{\mathrm{T}} (\tilde{x}^k - x^0) > \frac{\varepsilon^2}{2} \qquad \forall k \geqq k'' \tag{3.2.25}$$

also holds, and we obtain, if (3.2.23) is taken into consideration (for all k)

$$\begin{aligned} \|x^0 - x^k\|^2 &= \|x^0 - y^k\|^2 + \|y^k - x^k\|^2 + 2(y^k - x^k)^{\mathrm{T}} (x^0 - y^k) \\ &= \|x^0 - y^k\|^2 + \varepsilon^2 - 2\|y^k - x^k\|^2 - 2(y^k - x^k)^{\mathrm{T}} (x^k - x^0) \\ &= \|x^0 - y^k\|^2 - \varepsilon^2 + 2\left(\sum_{j=1}^{d} \beta_j u^{jk}\right)^{\mathrm{T}} (x^k - x^0) > \|x^0 - y^k\|^2 \end{aligned}$$

which contradicts definition (3.2.14) because of $y^k \in M^I(\lambda^k)$. In other words we have also arrived at a contradiction in the case $d \geqq 1$, (3.2.15) is therefore not possible. Hence $x^k \to x^0$ and as x^0 was arbitrary we have completed the proof. □

Conditions (2) and (3) of the last theorem may be expressed in simplified forms. From (3.2.12) and (3.2.13) we immediately find that (3) is automatically satisfied in the case $g_i(x, \lambda) =_{\mathrm{Df}} h_i(x) + \tau_i(\lambda)\ \forall i$ where h_i is convex on E_n and τ_i is continuous at λ^0 for all $i \in I$. Further simplifications of conditions (2) and (3) may be found and which are easier to verify than the original versions, they however refer to a certain subclass of convex functions which we will first define.

A function $f\colon E_n \to R$ is said to be *weakly analytic* if the following holds for any two vectors $x \in E_n$ and $u \in E_n$: If the function $f_{x,u}(\alpha) = f(x + \alpha u)$ is constant on an open nterval $(\underline{\alpha}, \bar{\alpha})$ then $f_{x,u}(\alpha) = f(x)$ for all $\alpha \in R$.

Corollary 3.2.2.1. *The proposition of Theorem* 3.2.2 *still holds if condition* (2) *is replaced by the condition that all functions* $g_i(\cdot, \lambda^0)$, $i \in I = \operatorname{ch} M(\lambda^0)$, *are convex and weakly analytic on* E_n.

Proof. One must show that

$$x(\alpha) =_{\mathrm{Df}} \alpha x + (1 - \alpha)\, y \in M^I(\lambda^0)$$

holds for any two points $x \in M^I(\lambda^0)$ and $y \in M^I(\lambda^0)$ and for all $\alpha \in R$. Assuming then that $x, y \in E_n$ we have, since $M^I(\lambda^0)$ is convex $x(\alpha) \in M^I(\lambda^0)$ for all $\alpha \in [0, 1]$, and from (3.2.10) we obtain

$$g_i\big(x(\alpha), \lambda^0\big) = 0 \qquad \forall i \in I \qquad \forall \alpha \in [0, 1]\,.$$

As $g_i(\cdot, \lambda^0)$ is weakly analytic for each $i \in I$ it follows that $g_i\big(x(\alpha), \lambda^0\big) = 0$ for all $\alpha \in R$ and all $i \in I$. □

It is evident that all convex analytic functions and all strictly convex functions are also weakly analytic. R. T. Rockafellar [3], [4] introduced the class of faithfully convex functions, which also contains all convex analytic functions and strictly convex functions and for which quite simple representations of the sets $L^+N_f(\alpha)$ (defined by (3.2.6) and (3.2.11)) may be found. This last may be utilized in reformulating condition (3) of Theorem 3.2.2.

A function $f\colon E_n \to R$ which is convex on E_n is said to be *faithfully convex* if there exist a function $h\colon E_r \to R$ which is strictly convex on E_r, a (r, n)-matrix A, vectors $b \in E_r$ and $p \in E_n$, and finally a $\gamma \in R$ such that f may be represented by

$$f(x) = h(Ax + b) + p^{\mathsf{T}}x + \gamma \qquad \forall x \in E_n\,. \tag{3.2.26}$$

If f is an arbitrary faithfully convex function, x any point of E_n, and u any vector of W, where

$$W = \{u \in E_n \mid Au = 0\} \tag{3.2.27}$$

then the function

$$f_{x,u}(\alpha) = f(x + \alpha u) \qquad \alpha \in R \tag{3.2.28}$$

is affine-linear on R. If $u \notin W$ it follows from the strict convexity of h that $f_{x,u}$ is strictly convex on R for all $x \in E_n$. A faithfully convex function is therefore either affine-linear or strictly convex along an arbitrary line. In fact a function is a faithfully convex function if and only if it exhibits this property.

Clearly the class of faithfully convex functions is a proper subclass of the convex weakly analytic functions.

Lemma 3.2.2. *If $f\colon \mathbb{E}_n \to \mathbb{R}$ is a faithfully convex function with the representation* (3.3.26) *then*

$$L^+N_f(\alpha) = \widetilde{W} =_{\mathrm{Df}} \{u \in \mathbb{E}_n \mid Au = 0, p^\mathsf{T}u = 0\} \tag{3.2.29}$$

holds for all α satisfying $N_f(\alpha) \neq \emptyset$.

Proof. If $u \in \widetilde{W}$ we immediately get from (3.2.26)

$$f(x + \theta u) = f(x) \qquad \forall \theta \in \mathbb{R} \qquad \forall x \in \mathbb{E}_n$$

from which together with (3.2.12) it then follows that $\widetilde{W} \subset L^+N_f(\alpha)$ if $N_f(\alpha) \neq \emptyset$. If on the other hand $\alpha \in \mathbb{R}$, $x \in N_f(\alpha)$, and $u \in L^+N_f(\alpha)$ then $f_{x,u}$ is constant on $\mathbb{R}$ and we therefore have $u \in W$. From

$$\begin{aligned} f(x) = f(x + \alpha u) &= h(Ax + \alpha Au + b) + p^\mathsf{T}(x + \alpha u) + \gamma \\ &= h(Ax + b) + p^\mathsf{T}x + \gamma + \alpha p^\mathsf{T}u \qquad \forall \alpha \in \mathbb{R} \end{aligned}$$

we obtain $p^\mathsf{T}u = 0$ and thus $u \in \widetilde{W}$. We consequently have $L^+N_f(\alpha) = \widetilde{W}$ for all α for which $L^+N_f(\alpha) \neq \emptyset$. □

We again consider a point-to-set mapping M which satisfies the conditions (3.2.1) to (3.2.4). We are now only interested in parameter points $\lambda^0 \in \Lambda$ for which $\operatorname{ch} M(\lambda^0)$ is non-empty since the lower semicontinuity of M follows from Theorem 3.2.1 if this is not the case.

Theorem 3.2.3. *If $\lambda^0 \in \Lambda$, $I =_{\mathrm{Df}} \operatorname{ch} M(\lambda^0) \neq \emptyset$, and the functions $g_i(\cdot, \lambda)$ are faithfully convex with the representation*

$$g_i(x, \lambda) = h\big(A^i(\lambda)\, x + b^i(\lambda), \lambda\big) + p^i(\lambda)^\mathsf{T}\, x + \gamma_i(\lambda) \tag{3.2.30}$$

where $A^i(\lambda) \in \mathbb{H}^{r_i,n}$, $b^i(\lambda) \in \mathbb{E}_{r_i}$, $p^i(\lambda) \in \mathbb{E}_n$, $\gamma_i(\lambda) \in \mathbb{R}$, and if further $M^I(\lambda) \neq \emptyset$ for all $\lambda \in \Lambda$, then the mappings M and M^I are l.s.c.-B at λ^0 providing

$$\operatorname{rg} R^I(\lambda) = \operatorname{rg} R^I(\lambda^0) \qquad \forall \lambda \in \Lambda \tag{3.2.31}$$

holds. $R^I(\lambda)$ $\forall \lambda \in \Lambda$ is a matrix with n columns formed from the row vectors $p_i(\lambda)^\mathsf{T}$, $i \in I$, and all rows of the matrices $A^i(\lambda)$, $i \in I$.

Proof. In order to apply Theorem 3.2.2 one must first verify that conditions (2) and (3) of that theorem are satisfied. We find that by Corollary 3.2.2.1 this is the case for condition (2) since the functions $g_i(\cdot, \lambda^0)$, $i \in I$, are faithfully convex and are therefore contained in the class of convex weakly analytic functions. It follows from (3.2.13), Lemma 3.2.2, and the representation (3.2.30) that for all $\lambda \in \Lambda$

$$L^+M^I(\lambda) = \{u \in \mathbb{E}_n \mid A^i(\lambda)\, u = 0, i \in I, p^i(\lambda)^\mathsf{T}\, u = 0, i \in I\}$$

so that by (3.2.31) condition (3) is likewise satisfied, q.e.d. □

The following corollaries are special forms of the last theorem as applied to parametric systems of quadratic and linear inequalities.

Corollary 3.2.3.1. *Let $\lambda^0 \in \Lambda$ and $I =_{\mathrm{Df}} \operatorname{ch} M(\lambda^0) \neq \emptyset$ be such that $M^I(\lambda)$ is non-empty for all $\lambda \in \Lambda$. Let the functions $g_i(\cdot, \lambda)$ for all $i \in I$ and all $\lambda \in \Lambda$ be given as follows*

$$g_i(x, \lambda) = x^\mathsf{T}C^i(\lambda)\, x + p^i(\lambda)^\mathsf{T}\, x + b_i(\lambda) \tag{3.2.32}$$

where $C^i(\lambda)$ are symmetric positive semi-definite $[n, n]$-matrices, $p^i(\lambda) \in \mathbb{E}_n$, and $b_i(\lambda) \in \mathbb{R}$ $\forall i \in I$ $\forall \lambda \in \Lambda$ and all C^i, p^i, b^i are continuous on Λ. *Further, define*

$$C^I(\lambda) =_{\mathrm{Df}} \big(C^{\alpha_1}(\lambda) \ldots C^{\alpha_r}(\lambda)\; p^{\alpha_1}(\lambda) \ldots p^{\alpha_r}(\lambda)\big)$$

where $I = \{\alpha_1, \dots, \alpha_r\}$. *The mappings* M *and* M^I *are then l.s.c.-B at* λ^0 *if*

$$\operatorname{rg} C^I(\lambda) = \operatorname{rg} C^I(\lambda^0) \qquad \forall \lambda \in \Lambda \tag{3.2.33}$$

holds.

Proof. There exists for each symmetric positive semi-definite $[n, n]$-matrix C of rank s a regular $[n, n]$-matrix Q such that $C = Q^{\mathsf T} D^s Q$ where the $[n, n]$-matrix $D^s = (d^s_{ij})$ is defined by

$$d^s_{ij} = \begin{cases} 1 & \text{if } \ 1 \leqq i = j \leqq s\,, \\ 0 & \text{in all other cases}\,. \end{cases}$$

We thus have $x^{\mathsf T}Cx = (D^sQx)^{\mathsf T}(D^sQx)$, i.e. the functions defined by (3.2.32) may be represented as in (3.2.30) (Theorem 3.2.3). If we note that

$$\{u \in E_n \mid D^sQu = 0\} = \{u \in E_n \mid Cu = 0\}$$

holds, the corollary then follows from the theorem. □

Taking $a^1(\lambda), \dots, a^r(\lambda)$ to be vectors of E_n, $\lambda \in \Lambda$, and $J = \{\alpha_1, \dots, \alpha_l\}$ to be a subset of $\{1, \dots, r\}$ we define the matrix $A^J(\lambda)$ by

$$A^J(\lambda) =_{\mathrm{Df}} (a^{\alpha_1}(\lambda) \dots a^{\alpha_l}(\lambda))^{\mathsf T}\,. \tag{3.2.34}$$

Corollary 3.2.3.2. *Let* Λ *be a metric space and let* $a^i\colon \Lambda \to E_n$ *and* $b_i\colon \Lambda \to R$ *be continuous functions on* Λ *for all* $i \in \{1, \dots, m\}$. *Let the point-to-set mapping* $M\colon \Lambda \to E_n$ *be defined by*

$$M(\lambda) = \{x \in E_n \mid a^i(\lambda)^{\mathsf T} x \leqq b_i(\lambda),\ i = 1, \dots, s,\ a^j(\lambda)^{\mathsf T} x = b_j(\lambda),\ j = s+1, \dots m\},\quad \lambda \in \Lambda \tag{3.2.35}$$

where $1 \leqq s \leqq m$. *Further let* $M^J(\lambda)$ *for all* $J \subset \{1, \dots, m\}$ *and* $\operatorname{ch} M(\lambda)$ *be defined in the same manner as in* (3.2.7) *and* (3.2.8).[1]) *Finally, let* λ^0 *be an element of* Λ *such that* $M(\lambda^0)$ *is non-empty and let there exist a* $\delta > 0$ *such that*

$$M^I(\lambda) \neq \emptyset \qquad \forall \lambda \in V_\delta\{\lambda^0\} \tag{3.2.36}$$

where $I = \operatorname{ch} M(\lambda^0)$.

M is then closed and l.s.c.-B at λ^0 if one of the following conditions is satisfied:

(1) $\operatorname{rg} A^I(\lambda) = \operatorname{rg} A^I(\lambda^0) \qquad \forall \lambda \in V_\delta\{\lambda^0\}$;

(2) $A^I(\lambda^0)$ *has maximal rank;*

(3) *if* $s = n$, $a^i(\lambda)^{\mathsf T} x = -x_i$, $b_i(\lambda) = 0\ \forall i \in \{1, \dots, s\}\ \forall \lambda \in \Lambda$, *and* $\tilde A^I(\lambda^0)$ *is the submatrix of* $A^{\{s+1, \dots, m\}}(\lambda^0)$ *formed by omitting all columns with the indices* $i \in I$, *then* $\tilde A^I(\lambda^0)$ *has maximal rank.*

Proof. By Theorem 3.2.1 M is closed (in fact even if the conditions (3.2.36) and (1) to (3) are not satisfied). If (1) is satisfied the lower semicontinuity (B) of M follows directly from Theorem 3.2.3. Clearly (1) follows from (2), and (2) results from (3) by simple matrix transformations, thus proving the corollary. □

The linear case considered in the last corollary was given by G. B. Dantzig, J. Folkman, and N. Shapiro [1] together with some further conclusions which may be drawn from that corollary.

---

[1]) i.e.

$$M^J(\lambda) = \{x \in E_n \mid a^i(\lambda)^{\mathsf T} x \leqq b_i(\lambda),\ i \in \{1, \dots, s \cap J,\ a^j(\lambda)^{\mathsf T} x = b_j(\lambda),\ j \in \{s+1, \dots, m\} \cap J\}$$

and

$$\operatorname{ch} M(\lambda) = \{i \in \{1, \dots, m\} \mid a^i(\lambda)^{\mathsf T} x = b_i(\lambda)\ \forall x \in M(\lambda)\}\,.$$

Note. The subclass of convex functions considered in this section enables one to infer continuity properties of solution sets of parametric convex inequalities without imposing a Slater condition. They also play an important role with respect to duality theorems without constraint qualifications and were indeed originally introduced for this reason. In work on optimization problems N. N. ASTAFJEV [1] demonstrated such duality theorems including inverse duality theorems using certain convex "semi-analytic" functions. His class of convex functions contains in particular the class of convex weakly analytic functions.

R. T. ROCKAFELLAR [3], [4] has shown that no duality gaps occur in ordinary convex programs with faithfully convex functions in the primal problem (if the feasible set of the primal problem is non-empty). Recently optimality criteria and solution procedures have been given by A. BEN-ISRAEL, A. BEN-TAL, S. ZLOBEC, H. WOLKOWICZ, and others (cf. A. BEN-ISRAEL, A. BEN-TAL, and S. ZLOBEC [1], A. BEN-TAL, A. BEN-ISRAEL, and S. ZLOBEC [1], A. BEN-TAL and S. ZLOBEC [1], S. ZLOBEC [1], H. WOLKOWICZ [1]).

B. KUMMER [5] has found a duality theorem without an additional constraint qualification for problems with convex weakly analytic constraint functions and to which we will turn our attention in Chapter 4.

Of the convex analytic functions the polynomial convex functions are of special interest and have been extensively studied by E. G. BELOUSOV [1].

If $M_0 = \{x \in E_n \mid g_i(x) \leqq 0, i = 1, \dots, m\}$ is described by convex polynomials then the recession cone O^+M_0 is polyhedral and the vertex set thereof L^+M_0 may be described in a relatively simple manner, which fact may then be used to simplify condition (3) of Theorem 3.2.2 in a fashion similar to that of Theorem 3.2.3.

Finally, we refer to the subclass of convex weakly analytic functions introduced by V. A. BULAVSKIJ and G. S. RUBINSTEIN [1] which we here denote as „quadratic-like": If $\nabla^2 f(x^0)$ is the Hessian matrix at $x^0 \in E_n$ for a convex, twice differentiable function $f\colon E_n \to R$ then f is quadratic-like if $h^\mathsf{T}\nabla^2 f(x)\, h = 0$ for all $x \in E_n$ follows from $h^\mathsf{T}\nabla^2 f(x^0)\, h = 0$. One then has for all $x \in E_n$

$$L^+N_f(f(x)) = \{u \in E_n \mid \nabla f(x^0)^\mathsf{T}\, h = 0,\ \nabla^2 f(x^0)\, h = 0\}$$

where $x^0 \in E_n$ is fixed, a further simplification of condition (3) of Theorem 3.2.2 is thus possible.

3.3. Perturbations of the Right-Hand Sides under Convex Constraints

In this section we consider point-to-set mappings which have the special form

$$M(\lambda) =_{\mathrm{Df}} \{x \in X \mid g_i(x) \leqq \lambda_i;\ i = 1, 2, \dots, m\}\,, \qquad \Lambda \subset E_m\,. \tag{3.3.1}$$

If not otherwise specified X is a (real) normed space and $g_i\colon X \to R$ are convex functions on X, $i = 1, 2, \dots, m$. The results of the previous section may obviously be employed if $X = E_n$; this will carried out in what follows and further results obtainable in the special case (3.3.1) and $X = E_n$ will also be included. First however we present a theorem due originally to S. DOLECKI [3] and which will be applied to spaces of infinite dimension.

Theorem 3.3.1 (S. Dolecki). *M is u.s.c.-H at λ^0 if $M(\lambda^0)$ is bounded and an $x^0 \in X$ exists such that $g_i(x^0) < \lambda_i^0$ for all i.*

Proof. We consider a sequence $\{\lambda^t\} \subset \Lambda$, $\lambda^t \to \lambda^0$, and assume that, contrarrily to the proposition, an $\varepsilon > 0$ and points $x^t \in M(\lambda^t)$ exist such that

$$d(x^t, M(\lambda^0)) \geqq \varepsilon \quad \forall t\,. \tag{3.3.2}$$

There then exist uniquely determined $\alpha_t \in (0, 1)$ such that

$$y^t \in \operatorname{bd} M(\lambda^0) \tag{3.3.3}$$

for $y^t = \alpha_t x^0 + (1 - \alpha_t)\, x^t$. Because of the continuity of the g_i the following must also hold on the line through x^0 and x^t for at least one $i = i(t)$:

$$g_i(y^t) = \lambda_i^0\,, \qquad g_i(x^t) > \lambda_i^0\,.$$

We may in addition assume without loss of generality * that $i = i_0$ for all t, hence

$$\lambda_{i_0}^0 \leqq \alpha_t g_{i_0}(x^0) + (1 - \alpha_t)\, g_{i_0}(x^t) \quad \forall t\,.$$

Because of $g_{i_0}(x^0) < \lambda_{i_0}^0$ and $g_{i_0}(x^t) \leqq \lambda_{i_0}^t \to \lambda_{i_0}^0$ this inequality implies $\alpha_t \to 0$, $t \to \infty$. We then obtain from this and

$$\|y^t - x^0\| = (1 - \alpha_t)\, \|x^t - x^0\|$$

as well as the fact that $M(\lambda^0)$ is bounded the conclusion that $\|x^t - x^0\|$ is also bounded. Thus we have

$$\|y^t - x^t\| = \alpha_t \|x^t - x^0\| \to 0$$

which contradicts (3.3.2) and (3.3.3). □

We wish to note, somewhat prematurely, that the last theorem is of particular interest in relation to Theorem 4.3.2.

If in addition the boundedness condition imposed on $M(\lambda^0)$ is relaxed, we are confined to finite-dimensional spaces.

Theorem 3.3.2. *Let $X = \mathbb{E}_n$, $M(\lambda)$ be non-empty for all $\lambda \in \Lambda$, and $M(\lambda^0)$ be an affine subspace. M is then Hausdorff-continuous at λ^0.*

Proof. Assume $M(\lambda^0) = x^0 + U$, $x^0 \in M(\lambda^0)$. Since U is a linear subspace and all functions g_i of (3.3.1) are convex it follows that

$$g_i(x + \alpha u) = g_i(x) \qquad \forall i \qquad \forall \alpha \in \mathbb{R} \qquad \forall x \in \mathbb{E}_n \qquad \forall u \in U \tag{3.3.4}$$

holds. If $\dim U = 0$, i.e. $U = \{0\}$, the proposition follows from Theorem 3.1.3. We therefore assume that the theorem is true for $\dim U < k$ and consider the case $\dim U = k$. Fixing $v \in U$, $v \neq 0$, we define $M_v(\lambda) = \{x \in M(\lambda) \mid v^\mathsf{T} x = 0\}$. It then follows from (3.3.4) that

$$M(\lambda) = M_v(\lambda) + L_v \tag{3.3.5}$$

where $L_v = \{\alpha v \mid \alpha \in \mathbb{R}\}$. Since $M_v(\lambda^0)$ is an affine subspace of dimension smaller than k the theorem is valid for M_v and thus because of (3.3.5) also for M. □

Although the next theorem is actually a consequence of Theorem 3.2.1 we present it here as a separate theorem because it represents an important special case.

Theorem 3.3.3. *Let $X = \mathbb{E}_n$, $s \in \{1, \dots, m\}$, and further, g_i be weakly analytic for $i = 1, \dots, s$. If there exists an $x^0 \in M(\lambda^0)$ such that*

$$g_i(x^0) < \lambda_i^0\,, \qquad i = s + 1, \dots, m\,,$$

and if $M(\lambda)$ is non-empty for all $\lambda \in \Lambda$ then M is l.s.c.-B at λ^0.

Proof. In this case we have for the sets I and $M^I(\lambda)$ of Theorem 3.2.1 (3)

$$I = \{i \mid g_i(x) = \lambda_i^0 \ \forall x \in M(\lambda^0)\} \subset \{1, \dots, m\},$$
$$M^I(\lambda) = \{x \in E_n \mid g_i(x) \leqq \lambda_i \ \forall i \in I\} \neq \emptyset .$$

Since the functions g_i are weakly analytic for $i \in I$, $M^I(\lambda^0)$ is an affine subspace. By Theorem 3.3.2 M^I is thus l.s.c.-B at λ^0 and the present theorem follows from Theorem 3.2.1 (3). □

Whereas Theorems 3.2.1 and 3.2.2 yield sufficient conditions for the lower semicontinuity of M which are in general difficult to confirm, the last theorem only uses the functions employed in the description of $M(\lambda)$. If these are all weakly analytic (and convex) then the l.s.c.-B property follows immediately for M (as a mapping of the feasible parameter set). If on the other hand one of the functions g_i is not weakly analytic we also need an interior point condition. Before we demonstrate that this is essential for the validity of the theorem we first give a somewhat more general version in which piece-wise weakly analytic functions are considered.

Corollary 3.3.3.1. *Let the hypotheses of Theorem* 3.3.3 *be weakened as follows: For each* $i = 1, \dots, s$ *there exists a finite number of convex weakly analytic functions*

$$h_{k_i}: E_n \to \mathbb{R}, \qquad k_i = 1, \dots, p_i,$$

and g_i *is given by*

$$g_i(x) = \max\{h_{k_i}(x) \mid k_i = 1, \dots, p_i\} \qquad \forall x \in E_n .$$

The proposition of Theorem 3.3.3 *then remains valid.*

Proof. $M(\lambda)$ may be written as

$$M(\lambda) = \{x \in E_n \mid h_{k_i}(x) \leqq \lambda_i, i = 1, \dots, s, k_i = 1, \dots, p_i; g_j(x) \leqq \lambda_j, \ j = s+1, \dots, m\} .$$

The corollary then follows from the theorem. □

We will now show that one cannot substitute the weaker condition $M(\lambda^0) \neq \emptyset$ for the existence of a "Slater-point" x^0 in Theorem 3.3.3, and that the hypotheses of the theorem do not imply that M is upper semicontinuous (H).

Consider the following function

$$g(x_1, x_2) = \begin{cases} |x_2| \, e^{-(x_1/|x_2|)} & \text{if } x_1 \geqq 0, \quad x_2 \neq 0, \\ 0 & \text{if } x_1 \geqq 0, \quad x_2 = 0, \\ |x_2| - x_1 & \text{if } x_1 \leqq 0. \end{cases} \tag{3.3.6}$$

One easily confirms that g has the following properties:

(P1) g is continuous,

(P2) $g(x_1, \pm 1) = \begin{cases} e^{-x_1} & \text{if } x_1 \geqq 0, \\ 1 - x_1 & \text{if } x_1 \leqq 0, \end{cases}$

(P3) $g(\cdot, 1)$ is convex,

(P4) $g(z) = \alpha g(x_1, 1)$ where $z = \alpha(x_1, 1)$ and $\alpha > 0$,

(P5) $g(\alpha z) = \alpha g(z) \quad \forall \alpha > 0 \quad \forall z \in E_2$,

(P6) $g(x_1, x_2) \leqq g(x_1, y_2)$ if $|x_2| < |y_2|$, and

(P7) $g(x_1, x_2) \geqq g(y_1, x_2)$ if $x_1 < y_1$.

Because of (P1) g is convex if the inequality

(P8) $g(\frac{1}{2}(u + v)) \leqq \frac{1}{2} g(u) + \frac{1}{2} g(v)$

holds for arbitrary $u, v \in E_2$. As a preliminary step we will prove the inequality under the conditions

$$u_2 + v_2 \neq 0, \qquad u_2 \neq 0, \qquad v_2 \neq 0.$$

Writing

$$(u_1, |u_2|) = \alpha(x_1, 1); \; (v_1, |v_2|) = \beta(v_1, 1)$$

we have, using the properties (P6), (P5), (P4), (P3), and again (P4) in that order

$$\begin{aligned} g\left(\tfrac{1}{2}(u+v)\right) &= g\left(\tfrac{1}{2}(u_1+v_1, u_2+v_2)\right) \\ &\leqq g\left(\tfrac{1}{2}(u_1+v_1, |u_2|+|v_2|)\right) \\ &= \tfrac{1}{2}\, g(\alpha x_1 + \beta y_1, \alpha+\beta) \\ &= \tfrac{1}{2}(\alpha+\beta)\, g(\alpha(\alpha+\beta)^{-1} x_1 + \beta(\alpha+\beta)^{-1} y_1, 1) \\ &\leqq \tfrac{1}{2}\alpha g(x_1, 1) + \tfrac{1}{2}\beta g(y_1, 1) \\ &= \tfrac{1}{2}\, g(u) + \tfrac{1}{2}\, g(v). \end{aligned}$$

We now assume that u and v are indeed arbitrary; there then exists an open interval $(0, \varepsilon_0)$ such that for all ε in this interval the inequalities

$$u_2 + v_2 + 2\varepsilon \neq 0, \qquad u_2 + \varepsilon \neq 0, \qquad v_2 + \varepsilon \neq 0$$

hold. From the above (P8) is valid for the points

$$u^\varepsilon = (u_1, u_2 + \varepsilon), \qquad v^\varepsilon = (v_4, v_2 + \varepsilon),$$

and in the limit $\varepsilon \to 0$ thus also for u and v because of the continuity of g (P1).

We can therefore say that g is in fact convex, it is however not weakly analytic (one need only consider the line $\{x \in E_2 \mid x_2 = 0\}$).

Example 3.3.1. M is not l.s.c.-B.

Assume

$$M(\lambda) = \{x \in E_2 \mid g(x) \leqq \lambda_1, -x_2 \leqq \lambda_2\},$$

$$\lambda^0 = (0, 0), \qquad \lambda(\varepsilon) = (\varepsilon\, e^{-1/\varepsilon}, -\varepsilon), \qquad \varepsilon > 0.$$

For small ε we have $\varepsilon\, e^{-1/\varepsilon} < \varepsilon$. The inequality $g(x) \leqq \varepsilon\, e^{-1/\varepsilon}$ thus cannot be satisfied for $x_1 \leqq 0$ and $-x_2 \leqq -\varepsilon$, hence $x \in M(\lambda(\varepsilon))$ implies $x_1 > 0$ for small ε. One thus obtains

$$x \in M(\lambda(\varepsilon)) \Rightarrow (x_1, -\varepsilon) \in M(\lambda(\varepsilon)) \Leftrightarrow \varepsilon\, e^{-x_1/\varepsilon} \leqq \varepsilon\, e^{-1/\varepsilon} \Leftrightarrow x_1 \geqq 1.$$

Since $x = (0, 0)$ is contained in $M(\lambda^0)$, M cannot be l.s.c.-B at λ^0.

Example 3.3.2. M is not u.s.c.-H under the conditions of Theorem 3.3.3.

Using the convex function g (3.3.6) we obtain a weakly analytic function h by defining

$$h(x) = g(x) + e^{-x_1}, \qquad x \in E_2,$$

and set

$$M(\lambda) = \{x \in E_2 \mid h(x) \leqq \lambda\}, \qquad \lambda \in R.$$

In order to confirm that M is not u.s.c.-H at any point $\lambda^0 = e^c$, $c \in R$, we proceed as follows: Consider a suitable sequence $\{x^t\}_{t=1,2,\dots} \subset E_2$ with $h(x^t) \to e^c$ and set $\lambda^t = h(x^t)$. We then have $x^t \in M(\lambda^t)$. Defining $z^t = (x_1^t + 1, x_2^t - 1)$ we find that $z^t \in M(\lambda^0)$ follows from $d(x^t, M(\lambda^0)) < 1$ because of the monotonicity properties (P6) and (P7) which are also valid for h. One verifies that $h(z^t) > e^c$, thus z^t is, for all t, not contained in $M(\lambda^0)$ and we therefore have $d(x^t, M(\lambda^0)) \geqq 1$ for sufficiently large t.

The remainder is largely of technical nature. If for $t > 1$ the points x^t and z^t are assumed to have the form

$$x_1^t = (t-1)\ln(t-1) - c(t-1) - 1\,, \qquad x_2^t = t\,,$$
$$z_1^t = (t-1)\ln(t-1) - c(t-1)\,, \qquad z_2^t = t-1\,,$$

for sufficiently large t, $t > t(c)$, one then has $x_1^t > 0$ and it is easily confirmed that $g(z^t) = e^c$ and $h(z^t) > e^c$ hold. In order to show that $\lim_{t\to\infty} g(x^t) = e^c$, from which the above assertion $h(x^t) \to e^c$ obviously follows, we employ the second mean value theorem of elementary analysis:

$$t \ln t - (t-1)\ln(t-1) = 1 + \ln \tau\,, \qquad \tau \in (t-1, t)\,.$$

It follows that

$$g(x^t) = e^{c(t-1)/t + 1/t(2+\ln\tau)}\,.$$

Because of $\tau < t$ the limit of this product is in fact e^c. The mapping M is not u.s.c.-H at any λ for which $M(\lambda)$ is non-empty.

3.4. Solution Sets of Parametric Linear Systems of Inequalities with or without Integer Conditions

In this section we examine the conditions under which the solution set of a linear system of inequalities with variable right-hand sides is a continuous point-to-set mapping under integer conditions imposed on the variables. To this purpose we define the set

$$G(\lambda) =_{\mathrm{Df}} \{x \in E_n \mid Ax \leqq \lambda,\, x_1, \ldots, x_s \in Z\} \tag{3.4.1}$$

where A is a fixed $(m \times n)$-matrix, Z the set of all integers, s a natural number satisfying $0 < s \leqq n$, and λ a vector of the set $\Lambda \subset E_m$. $G(\lambda)$ is imbedded in the convex polyhedron

$$M(\lambda) =_{\mathrm{Df}} \{x \in E_n \mid Ax \leqq \lambda\} \tag{3.4.2}$$

and with $G(\lambda)$ we associate the set U_G of "half lattice points" of the recession cone of $M(\lambda)$

$$U =_{\mathrm{Df}} O^+M(\lambda) = \{u \in E_n \mid Au \leqq 0\}\,, \tag{3.4.3}$$

i.e. we set

$$U_G =_{\mathrm{Df}} \{u \in E_n \mid Au \leqq 0,\, u_1, \ldots, u_s \in Z\}\,. \tag{3.4.4}$$

Theorem 3.4.1. *If the convex polyhedron $M(\lambda)$ (3.4.2) is non-empty for all $\lambda \in \Lambda$ then there exists an H-continuous mapping $K\colon \Lambda \to 2^{E_n}$ whose images are convex compact polyhedrons, such that $M(\lambda)$ may be represented by*

$$M(\lambda) = K(\lambda) + U \qquad \forall \lambda \in \Lambda\,. \tag{3.4.5}$$

Proof. It suffices to consider the case that $\{0\}$ is the only linear subspace of the polyhedral cone U, since in the general case one may reason as follows: if H is a subspace of maximal dimension of U, and if H_0 is its orthogonal complement we then have

$$M(\lambda) = \big(M(\lambda) \cap H_0\big) + H\,,$$

and the recession cone U_0 of $M(\lambda) \cap H_0$ only contains the subspace $\{0\}$; from $M(\lambda) \cap H_0 = K(\lambda) + U_0$ we then have

$$M(\lambda) = \big(K(\lambda) + U_0\big) + H = K(\lambda) + (U_0 + H) = K(\lambda) + U$$

and therefore we may indeed assume that U is a pointed cone.

We denote the components of λ by λ_i and the rows of A by a^i (and consider them to be column vectors), further we set

$$a = \sum_{i=1}^{m} a^i .$$

For each $\lambda \in \Lambda$ and each $x \in M(\lambda)$ consider the following linear optimization problem:

$$(\mathrm{P}_{\lambda, x}) \qquad \max \{\alpha \mid (\alpha, u) \in V(\lambda, x)\}$$

and the feasible set

$$V(\lambda, x) =_{\mathrm{Df}} \{(\alpha, u) \in E_{1+n} \mid Au \leqq 0,\, Au \geqq Ax - \lambda,\, a^{\mathsf{T}}u + \alpha \leqq a^{\mathsf{T}}x\} .$$

The problem $(\mathrm{P}_{\lambda, x})$ always possesses a solution since $(a^{\mathsf{T}}x, 0)$ lies in $V(\lambda, x)$ and the objective function is upper bounded on $V(\lambda, x)$ because of

$$\alpha \leqq a^{\mathsf{T}}x - a^{\mathsf{T}}u \leqq \left(\sum_{i=1}^{m} a^{i\mathsf{T}}u + \sum_{i=1}^{m} \lambda_i\right) - a^{\mathsf{T}}u = \sum_{i=1}^{m} \lambda_i .$$

Hence the dual problem

$$(\mathrm{D}_{\lambda, x}) \qquad \min \{-(Ax - \lambda)^{\mathsf{T}} w + a^{\mathsf{T}}x \mid (v, w) \in W\}$$

with the constraint polyhedron

$$W =_{\mathrm{Df}} \{(v, w) \in E_{m+m} \mid A^{\mathsf{T}}v - A^{\mathsf{T}}w = -a,\, v \geqq 0,\, w \geqq 0\}$$

also always possesses a solution. We denote the common optimal value of the objective function of both problems by $h(\lambda, x)$ and demonstrate that the function φ defined by

$$\varphi(\lambda) =_{\mathrm{Df}} \inf \{h(\lambda, x) \mid x \in M(\lambda)\} \tag{3.4.6}$$

possesses a continuous minorant $\alpha(\cdot)$ which is suitable for being employed in the construction of the desired mapping $K\colon \Lambda \to 2^{E_n}$.

For each $\lambda \in \Lambda$ and each $x \in M(\lambda)$ there exists a vertex $(v, w) \in W$ such that

$$\begin{aligned} h(\lambda, x) &= -(Ax - \lambda)^{\mathsf{T}} w + a^{\mathsf{T}}x , \\ a^{\mathsf{T}}x &= -v^{\mathsf{T}}Ax + w^{\mathsf{T}}Ax , \qquad v \geqq 0 , \qquad w \geqq 0 \end{aligned} \tag{3.4.7}$$

holds, from which

$$h(\lambda, x) \geqq \lambda^{\mathsf{T}}(w - v) \tag{3.4.8}$$

follows. Now, if $\{(v^1, w^1), \dots, (v^r, w^r)\} \subset W$ is the set of all vertices of W then

$$\alpha(\lambda) =_{\mathrm{Df}} \min \{\lambda^{\mathsf{T}}(w^i - v^i) \mid i = 1, \dots, r\} \tag{3.4.9}$$

is a continuous minorant, since the continuity of $\alpha(\cdot)$ is obvious, and

$$\varphi(\lambda) \geqq \alpha(\lambda) \qquad \forall \lambda \in \Lambda \tag{3.4.10}$$

follows from (3.4.6) and (3.4.8). For each fixed $\lambda \in \Lambda$ we now define

$$K(\lambda) =_{\mathrm{Df}} M(\lambda) \cap \{y \in E_n \mid a^{\mathsf{T}}y \geqq \alpha(\lambda)\} . \tag{3.4.11}$$

First observe that $K(\lambda) \neq \emptyset$ for each $\lambda \in \Lambda$, since $h(\lambda, x) \geqq \alpha(\lambda)$ follows for each $x \in M(\lambda)$ from (3.4.8) and (3.4.9), hence there exists a $u \in E_n$ such that $(\alpha(\lambda), u) \in V(\lambda, x)$ and thus $y = x - a \in K(\lambda)$ follows immediately. The representation (3.4.5) is thus established. $K(\lambda)$ is a bounded polyhedron for each $\lambda \in \Lambda$, since if u is an unbounded direction of $K(\lambda)$ then $Au \leqq 0$ and $a^{\mathrm{T}}u \geqq 0$ must hold, from which, taking the definition of a into account, the equality $Au = 0$ follows. As the cone U is pointed the only possibility that remains is $u = 0$.

The continuity of the mapping K may now be shown as follows: Theorem 3.1.1 implies the closedness of the mapping K (3.4.11) on Λ, likewise K is l.s.c.-B on Λ by Theorem 3.2.3. Since the image sets $K(\lambda)$ are furthermore compact and convex for all $\lambda \in \Lambda$ the proof may be completed by applying Corollary 3.1.3.2. □

An immediate consequence of the foregoing theorem is given by the following corollary.

Corollary 3.4.1.1. *If $M(\lambda)$ (3.4.2) is non-empty for all $\lambda \in \Lambda$ then M is continuous (H) on Λ.*

In the remainder of this section we examine additional hypotheses which may be used to establish the upper semicontinuity (H) or the continuity of the mapping G defined by (3.4.1).

Theorem 3.4.2. *Let the matrix A have rational elements only. If $G(\lambda)$ (3.4.1) is non-empty for all $\lambda \in \Lambda$ then there exists an u.s.c.-H mapping $K_G\colon \Lambda \to 2^{E_n}$ with compact image sets and which has the representation*

$$G(\lambda) = K_G(\lambda) + U_G \tag{3.4.12}$$

for all $\lambda \in \Lambda$ if U_G is the set defined by (3.4.4).

Proof. Since A is rational the existence of integer-valued vectors $\bar{u}^i$, $i = 1, \dots, \tau$, which are generators of the recession cone U (3.4.3), is assured, thus

$$U = \operatorname{con}\{\bar{u}^1, \dots, \bar{u}^\tau\}\,. \tag{3.4.13}$$

If λ is an arbitrary fixed element of Λ then we also have $M(\lambda) \neq \emptyset$ because of (3.4.2), and there exists a bounded polyhedron $K(\lambda)$ such that

$$M(\lambda) = K(\lambda) + U$$

holds. We then define the set

$$K_G(\lambda) =_{\mathrm{Df}} \{x \in E_n \mid x = v + \sum_{i=1}^{\tau} \alpha_i \bar{u}^i,\ v \in K(\lambda), \alpha_i \in [0, 1],$$
$$i = 1, \dots, \tau, x_1, \dots, x_s \in \mathbb{Z}\} \tag{3.4.14}$$

and prove that

$$K_G(\lambda) \neq \emptyset$$

is true. From $x \in G(\lambda) \subset M(\lambda)$ it follows that there exist points $v \in K(\lambda)$ and $u \in U$ such that $x = v + u$. Further, because of (3.4.13) we can find numbers $\varrho_i \geqq 0$, $i = 1, \dots, \tau$, such that $u = \sum_{i=1}^{\tau} \varrho_i \bar{u}_i$. We may therefore write

$$x = v + u = v + \sum_{i=1}^{\tau} (\varrho_i - [\varrho_i])\, \bar{u}^i + \sum_{i=1}^{\tau} [\varrho_i]\, \bar{u}^i$$

where $[a]$ denotes the largest integer which is not greater than a. The above assertion follows from

(i) $x_1, \dots, x_s \in \mathbb{Z}$,

(ii) all components of $\sum_{i=1}^{\tau} [\varrho_i] \bar{u}^i$ are contained in $\mathbb{Z}$, and

(iii) $v + \sum_{i=1}^{\tau} (\varrho_i - [\varrho_i]) \bar{u}^i = x - \sum_{i=1}^{\tau} [\varrho_i] \bar{u}^i$.

We thus also have

$$G(\lambda) \subset K_G(\lambda) + U_G$$

and (3.4.12) follows since one easily confirms that the inclusion relation may be reversed.

The compactness of $K_G(\lambda)$ is obvious, all that remains is to show that the point-to-set mapping $K_G: \Lambda \to 2^{\mathbb{E}_n}$ defined by (3.4.14) is upper semicontinuous (H). The mapping will exhibit this property if we employ the images of the continuous mapping K given by Theorem 3.4.1 in the definition of K_G. Let $K(\lambda)$ be chosen in this sense, $\lambda^0 \in \Lambda$ be arbitrary and fixed, and $\{\lambda^t\} \subset \Lambda: \lambda^t \to \lambda^0$ be an arbitrary sequence. We assign to each t an $x^t \in K_G(\lambda^t)$ to which there correspond points $v^t \in K(\lambda^t)$ and numbers $\alpha_{it} \in [0, 1]$, $i = 1, \dots \tau$, , such that

$$x^t = v^t + \sum_{i=1}^{\tau} \alpha_{it} \bar{u}^i , \qquad t = 1, 2, \dots ,$$

holds. The sequence $\{v^t\}$ so defined has, by Theorem 3.4.1 an accumulation point $v^0 \in K(\lambda^0)$. Since the sequences $\{\alpha_{it}\}$, $i = 1, \dots, \tau$, also possess accumulation points α_{i0} the sequence $\{x^t\}$ likewise has an accumulation point x^0 such that without loss of generality *

$$x^t \to x^0 = v^0 + \sum_{i=1}^{\tau} \alpha_{i0} \bar{u}^i$$

holds. Because the components $x_1^t, \dots, x_s^t$ are contained in $\mathbb{Z}$ and $\mathbb{Z}$ is closed it follows that $x^0 \in K_G(\lambda^0)$. □

Corollary 3.4.2.1. *Let the matrix A have rational elements. If $G(\lambda)$ is non-empty for all $\lambda \in \Lambda$ then $G: \Lambda \to 2^{\mathbb{E}_n}$ is upper semicontinuous (H).*

Proof. If $\lambda^0 \in \Lambda$ and $\{\lambda^t\} \subset \Lambda: \lambda^t \to \lambda^0$ are chosen arbitrarily then any $x^t \in G(\lambda^t)$ may be represented as

$$x^t = v^t + u^t , \qquad v^t \in K_G(\lambda^t) , \qquad u^t \in U_G ,$$

for all t because of (3.4.12). The existence of an accumulation point $v^0 \in K_G(\lambda^0)$ may be demonstrated as in the previous proof and one then has

$$\forall \varepsilon > 0 \ \exists t': \ v^t + u^t \in U_\varepsilon K_G(\lambda^0) + U_G \qquad \forall t \geqq t' .$$

The relation

$$U_\varepsilon K_G(\lambda^0) + U_G \subset U_\varepsilon \big(K_G(\lambda^0) + U_G\big)$$

completes the proof. □

The condition that the matrix A be rational is necessary if one wishes to ensure that the mapping is upper semicontinuous (H), as is shown by the following example using the irrational number

$$\alpha = 1.\ 010010001 \dots$$

The set

$$G(\lambda) = \{(x, y) \in E_2 \mid -\alpha x + y \leqq -\lambda, \alpha x - y \leqq \lambda, x, y \in Z\}$$

obviously only contains the origin of E_2 if $\lambda_0 = 0$. Consider the sequence

$$\{\lambda_t\} = \{0.01001 \ldots; 0.0010001 \ldots; \ldots\}$$

which converges to zero. The sequence $\{(x_t, y_t)\} \subset E_2$ of integer points defined by

$$\begin{aligned} (x_1, y_1) &= (1; 1)\,, \\ (x_2, y_2) &= (100; 101)\,, \\ (x_3, y_3) &= (100000; 101001)\,, \\ &\vdots \qquad\quad \vdots \end{aligned}$$

has the property $(x_t, y_t) \in G(\lambda_t)$ for $t = 1, 2, \ldots$, however we find

$$d\big((x_t, y_t), G(\lambda^0)\big) > \varepsilon\,, \qquad t = 1, 2, \ldots,$$

for each ε satisfying $0 < \varepsilon < 1$.

Assume $S \subset E_n$, if $z \in E_n$ is partitioned into partial vectors x and y then $P_x S$ denotes the projection of the set S onto the subspace spanned by the unit vectors corresponding to x.

Theorem 3.4.3. *Let the matrix A have rational elements and $G(\lambda)$ (3.4.1) be non-empty for all $\lambda \in \Lambda$. Further, let*

$$P_{x_1,\ldots,x_s} G(\lambda) = P_{x_1,\ldots,x_s} G(\lambda^0) \tag{3.4.15}$$

hold for $\lambda \in \Lambda$. The mapping $G: \Lambda \to 2^{E_n}$ defined by (3.4.1) is then continuous at λ^0.

Proof. By Corollary 3.4.2.1 G is u.s.c.-H at λ^0 and in order to establish the lower semicontinuity (B) we then need only show that the following holds for almost all t

$$\forall x \in G(\lambda^0)\ \exists \{x^t\} \subset E_n : x^t \in G(\lambda^t)$$

where $\{\lambda^t\} \subset \Lambda$ is an arbitrary sequence converging to λ^0 and that $x^t \to x^0$ is satisfied. If $\bar{x} \in G(\lambda^0)$ is chosen arbitrarily then $\bar{x}$ is contained in the set

$$G(\bar{x}_1, \ldots, \bar{x}_s, \lambda^0) =_{\mathrm{Df}} \left\{ x \in E_n \,\middle|\, \sum_{i=s+1}^{n} A^i x_i \leqq \lambda^0 - \sum_{i=1}^{s} A^i \bar{x}_i \right\} \tag{3.4.16}$$

where A^i are the columns of the matrix A. Because of (3.4.15) we have

$$G(\bar{x}_1, \ldots, \bar{x}_s, \lambda^t) \neq \emptyset \quad \forall t\,.$$

By Theorem 3.2.3 the mapping $G(\bar{x}_1, \ldots, \bar{x}_s): \Lambda \to 2^{E_n}$ with images given by (3.4.16) is l.s.c.-B at λ^0. Therefore there exists a sequence $\{\bar{x}^t\} \subset E_n$ such that

$$\bar{x}^t \in G(\bar{x}_1, \ldots, \bar{x}_s, \lambda^t) \quad \forall t \quad \text{and} \quad \bar{x}^t \to \bar{x}\,.$$

Since $G(\lambda^t) \supset G(\bar{x}_1, \ldots, \bar{x}_s, \lambda^t)$ for all t, $\bar{x} \in G(\lambda^0)$ and finally the sequence $\{\lambda^t\}, \lambda^t \to \lambda^0$ was assumed to be arbitrary we find that the proof is completed. □

4. Stability Considerations

4.1. Introduction

In this chapter we present conditions for the *stability* of the parameter-dependent problem

(P_λ) $\quad \inf \{f(x, \lambda) \mid x \in M(\lambda)\}\ , \qquad \lambda \in \Lambda\ ,$

under perturbations of the parameter λ. The word *stability* does not stand for some well-defined property but rather for various types of semicontinuity of the extre me value function

$$\varphi = \varphi(\lambda) =_{\mathrm{Df}} \inf \{f(x, \lambda) \mid x \in M(\lambda)\}\ ,$$

the optimal set mapping

$$\psi = \psi(\lambda) =_{\mathrm{Df}} \{x \in M(\lambda) \mid f(x, \lambda) = \varphi(\lambda)\}\ ,$$

or of the *ε-optimal set mappings*

$$\psi_\varepsilon = \psi_\varepsilon(\lambda) =_{\mathrm{Df}} \{x \in M(\lambda) \mid f(x, \lambda) < \varphi(\lambda) + \varepsilon\}\ ,$$

$$\tilde{\psi} = \tilde{\psi}(\lambda, \varepsilon) =_{\mathrm{Df}} \{x \in M(\lambda) \mid f(x, \lambda) \leqq \varphi(\lambda) + \varepsilon\}\ .$$

For the main part we consider conditions under which the extreme value function φ is upper or lower semicontinuous or the optimal set mapping ψ is u.s.c.-H or l.s.c.-B.

The chapter is sectioned according to the hypotheses employed. Thus in Section 4.2, in which the parameter space Λ and the inverse image space X are metric spaces, certain continuity properties of the objective function f and the constraint set mapping M are assumed, and the relations between the semicontinuity of φ and ψ are then examined. In order to ensure that φ is lower semicontinuous (or ψ is u.s.c.-H) in the general problems considered in that section we find that certain compactness conditions are necessary (see for instance Theorems 4.2.1 and 4.2.2), however these conditions are in general difficult to verify if X itself is not compact.

Section 4.3 is devoted to parametric problems which satisfy certain convexity conditions (parameter-dependent problems described by convex or quasiconvex functions or those problems for which φ turns out to be convex). In this case the compactness conditions mentioned above may be weakened, for various classes of problems they may dispensed with altogether. If convexity conditions are imposed one may furthermore find conditions for the lower semicontinuity (B) of the mapping ψ; these are based on an explicit representation of the mapping M in the form

$$M(\lambda) = \{x \in E_n \mid g_i(x, \lambda) \leqq 0, i = 1, \dots, m\}$$

and, apart from certain special cases, consist of assuming that the index set $I(\lambda)$ of all constraints active at all optimal points $x \in \psi(\lambda)$ is constant on Λ. The latter condition will obviously be seldom fulfilled if Λ is an arbitrary parameter set, together with the other l.s.c.-B conditions for ψ it however enables one to find a finite or infinite partitioning of Λ into subsets Λ_ν, $\nu = 1, 2, \dots$, on which ψ (or to be exact ψ restricted to Λ_ν) is l.s.c.-B. We call these subsets "local stability sets" and discuss their properties in Chapter 5.

In Section 4.4 we give a survey of problems for which the extreme values of the primal and (Lagrangian) dual problems are identical; this is based on well-known relations between duality and stability of perturbed convex optimization problems and also uses results on the lower semicontinuity of φ presented in Section 4.3. We also study the behaviour of the primal and dual optimal set mappings in certain particular primal and associated dual parametric problems.

The remainder of the chapter consists of studies devoted to mixed-integer quadratic optimization problems in which the linear parts of the objective function and of the constaints are parameter-dependent. The results of this closing section are based on work by R. Hansel [1] and constitute a generalization of corresponding studies by B. Bank [1], B. C. Eaves [3], and B. Kummer [3] on parametric linear mixed-integer optimization and parametric quadratic optimization respectively.

We now list the classes of optimization problems examined in this chapter.

(1) $\inf \{f(x, \lambda) \mid x \in M(\lambda)\}$, $M: \Lambda \to 2^X$, Λ and X are metric spaces (Section 4.2);

(2) $\inf \{f(x) \mid x \in M(\lambda)\}$, $M: \Lambda \to 2^X$, X is a normed space, $M(\lambda)$ and $f: X \to \mathbb{R}$ are convex (Section 4.3);

(3) $\inf \{f(x, \lambda) \mid x \in M(\lambda)\}$, $M: \Lambda \to 2^{\mathbb{E}_n}$, $M(\lambda)$ is convex, $f(\cdot, \lambda)$ is quasiconvex (Section 4.3);

(4) as under (3), $M(\lambda) = \{x \in \mathbb{E}_n \mid g_i(x, \lambda) \leqq 0, i \in I\}$, $f(\cdot, \lambda)$ and $g_i(\cdot, \lambda)$ are quasiconvex (Section 4.3);

(5) as under (4), $f(\cdot, \lambda)$ and $g_i(\cdot, \lambda)$ are convex, I is finite (Section 4.3);

(6) $\inf \{f(x) \mid g_i(x) \leqq \lambda_i, i = 1, \dots, m\}$, $f, g_i: \mathbb{E}_n \to \mathbb{R}$ are convex (Section 4.3);

(7) $\inf \{f(x, \lambda) \mid x \in M\}$, $\lambda \in \Lambda \subset \mathbb{E}_m$, Λ is a polyhedron, $f(x, \cdot)$ is concave (Section 4.3);

(8) $\inf \{x^{\mathsf{T}}Dx + p^{\mathsf{T}}x \mid Ax \leqq b, x \in \mathbb{E}_n, x_i \text{ integer}, i = 1, \dots, s\}$, $\lambda = (p, b) \in \mathbb{E}_{n+m}$, $0 \leqq s \leqq n$ (Section 4.5).

4.2. Relations between Continuity Properties of the Extreme Value Function and the Optimal Set and Constraint Set Mappings

The present section is concerned with the relations existing between the continuity properties of the mappings of interest. The results obtained enable us to give the investigations which follow later a unified and concise form. The theorems thus have an auxiliary function with respect to the objective of establishing the stability of a problem by examining the description of the same.

We consider the problem

$$(\mathrm{P}_\lambda) \qquad \inf \{f(x, \lambda) \mid x \in M(\lambda)\}\,, \qquad \lambda \in \Lambda\,,$$

subject to the general conditions

(i) X and Λ are metric spaces;

(ii) $f: X \times \Lambda \to \mathbb{R}$, $M: \Lambda \to 2^X$.

As usual λ^0 denotes an arbitrary fixed parameter point of Λ. In all sequences $\{\lambda^t\}$, $\{x^t\}$ the index takes on all positive integer values $t \geqq 1$, $\varphi(\lambda)$ and $\psi(\lambda)$ denote the infimum and the optimal set respectively of (P_λ). The values $\varphi(\lambda) = \pm\infty$ are allowed.

Theorem 4.2.1. *Let the mapping M be closed at λ^0 and K be a non-empty compact subset of X. Then*

(1) *φ is lower semicontinuous at λ^0 if f is lower semicontinuous on $(M(\lambda^0) \cap K) \times \{\lambda^0\}$ and $\psi(\lambda) \cap K \neq \emptyset$ holds for all $\lambda \in \Lambda$;*

(2) *φ is upper semicontinuous at λ^0 if f is upper semicontinuous on $(\psi(\lambda^0) \cap K) \times \{\lambda^0\}$, $\psi(\lambda) \cap K \neq \emptyset$ holds for all $\lambda \in \Lambda$, and the mapping ψ is closed at λ^0;*

(3) *ψ is closed at λ^0 if φ is upper semicontinuous at λ^0 and f is lower semicontinuous on $X \times \{\lambda^0\}$;*

(4) *ψ is u.s.c.-B at λ^0 if the hypotheses of (3) are fulfilled and $\psi(\lambda) \subset K$ holds for all $\lambda \in \Lambda$.*

Proof. (1) Assume $\{\lambda^t\} \subset \Lambda$, $\lambda^t \to \lambda^0$. Choosing x^t from $\psi(\lambda^t) \cap K$ we find that a convergent subsequence $\{x^{t_k}\}$ of $\{x^t\}$ exists such that $x^{t_k} \to x^0$, $k \to \infty$. Since M is closed at λ^0 the point x^0 is contained in $M(\lambda^0) \cap K$. The lower semicontinuity of f at (x^0, λ^0) then implies

$$\begin{aligned}\overline{\lim}\, \varphi(\lambda^t) &= \overline{\lim}\, f(x^t, \lambda^t) \\ &\geqq \underline{\lim}_{k\to\infty} f(x^{t_k}, \lambda^{t_k}) \geqq f(x^0, \lambda^0) \geqq \varphi(\lambda^0)\,.\end{aligned}$$

(2) We proceed as in (1). The point x^0 is contained in $\psi(\lambda^0) \cap K$ since ψ is closed at λ^0. Using the upper semicontinuity of f at (x^0, λ^0) we find in the same manner as in (1)

$$\underline{\lim}\, \varphi(\lambda^t) \leqq \overline{\lim}_{k\to\infty} f(x^{t_k}, \lambda^{t_k}) \leqq f(x^0, \lambda^0) = \varphi(\lambda^0)\,.$$

(3) This follows immediately from Theorem 3.1.1 because of

$$\psi(\lambda) = M(\lambda) \cap \{x \in X \mid f(x, \lambda) \leqq \varphi(\lambda)\}\,.$$

(4) See (3) and Lemma 2.2.2. □

One may easily find counter-examples to confirm that none of the hypotheses of the last theorem may be relaxed — with the obvious exception of the closedness of M at λ^0 assumed in proposition (2). Using stronger conditions one may express the propositions of the theorem in a considerably clearer form.

Corollary 4.2.1.1. *Let M be closed at λ^0, $M(\lambda^0)$ be non-empty, f be continuous, and the metric space X be compact. Then φ is lower semicontinuous at λ^0; φ is also upper semicontinuous at λ^0 if and only if ψ is u.s.c.-B at λ^0.*

Proof. This follows immediately from Theorem 4.2.1 and Lemma 2.2.2. □

The next theorem, due originally to C. Berge [2] can be said to be part of the standard equipment used in stability investigations relating to parametric optimization problems, it may be found in slightly modified form in work by W. W. Hogan [3], R. R. Meyer [1] and others; our particular version contains somewhat stronger statements as it refers to metric and not just topological spaces.

Theorem 4.2.2. (1) *φ is upper semicontinuous at λ^0 if M is l.s.c.-B at λ^0 and f is upper semicontinuous on $M(\lambda^0) \times \{\lambda^0\}$.*

(1') *φ is upper semicontinuous at λ^0 if M is l.s.c.-B at λ^0 and an $x^0 \in \psi(\lambda^0)$ exists such that f is upper semicontinuous at (x^0, λ^0).*

(2) *φ is lower semicontinuous at λ^0 if M is u.s.c.-H at λ^0, $M(\lambda^0)$ is compact and f is lower semicontinuous on $M(\lambda^0) \times \{\lambda^0\}$.*

(3) *ψ is u.s.c.-B at λ^0 if φ is upper semicontinuous at λ^0 and the hypotheses of* (2) *hold.*

Proof. (1), (1′) Assume $\varepsilon > 0$ and that x^0 is the point hypothesized in (1′) or a point contained in $M(\lambda^0)$ which satisfies[1])

$$f(x^0, \lambda^0) < \varphi(\lambda^0) + \varepsilon .$$

Assume further $\{\lambda^t\} \subset \Lambda$, $\lambda^t \to \lambda^0$. Since M is l.s.c.-B at λ^0 there exist for large t points $x^t \in M(\lambda^t)$ such that $x^t \to x^0$. The assertion then follows from the upper semicontinuity of f at (x^0, λ^0):

$$\overline{\lim}\, \varphi(\lambda^t) \leqq \overline{\lim}\, f(x^t, \lambda^t) \leqq f(x^0, \lambda^0) < \varphi(\lambda^0) + \varepsilon .$$

(2) Assume $\lambda^t \in \Lambda$, $\lambda^t \to \lambda^0$, and that x^t is contained in $M(\lambda^t)$. It suffices to show that

$$\overline{\lim}\, f(x^t, \lambda^t) \geqq \varphi(\lambda^0) .$$

According to Theorem 3.1.2 (2) M is in fact u.s.c.-B at λ^0, together with the compactness of $M(\lambda^0)$ Lemma 2.2.2 then implies the existence of an accumulation point x^0 of the sequence $\{x^t\}$ and which is contained in $M(\lambda^0)$. From the lower semicontinuity of f at (x^0, λ^0) we finally obtain from $x^{t_k} \to x^0$

$$\overline{\lim}\, f(x^t, \lambda^t) \geqq \underline{\lim}\, f(x^{t_k}, \lambda^{t_k}) \geqq f(x^0, \lambda^0) \geqq \varphi(\lambda^0) .$$

(3) The proof follows from Theorem 3.1.2 (1) and

$$\psi(\lambda) = M(\lambda) \cap \{x \in X \mid f(x, \lambda) \leqq \varphi(\lambda)\}$$

if $M(\lambda^0)$ is non-empty, and is trivial otherwise. □

The compactness needed in (2) and (3) has drastic consequences for many applications, this condition however cannot be simply dispensed with as was illustrated by Example 1.1 in the General Introduction (φ was not lower semicontinuous). A further unpleasant example will now be given; in this case the constraint set is fixed, however neither φ is lower semicontinuous nor is ψ u.s.c.-H.

Example 4.2.1

$$f(x, \lambda) = -x_1 + x_1 x_2 + 2[\lambda^2 x_2^2 - 2\lambda(1-\lambda)\, x_2] ,$$
$$M = \{x \in E_2 \mid 0 \leqq x_1 \leqq 1, x_2 \geqq 0\} , \qquad \lambda \in R .$$

The function in square brackets is the same as that employed in Example 1.1, it is lower bounded by -1 if $x_2 \geqq 0$ and this is also the case for $x_1 x_2 - x_1$ defined on M. We therefore have $\inf\{f(x, \lambda) \mid x \in M\} \geqq -3$ for all λ. Since M is a convex polyhedron and $f(\cdot, \lambda)$ is a quadratic function it follows from the existence theorem of quadratic optimization that $\psi(\lambda)$ is non-empty. The sets $\psi(\lambda)$ are even compact if $\lambda \neq 0$ since in the limit $\|x\| \to \infty$ we immediately find $f(x, \lambda) \to \infty$ if $x \in M$.

We now consider φ and ψ. Obviously $\psi(0) = \{(1, 0)\}$, $\varphi(0) = -1$. Setting

$$\bar{x}(\lambda) = (0, (1-\lambda)\, \lambda^{-1})$$

where $\lambda \in (0, 1)$ we obtain

$$\varphi(\lambda) \leqq f(\bar{x}(\lambda), \lambda) = -2(1-\lambda)^2$$

[1]) One easily confirms that the proof is trivial if $\varphi(\lambda^0) = +\infty$, and that it may be carried out in similar fashion if $\varphi(\lambda^0) = -\infty$ by using the convention $-\infty + \varepsilon = -1/\varepsilon$.

from which

$$\varliminf_{\lambda \to +0} \varphi(\lambda) \leqq -2 \tag{4.2.1}$$

follows, i.e. φ is not lower semicontinuous at $\lambda = 0$. Finally, assuming that ψ were u.s.c.-H at $\lambda = 0$ we would have $\bar{x}(\lambda) \to (1, 0)$ if $\lambda \to +0$ and $\bar{x}(\lambda) \in \psi(\lambda)$ and thus, since f is continuous

$$\varlimsup_{\lambda \to +0} \varphi(\lambda) = \varlimsup_{\lambda \to +0} f(\bar{x}(\lambda), \lambda) = f((1, 0), 0) = -1$$

which contradicts (4.2.1).

As was shown by S. Dolecki [3] the compactness condition contained in proposition (2) of Theorem 4.2.2 may be avoided as follows:

Theorem 4.2.3 (S. Dolecki)[1]. *Let $M(\lambda^0)$ be closed. Then*

(1) *φ is lower semicontinuous at λ^0 if the function $f(x, \lambda) = f(x)$ does not depend on λ and is lower semicontinuous on $M(\lambda^0)$ and if M is u.s.c.-B at λ^0;*

(2) *if on the other hand M is not u.s.c.-B at λ^0 then there exists a continuous function $f: X \to \mathbb{R}$ such that*

$$\varphi(\lambda) = \inf \{f(x) \mid x \in M(\lambda)\}$$

is not lower semicontinuous at λ^0.

Proof. (1) If φ were not lower semicontinuous then there would exist sequences $\{\lambda^t\} \subset \Lambda$, $\{x^t\} \subset X$ and an $\varepsilon > 0$ such that $\lambda^t \to \lambda^0$, $x^t \in M(\lambda^t)$, $f(x^t) < \varphi(\lambda^0) - \varepsilon$. Obviously this implies $x^t \in M(\lambda^t) \setminus M(\lambda^0)$ for all t. Since M is u.s.c.-B there exists by Lemma 2.2.2 an accumulation point x^0 of the sequence $\{x^t\}$ and which is contained in $M(\lambda^0)$. The lower semicontinuity of f however leads to a contradiction:

$$\varphi(\lambda^0) \leqq f(x^0) \leqq \varlimsup f(x^t) \leqq \varphi(\lambda^0) - \varepsilon .$$

(2) Since M is not u.s.c.-B there exists a sequence $\{\lambda^t\}$, $\lambda^t \to \lambda^0$, and an open set $\Omega \supset M(\lambda^0)$ such that

$$M(\lambda^t) \setminus \Omega \neq \emptyset , \qquad t = 1, 2, \ldots$$

Defining $A = X \setminus \Omega$ and setting

$$s(x) = d(x, A) + d(x, M(\lambda^0)) , \qquad f(x) = d(x, A) \cdot s(x)^{-1}$$

where $x \in X$, we find that f is well-defined and continuous since A and $M(\lambda^0)$ are disjoint closed sets and thus $s(x) > 0$. Clearly

$$f(x) = \begin{cases} 0 & \text{if} \quad x \in A , \\ 1 & \text{if} \quad x \in M(\lambda^0) \end{cases}$$

and we thus conclude that the function

$$\varphi(\lambda) = \inf \{f(x) \mid x \in M(\lambda)\}$$

is not lower semicontinuous at λ^0. □

We now turn our attention to the lower semicontinuity of M and ψ and additionally consider the mapping

$$\psi_\varepsilon(\lambda) = \{x \in M(\lambda) \mid f(x, \lambda) < \varphi(\lambda) + \varepsilon\}$$

---

[1]) We refer to the original paper cited above for the case that Λ and X are certain more general topological spaces.

where we as usual use the convention $-\infty + \varepsilon = -1/\varepsilon$, $\varepsilon > 0$. The points of $\psi_\varepsilon(\lambda)$ are said to be *ε-optimal* for the problem (P_λ).

Theorem 4.2.4. *Let f be continuous on $X \times \Lambda$. Then*

(1) *ψ_ε is l.s.c.-B at λ^0 for each $\varepsilon > 0$ if M is l.s.c.-B and φ is lower semicontinuous both at λ^0;*

(2) *φ is continuous at λ^0 if ψ is l.s.c.-B at λ^0 and $\psi(\lambda^0)$ is non-empty;*

(3) *φ is continuous at λ^0 if there exists an $\varepsilon_0 > 0$ such that ψ_ε is l.s.c.-B at λ^0 for all $\varepsilon \in (0, \varepsilon_0)$.*

Proof. (1) This follows from Lemma 2.2.4 because of $\psi_\varepsilon(\lambda) = \Gamma\lambda \cap M(\lambda)$ and since the mapping

$$\Gamma\lambda = \{x \in X \mid f(x, \lambda) < \varphi(\lambda) + \varepsilon\}$$

is strongly lower semicontinuous at λ^0 for fixed ε.

(2) Assume $\{\lambda^t\} \subset \Lambda$, $\lambda^t \to \lambda^0$, and $x^0 \in \psi(\lambda^0)$. Since ψ is l.s.c.-B at λ^0 there exist for large t points $x^t \in \psi(\lambda^t)$ such that $x^t \to x^0$, $t \to \infty$. The assertion

$$\lim \varphi(\lambda^t) = \lim f(x^t, \lambda^t) = f(x^0, \lambda^0) = \varphi(\lambda^0)$$

is a consequence of the continuity of f.

(3) In analogous fashion we find

$$\overline{\lim}\, \varphi(\lambda^t) \leqq \varphi(\lambda^0) + \varepsilon , \qquad \underline{\lim}\, \varphi(\lambda^t) \geqq \varphi(\lambda^0) - \varepsilon$$

if $x^t \in \psi_\varepsilon(\lambda^t)$ since ψ_ε is by hypothesis l.s.c.-B at λ^0 and the proof follows in the limit $\varepsilon \to 0$. □

Propositions (2) and (3) of the preceding theorem are only of theoretical interest in establishing the continuity of φ, in practical applications it is usually simpler to show that φ is continuous than to demonstrate that ψ or ψ_ε is lower semicontinuous. One reads from (1) that ψ_ε is l.s.c.-B under relatively weak conditions in contrast to ψ, in the following sections we will see just how strong the corresponding conditions for ψ must be in even quite simple problems.

From the results of Section 2.3 we can conclude that in the case of convex problems (it is sufficient if $\psi_\varepsilon(\lambda)$ is convex) a continuous mapping which assigns an ε-optimal point $x(\lambda)$ to each λ will occur more frequently than a continuous correspondence between parameter and (exact) solution. A typical example is given by linear parametric problems of the form (cf. Section 4.3)

$$\lambda_1^{\mathsf{T}} x \to \min , \qquad Ax = \lambda_2 , \qquad x \geqq 0$$

where ψ_ε is always l.s.c.-B, the same however only applies to the mapping ψ if additional conditions are met.

A simple consequence of Theorem 4.2.4 now follows.

Corollary 4.2.4.1. *If the mapping M is l.s.c.-B at λ^0 then the following statements are equivalent:*

(1) *φ is continuous at λ^0;*

(2) *ψ_ε is l.s.c.-B at λ^0 for each $\varepsilon > 0$.*

Proof. The equivalence follows immediately from Theorem 4.2.4 (1) and (3). □

As in Section 3.1 we also consider the possibility of reducing parametric problems to problems with perturbed right-hand sides. In this respect the following theorem is analogous to Theorem 3.1.4.

Theorem 4.2.5. *Let f_i, $i = 0, 1, \dots, m$, be arbitrary (not necessarily continuous) real functions defined on $X \times \Lambda$, and let $\delta_i: \Lambda \to \mathbb{R}$ be functions which are continuous at $\lambda^0 \in \Lambda$ and satisfy $\delta_i(\lambda^0) = 0\ \forall i$. Under the condition*

$$f_i(x, \lambda^0) - f_i(x, \lambda) \leqq \delta_i(\lambda) \quad \forall i \qquad \forall x \in X \tag{4.2.2}$$

the following then holds:

The function φ defined by

$$\varphi(\lambda) = \inf \{f_0(x, \lambda) \mid f_i(x, \lambda) \leqq 0, i = 1, \dots, m\}$$

is lower semicontinuous at λ^0 if the function

$$\varphi^0: \mathbb{E}_m \to \mathbb{R} \cup \{-\infty, +\infty\}$$

$$\varphi^0(y) = \inf \{f_0(x, \lambda^0) \mid f_i(x, \lambda^0) \leqq y_i, i = 1, \dots, m\}$$

is lower semicontinuous at $y = 0$.

Proof. For each sequence $\{\lambda^t\}$ with $\lambda^t \in \Lambda$, $\lambda^t \to \lambda^0$ and arbitrary x^t satisfying $f_i(x^t, \lambda^t) \leqq 0$, $i = 1, \dots, m$, we obtain from (4.2.2)

$$f_0(x^t, \lambda^t) \geqq f_0(x^t, \lambda_0) - \delta^0(\lambda^t),$$

$$f_i(x^t, \lambda^0) \leqq \delta_i(\lambda^t).$$

The conditions of the theorem then give

$$\begin{aligned} \underline{\lim}\, f_0(x^t, \lambda^t) &\geqq \underline{\lim}\, f_0(x^t, \lambda^0) \\ &\geqq \underline{\lim}\, \varphi^0(\delta_1(\lambda^t), \dots, \delta_m(\lambda^t)) \geqq \varphi^0(0) = \varphi(\lambda^0). \end{aligned}$$

By the choice of x^t this however signifies that φ is lower semicontinuous at λ^0. □

Finally, we remark that the condition (4.2.2) may be used to characterize a mapping $\Gamma: \Lambda \to 2^X$ as being u.s.c.-H. Writing

$$f_1(x, \lambda) = d(x, \Gamma\lambda), \qquad x \in X, \qquad \lambda \in \Lambda,$$

we find that the following is true:

Γ is u.s.c.-H at λ^0 if and only if there exists a real function δ_1 which is continuous at λ^0 and which satisfies $\delta_1(\lambda^0) = 0$ and (4.2.2).

We leave the simple proof to the reader.

4.3. Convex Parametric Problems

In this section we exploit convexity properties in examining the parametric problem

$$(\mathrm{P}_\lambda) \qquad \inf \{f(x, \lambda) \mid x \in M(\lambda)\}, \qquad \lambda \in \Lambda.$$

Besides the symbols φ and ψ for the extreme value function and the optimal set mapping of (P_λ) we employ $\mathfrak{A}$ and $\mathfrak{B}$ to denote the solubility set and the feasible parameter set respectively:

$$\mathfrak{A} =_{\mathrm{Df}} \{\lambda \in \Lambda \mid \psi(\lambda) \neq \emptyset\},$$

$$\mathfrak{B} =_{\mathrm{Df}} \{\lambda \in \Lambda \mid M(\lambda) \neq \emptyset\}.$$

Theorem 4.3.1. *Let X and Y be real linear normed spaces, Ω and Λ be open, non-empty and convex subsets of X and Y respectively with induced topologies and let*

$$\Gamma: \Lambda \to 2^X$$

be a mapping which is l.s.c.-B at $\lambda^0 \in \Lambda$ and which has the property

$$\tfrac{1}{2}\Gamma\lambda^1 + \tfrac{1}{2}\Gamma\lambda^2 \subset \Gamma(\tfrac{1}{2}\lambda^1 + \tfrac{1}{2}\lambda^2) \quad \forall \lambda^1, \lambda^2 \in \Lambda \,. \tag{4.3.1}$$

Further, let f be independent of λ, the restriction $f|_\Omega$ be convex and upper semicontinuous[1]), *and let the following hold on Λ*

$$M(\lambda) = \Gamma\lambda \cap \Omega \,, \qquad M(\lambda^0) \neq \emptyset \,.$$

Then φ is continuous at λ^0.

Note 4.3.1.1. The conditions of the theorem mean that the graph

$$G = \{(x, \lambda) \mid \lambda \in \Lambda, x \in M(\lambda)\}$$

contains the centre point of the line segment between any two arbitrary points contained in G, in other words G is "almost convex". An essential feature is that λ^0 is contained in the interior of the feasible parameter set $\mathfrak{B}$, this means that the theorem cannot be applied directly in proving Theorem 4.3.5 below.

Proof of Theorem 4.3.1. By Lemma 2.2.5 the mapping M is l.s.c.-B at λ^0 and thus it follows from Theorem 4.2.2 that φ is upper semicontinuous at λ^0.

If $\varepsilon > 0$ is arbitrary and fixed then one need only show that a $\delta > 0$ exists such that

$$\varphi(\lambda) \geqq \varphi(\lambda^0) - \varepsilon \qquad \forall \lambda \in V_\delta\{\lambda^0\} \,. \tag{4.3.2}$$

Since Λ is an open subset of Y the neighbourhoods $V_\delta\{\lambda^0\}$ are identical with the open balls $K_\delta = \{y \in Y \mid \|y - \lambda^0\| < \delta\}$ for $\delta < \delta_1$ and sufficiently small δ_1. There also exists a $\delta \in (0, \delta_1)$ such that

$$\varphi(\lambda) < \varphi(\lambda^0) + \varepsilon \qquad \forall \lambda \in V_\delta\{\lambda^0\} \tag{4.3.3}$$

holds (since φ is upper semicontinuous at λ^0). We now fix $\lambda \in V_\delta\{\lambda^0\}$ and an arbitrary element $x \in M(\lambda)$[2]). The choice of δ ensures that the point $\mu = \lambda^0 + (\lambda^0 - \lambda)$ is also contained in $V_\delta\{\lambda^0\}$. By (4.3.3) and because of $\varphi(\lambda^0) < +\infty$ there exists a $z \in M(\mu)$ such that $f(z) < \varphi(\lambda^0) + \varepsilon$. Since Ω is convex it then follows from (4.3.1) that $\frac{1}{2}x + \frac{1}{2}z \in M(\lambda^0)$ holds. Employing the convexity of f we obtain

$$\varphi(\lambda^0) \leqq f(\tfrac{1}{2}x + \tfrac{1}{2}z) < \tfrac{1}{2}f(x) + \tfrac{1}{2}(\varphi(\lambda^0) + \varepsilon) \,,$$

$$f(x) > \varphi(\lambda^0) - \varepsilon \,.$$

Since $x \in M(\lambda)$ was chosen arbitrarily we find that (4.3.2) holds. □

Note 4.3.1.2. The proof remains the same if X is only a real linear metric space. If Y also has this property then the proof follows in analogous fashion provided the metric of Y is invariant with respect to linear transformations:

$$d_Y(a + c, b + c) \equiv d_Y(a, b) \,.$$

If this is not the case the conclusion $\mu \in V_\delta\{\lambda^0\}$ is no longer valid.

Theorem 4.3.1 may in particular be applied to parametric problems of the type

$$\inf\{f(x) \mid x \in \Omega, g(x) - \lambda \in K\} \,, \qquad \lambda \in Y \,, \tag{4.3.4}$$

---

[1]) Setting $Y = X$, $\Lambda = \Omega$ and $\Gamma\lambda = \{\lambda\}$ the Theorem gives the well-know fact that a real-valued convex and upper semicontinuous function defined on an open convex subset of a normed space is continuous.

[2]) If $M(\lambda) = \emptyset$ there is nothing more to prove.

if Ω and f are defined as in the theorem, K is a convex cone contained in the normed space Y and $g: \Omega \to Y$ is a continuous operator satisfying the convexity condition

$$g(\tfrac{1}{2} x^1 + \tfrac{1}{2} x^2) - \tfrac{1}{2} [g(x^1) + g(x^2)] \in K \qquad \forall x^1, x^2 \in \Omega .$$

If $\Gamma\lambda = \{x \in \Omega \mid g(x) - \lambda \in K\}$ one finds that (4.3.1) holds. Further, if a point $x^0 \in \Omega$ exists such that

$$g(x^0) - \lambda^0 \in \text{int } K$$

then Γ is l.s.c.-B at λ^0. The mapping Γ^0 with

$$\Gamma^0\lambda = \{x \in \Omega \mid g(x) - \lambda \in \text{int } K\}$$

is then in fact strongly lower semicontinuous at λ^0 and $\Gamma\lambda^0 \subset \text{cl } \Gamma^0\lambda^0$ holds since for each $x \in \Gamma\lambda^0$ all points of the open line segment between x and x^0 are contained in $\Gamma^0\lambda^0$ and the lower semicontinuity of Γ follows from Corollary 2.2.5.1.

The Slater condition $\{x \in \Omega \mid g(x) - \lambda^0 \in \text{int } K\} \neq \emptyset$ thus establishes the continuity of the extreme value function corresponding to (4.3.4) at λ^0 if the above conditions are fulfilled. The next theorem constitutes a further application of the previous one.

Theorem 4.3.2. *Let X be a real linear normed space, $M: \Lambda \to 2^X$ be u.s.c.-H at $\lambda^0 \in \Lambda$ and the set $M(\lambda^0)$ be non-empty and convex. Let further the function f be independent of λ and convex and continuous on X.*

The function φ is then lower semicontinuous at λ^0.

Proof. The mapping

$$\Gamma z = z + M(\lambda^0) , \qquad z \in X ,$$

satisfies (4.3.1), the function

$$h(z) = \inf \{f(x) \mid x \in \Gamma z\}$$

is thus continuous. Let $\{\lambda^t\} \subset \Lambda$ be a sequence which converges to λ^0. Since M is u.s.c.-H at λ^0 one may find for large t positive ε_t such that

$$M(\lambda^t) \subset \bigcup_{||z||<\varepsilon_t} \Gamma_z \quad \text{and} \quad \varepsilon_t \to 0$$

hold. Thus

$$\underline{\lim}\, \varphi(\lambda^t) \geqq \underline{\lim} \inf \{h(z) \bigm| ||z|| < \varepsilon_t\} = h(0) = \varphi(\lambda^0)$$

and the proof is complete. □

Whereas upper semicontinuity (B) of the constraint set mapping M is by Theorem 4.2.3 sufficient to establish the lower semicontinuity of φ in problems with fixed continuous objective function and variable constraint sets, if additional convexity assumptions are made upper semicontinuity (H) of M suffices in this respect. In this sense the Theorems 4.2.3 and 4.3.2 are related to each other as sufficient continuity conditions. On the other hand the latter theorem may not be reversed in the same manner as was the case for the former, the function $\varphi(\lambda) = \inf \{f(x) \mid x \in M(\lambda)\}$ may be continuous for any convex objective function f even if the mapping M is nowhere u.s.c.-H. In this connection we refer to Theorem 4.3.5 and Example 3.3.2. One should also remember that Theorem 3.3.1 is a sufficient condition for the upper semicontinuity (H) of M.

We now consider the problem (P_λ) subject to the following conditions:

$$X = E_n; \tag{4.3.5}$$

$$M: \Lambda \to 2^X \text{ is l.s.c.-B at } \lambda^0 \in \Lambda; \tag{4.3.6}$$

$$\psi(\lambda^0) \text{ is non-empty and bounded;} \tag{4.3.7}$$

f is lower semicontinuous on $X \times \{\lambda^0\}$ and a point $x^0 \in \psi(\lambda^0)$ exists such that f is upper semicontinuous at (x^0, λ^0); (4.3.8)

$$f(\cdot, \lambda) \text{ is quasiconvex on } X \text{ for each fixed } \lambda \in \Lambda; \tag{4.3.9}$$

$$M(\lambda^0) \text{ is convex and closed.} \tag{4.3.10}$$

Theorem 4.3.3. *If the conditions* (4.3.5) *to* (4.3.10) *are fulfilled, all sets* $M(\lambda)$, $\lambda \in \Lambda$, *are convex and the mapping* M *is closed at* λ^0, *then* φ *is continuous at* λ^0, *and* ψ *is u.s.c.-B at* λ^0.

Note 4.3.3.1. The assumptions of the theorem are not just of a technical nature for establishing a proof. If any one is omitted this theorem no longer holds. (See the Examples 4.3.1 to 4.3.4.)

Proof of Theorem 4.3.3. By (4.3.6), (4.3.8) and Theorem 4.2.2 φ is upper semicontinuous at λ^0. The upper semicontinuity (B) of

$$\psi(\lambda) = M(\lambda) \cap \{x \in E_n \mid f(x, \lambda) - \varphi(\lambda) \leqq 0\}$$

at λ^0 follows from Theorem 3.1.3 if M is identified with Γ, $f_i = f$, and $\tau_i = \varphi$. We demonstrate the lower semicontinuity of φ at λ^0 by contradiction. Let $\{\lambda^t\} \subset \Lambda$, $\lambda^t \to \lambda^0$, be a sequence satisfying

$$\underline{\lim}\, \varphi(\lambda^t) < \varphi(\lambda^0) .$$

There then exist for large t points $x^t \in M(\lambda^t) \cup M(\lambda^0)$ such that

$$\underline{\lim}\, f(x^t, \lambda^t) < \varphi(\lambda^0) \tag{4.3.11}$$

also holds. If $G(\lambda) = M(\lambda) \cap \{x \in E_n \mid f(x, \lambda) \leqq \varphi(\lambda^0)\}$ then $x^t \in G(\lambda^t)$ obviously follows from $f(x^t, \lambda^t) < \varphi(\lambda^0)$. Because of (4.3.11) the function

$$h(\lambda) = \inf \{f(x, \lambda) \mid x \in G(\lambda)\}$$

is also not lower semicontinuous at λ^0. On the other hand $G(\lambda^0) = \psi(\lambda^0)$ is compact and G is u.s.c.-B at λ^0 by Theorem 3.1.3. It follows from Theorem 4.2.2 that h is lower semicontinuous at λ^0, this contradiction completes the proof. □

Theorem 4.3.3 may also be derived from Theorem 9 of W. W. HOGAN [3] if f is continuous. The basic idea of the proof (in the present text it is contained in hidden form in Lemma 3.1.1 and consists of constructing the accumulation point z^*) may be found in an earlier paper by E. G. GOL'ŠTEĬN [3] in that author's proof of Theorem 3.1, Chapter 5, from which the proposition of Theorem 4.3.3 for constant constraint sets follows.

The following corollaries are consequences of the preceding theorem.

Corollary 4.3.3.1. *Under the conditions of Theorem* 4.3.3 *the* ε*-optimal set mapping* ψ_ε *defined by*

$$\psi_\varepsilon(\lambda) = \{x \in M(\lambda) \mid f(x, \lambda) < \varphi(\lambda) + \varepsilon\}$$

is l.s.c.-B at λ^0 *for each* $\varepsilon > 0$.

Proof. This follows immediately if Theorem 4.2.4 is taken into account. □

Corollary 4.3.3.2. *If the conditions of Theorem* 4.3.3 *are fulfilled then the mapping* $\overline{\psi}$ *defined by*

$$\overline{\psi}(\lambda, \varepsilon) = \{x \in M(\lambda) \mid f(x, \lambda) \leqq \varphi(\lambda) + \varepsilon\}\,, \qquad \lambda \in \Lambda\,, \qquad \varepsilon \geqq 0\,,$$

is u.s.c.-B at $(\lambda^0, 0)$.

Proof. Follows from the continuity of φ at λ^0 and Theorem 3.1.3. □

The last corollary is of interest if the problems corresponding to the parameters λ^t, $t = 1, 2, \dots, \lambda^t \to \lambda^0$, constitute certain substitute problems with respect to the problem to be solved, (P_{λ^0}), and the problems (P_{λ^t}) are solved with increasing accuracy $\varepsilon_t \to 0$. By the last corollary and Lemma 2.2.2 the following holds for each sequence $\{x^t\}$ of ε_t-optimal solutions of (P_{λ^t}): The sequence $\{x^t\}$ possesses an accumulation point and each of its accumulation points is contained in the solution set $\psi(\lambda^0)$.

We now return to the hypotheses of Theorem 4.3.3 and consider the question of whether any of them may be omitted. Example 1.1 in the General Introduction illustrates the necessity of the boundedness condition of (4.3.7). If we admit the possibility $\psi(\lambda^0) = \emptyset$ then one obtains a counter-example as follows.

Example 4.3.1. $\Lambda = \mathbb{R}^+$, $\lambda^0 = 0$, $X = \mathbb{R}$, $M(\lambda) \equiv \mathbb{R}$, $f(x, \lambda) = \min\{e^x, 1 + \lambda x\}$. Here $\varphi(0) = 0$, $\varphi(\lambda) = -\infty\ \forall \lambda > 0$.

In the next example X is only a Banach space.

Example 4.3.2. See Example 3.1.1 and set $M = \Gamma$, $f(x, \lambda) = g(x)$. ψ is then not u.s.c.-H at λ^0.

The next two examples do not satisfy the lower semicontinuity condition (4.3.6) or alternatively the condition that all sets $M(\lambda)$, $\lambda \in \Lambda$, are convex.

Example 4.3.3. $X = \Lambda = \mathbb{R}$, $M(0) = \{0\}$, $M(\lambda) = \left\{\frac{1}{\lambda}\right\}$, $\lambda \neq 0$, $f(x, \lambda) = x$. φ is neither lower nor upper semicontinuous at $\lambda = 0$, ψ is not u.s.c.-H at $\lambda = 0$.

Example 4.3.4. As in Example 4.3.3 except for $M(\lambda) = \left\{0, \frac{1}{\lambda}\right\}$ for $\lambda \neq 0$.

Further examples corresponding to the remaining conditions are easily constructed, this however is left to the reader.

The following corollary is intended to show that in Theorem 4.3.3 the convexity of the sets $M(\lambda)$, $\lambda \neq \lambda^0$, may be dispensed with provided M satisfies a stronger continuity condition at λ^0.

Corollary 4.3.3.3. *If conditions* (4.3.5) *to* (4.3.10) *are satisfied and M is u.s.c.-H at* λ^0 *then* φ *is continuous at* λ^0 *and* ψ *is u.s.c.-B at* λ^0.

Proof. This is the same as the proof of Theorem 4.3.3 if Theorem 3.1.3 is replaced by Corollary 3.1.3. □

Obviously Theorem 4.3.3 may be applied to parametric problems where

$$M(\lambda) = \{x \in \mathbb{E}_n \mid g_i(x, \lambda) \leqq 0\ \forall i \in I\} \tag{4.3.12}$$

if f and all g_i are quasiconvex with respect to x, however condition (4.3.6) (M is l.s.c.-B at λ^0) would appear to present difficulties. Example 4.3.3 illustrates the fact that the condition is needed even just to establish the lower semicontinuity of φ. The next theorem does not use condition (4.3.6) and is thus a welcome supplement to the last theorem.

Theorem 4.3.4. *Let the conditions* (4.3.5) *and* (4.3.7) *be fulfilled for the problem* (P_λ) *and the sets* $M(\lambda)$ *be defined by* (4.3.12) *with arbitrary index set* I. *Further let the functions* f *and* g_i, $i \in I$, *be lower semicontinuous on* $\mathbb{E}_n \times \{\lambda^0\}$ *and quasiconvex for all* λ *with respect to* $x \in \mathbb{E}_n$. *If a point* $x^0 \in \psi(\lambda^0)$ *exists such that* $f(x^0, \cdot)$ *and* $g_i(x^0, \cdot)$ *are upper semicontinuous at* λ^0 *then* φ *is lower semicontinuous at* λ^0.

Proof. Define

$$G(\lambda) = M(\lambda) \cap \{x \in \mathbb{E}_n \mid f(x, \lambda) \leqq \varphi(\lambda^0)\} .$$

Because of the hypotheses of the theorem one may apply Theorem 3.1.3, thus G is u.s.c.-B at λ^0 and the function

$$h(\lambda) = \inf \{f(x, \lambda) \mid x \in G(\lambda)\}$$

is lower semicontinuous at λ^0 taking Theorem 4.2.2 into account. As was shown in the proof of Theorem 4.3.3 φ is then also lower semicontinuous at λ^0. □

Using the duality theorem of linear optimization D. H. Martin [1] proved the last theorem for the special case

$$\inf \{c(\lambda)^T x \mid A(\lambda) x \leqq b(\lambda), x \in \mathbb{E}_n\}$$

where the matrix and vector functions A, c, b are continuous. We again find that no conclusions can be drawn from the hypotheses regarding the upper semicontinuity of φ or ψ.

Example 4.3.5

$$(P_\lambda) \qquad \min \{x \mid x \in \mathbb{R}, \lambda x = 0, x \geqq -1\} , \qquad \lambda \in \mathbb{R} .$$

Here $\varphi(0) = -1$ and $\varphi(\lambda) = 0, \lambda \neq 0$. We also have $\psi(0) = \{-1\}, \psi(\lambda) = \{0\}, \lambda \neq 0$.

We now turn our attention to problems (P_λ) described by convex and weakly analytic functions. In contrast to the preceding theorems we no longer need interior point conditions, upper semicontinuity of M or compactness of the solution set $\psi(\lambda^0)$ in order to establish the continuity of the extreme value function.

Consider first the special case that (P_λ) is of the form

$$\inf \{f(x) \mid x \in \mathbb{E}_n, g_i(x) \leqq \lambda_i, i = 1, \dots, m\} , \qquad \lambda \in \Lambda , \tag{4.3.13}$$

where Λ is a subset of $\mathbb{E}_m$ (with induced topology) and

$$M(\lambda) = \{x \in \mathbb{E}_n \mid g_i(x) \leqq \lambda_i, i = 1, \dots, m\} .$$

Theorem 4.3.5. *Let the functions* f, g_i, $i = 1, \dots, m$, *be convex on* $\mathbb{E}_n$, $\lambda^0 \in \Lambda$ *and*

$$J =_{\mathrm{Df}} \{i \mid g_i \text{ is weakly analytic}\} .$$

Then,

(1) *the extreme value function* φ *corresponding to* (4.3.13) *is continuous at* λ^0 *and* ψ *is closed at* λ^0 *if* $M(\lambda) \neq \emptyset$ *for all* $\lambda \in \Lambda$ *and a point* $x^0 \in M(\lambda^0)$ *exists such that*

$$g_i(x^0) < \lambda_i^0 \qquad \forall i \in \{1, \dots, m\} \setminus J$$

holds;

(2) *the optimal set mapping* ψ *is l.s.c.-B at* λ^0 *if* $\psi(\lambda) \neq \emptyset$ *for all* $\lambda \in \Lambda$ *and* f *and all* g_i *are weakly analytic.*

Note 4.3.5.1. From the above we find that the mapping $\varphi|_{\mathfrak{B}}$ restricted to the feasible parameter set is continuous, and the mapping $\psi|_{\mathfrak{A}}$ restricted to the solubility set is l.s.c.-B, assuming that f is also weakly analytic, if $\Lambda = \mathbb{E}_m$ and $J = \{1, \dots, m\}$.

Proof of Theorem 4.3.5. (1) The mapping M is l.s.c.-B at λ^0 because of Theorem 3.4.3. The upper semicontinuity of φ at λ^0 follows from Theorem 4.2.2 and likewise the closedness of ψ from Theorem 4.2.1.

In order to show that φ is lower semicontinuous by using Theorem 4.3.1 we decompose the mapping M. The existence of an index set $I \subset \{1, \dots, m\}$ with the property

$$\text{ri}\, M(\lambda^0) = M^I\lambda^0 \cap \Omega \tag{4.3.14}$$

where $\Omega = \{x \in E_n \mid g_i(x) < \lambda_i^0, i \in \{1, \dots, m\} \setminus I\}$, follows from (3.2.9); the sets $M^I\lambda$, $\lambda \in \Lambda$, are defined according to (3.2.7). We further define a mapping Γ by

$$\Gamma u = u + M^I\lambda^0 , \qquad u \in E_n .$$

Theorem 4.3.1 may obviously be applied to the function

$$h(u) = \inf \{f(x) \mid x \in \Gamma u \cap \Omega\} ,$$

h is continuous at $u = 0$, and $h(0) = \varphi(\lambda^0)$ follows from (4.3.14). Let the following be arbitrary: a sequence $\{\lambda^t\} \subset \Lambda$, $\lambda^t \to \lambda^0$, a sequence $\{x^t\}$ where $x^t \in M(\lambda^t)$ and an $\varepsilon > 0$. We must then prove

$$f(x^t) > \varphi(\lambda^0) - \varepsilon \tag{4.3.15}$$

for sufficiently large t. From (3.2.10) we find

$$M^I\lambda^0 = \{x \in E_n \mid g_i(x) = \lambda_i^0, i \in I\} .$$

$I \subset J$ follows from the hypotheses, all g_i, $i \in I$, are therefore weakly analytic. $M^I\lambda^0$ is thus an affine subspace[1]) and the mapping M^I is u.s.c.-H at λ^0 because of Theorem 3.4.2, there then exists a sequence $\{u^t\} \subset E_n$ satisfying $x^t \in \Gamma u^t$, $\forall t$, and $u^t \to 0$. Because of (3.1.14) there further exists a point $x \in M^I\lambda^0 \cap \Omega$ satisfying

$$\varphi(\lambda^0) - f(x) > -\frac{\lambda}{3} .$$

Finally, a t_0 exists such that

$$g_i(x) < \lambda_i^0 - (\lambda_i^t - \lambda_i^0) \qquad \forall t > t_0 \qquad \forall i \in \{1, \dots, m\} \setminus I \tag{4.3.16}$$

since $\lambda^t \to \lambda^0$. Defining $z^t =_{\text{Df}} \frac{1}{2}(x^t + x) \in \Gamma(\frac{1}{2} u^t)$ $\forall t$ one finds from (4.3.16)

$$z \in \Gamma(\tfrac{1}{2} u^t) \cap \Omega \qquad \forall t > t_0$$

and thus because of the continuity of h at $u = 0$

$$f(z^t) \geqq h\left(\frac{1}{2} u^t\right) > h(0) - \frac{\varepsilon}{3} \qquad \forall t > t' .$$

Taking the convexity of f into account we then obtain for $t > t'$

$$f(x) \geqq 2f(z^t) - f(x) > \varphi(\lambda^0) + (\varphi(\lambda^0) - f(x)) - \tfrac{2}{3}\varepsilon > \varphi(\lambda^0) - \varepsilon .$$

Hence (4.3.15) holds, which proves (1).

Proposition (2) follows immediately from the representation

$$\psi(\lambda) = \{x \in E_n \mid f(x) \leqq \varphi(\lambda), g_i(x) \leqq \lambda_i, i = 1, \dots, m\} ,$$

the continuity of φ, and Theorem 3.4.3. □

Theorem 4.3.5. (1) was first proved by R. T. Rockafellar [4] in a weaker version: φ is lower semicontinuous at λ^0 if $M(\lambda^0) \neq \emptyset$ and all functions f, g_i are "faithfully con-

[1]) See Corollary 3.2.2.1.

vex" (see (3.3.26)). As we have seen in Theorem 4.3.5 f then need only be convex assuming that the other conditions are still satisfied. It is of no significance for the continuity properties considered which of the two conditions is fulfilled: the function g_i is faithfully convex or alternatively convex and weakly analytic; instead of the constraint $g_i(x) \leqq \lambda_i$ referring to a weakly analytic convex function one need only write

$$e^{g_i(x)} \leqq e^{\lambda_i} ,$$

the function on the left-hand side is then faithfully convex.[1])

The mapping ψ is not necessarily u.s.c.-H at λ^0 under the conditions of Theorem 4.3.5 (b). A counter-example follows if $f \equiv 0$ and M is defined as in Example 3.3.2; from the same example one may conclude that ψ is in general not l.s.c.-H, this however involves some effort. Theorem 4.3.5 remains valid in all parts if "weakly analytic" is replaced everywhere by "piece-wise weakly analytic" in the sense of Corollary 3.3.3.1 as may immediately be shown using the representation for $M(\lambda)$ (and analogously for $\psi(\lambda)$) given in that corollary.

We now consider some further consequences of the last theorem.

Corollary 4.3.5.1. *Under the conditions of Theorem* 4.3.5 (2) *there exists a continuous function* $x\colon \Lambda \to \mathbb{E}_n$ *satisfying*

$$x(\lambda) \in \psi(\lambda) \qquad \forall \lambda \in \Lambda .$$

Proof. Since the Euclidean norm constitutes a convex and weakly analytic function Theorem 4.3.5 may be applied to the problem

$$\inf \{||x|| \mid x \in \psi(\lambda)\} , \qquad \lambda \in \Lambda .$$

By hypothesis this problem possesses uniquely determined solutions $x(\lambda)$ which are continuously dependent on λ. □

Corollary 4.3.5.2. *Let the conditions of Theorem* 4.3.5 (2) *be fulfilled. Further, let vectors* $a^k \in \mathbb{E}_n$, $k = 1, \dots, p$, *and numbers* b_k *be given such that*

$$\psi(\lambda^0) \subset \{x \in \mathbb{E}_n \mid a^{k^{\mathsf{T}}} x \leqq b_k, k = 1, \dots, p\} .$$

There then exists a $\delta > 0$ *for each* $\varepsilon > 0$ *such that*

$$\psi(\lambda) \subset \{x \in \mathbb{E}_n \mid a^{k^{\mathsf{T}}} x \leqq b_k + \varepsilon, k = 1, \dots, p\}$$

holds for all $\lambda \in V_\delta\{\lambda^0\}$.

Proof. By Theorem 4.3.5 the functions

$$\varphi_k(\lambda) = \sup \{a^{k^{\mathsf{T}}} x \mid x \in \psi(\lambda)\}$$

are continuous on Λ, the proof follows immediately. □

In the second corollary a certain type of upper semicontinuity of ψ with respect to polyhedral neighbourhoods of $\varphi(\lambda^0)$ is established: If $P \subset \mathbb{E}_n$ is a convex polyhedron containing $\psi(\lambda^0)$ then there exists a $\delta > 0$ for each ε <0 such that

$$\psi(\lambda) \subset U_\varepsilon P \qquad \forall \lambda \in V_\delta\{\lambda^0\}$$

holds. This may be deduced from Corollary 4.3.5.2 and Theorem 3.4.1.

If one now wishes to consider parametric dependence of the objective function and/or an arbitrary continuous parametric dependence of the constraint functions

---

[1]) In fact, using the properties of faithfully convex functions one may without difficulty show that a convex function $g\colon \mathbb{E}_n \to \mathbb{R}$ is weakly analytic iff e^g is faithfully convex (see the remarks following definition (3.2.26)).

then one finds that the lower semicontinuity of ψ may only be demonstrated under highly restrictive conditions. Thus even linear parametric optimization problems with a parameter-dependent objective function exemplify the fact that the optimal set mapping ψ is in general not necessarily l.s.c.-B on the solubility set $\mathfrak{A}$. The possibility offers itself of showing — as in linear parametric optimization — that ψ is l.s.c.-B on certain local stability sets, and the lemma which follows below plays a central role in such investigations.

We consider a parametric problem

$$(\mathrm{P}_\lambda) \qquad \min \{f(x, \lambda) \mid g_i(x, \lambda) \leqq 0, i = 1, \dots, m\}, \qquad \lambda \in \Lambda, \tag{4.3.17}$$

where Λ is a metric space f, g_i are continuous on $E_n \times \Lambda$ and $f(\cdot, \lambda), g_i(\cdot, \lambda), i = 1, \dots, m$, are convex on E_n for each $\lambda \in \Lambda$.

According to the definition (3.2.8) the characteristic index set of $\psi(\lambda)$ has the representation

$$\operatorname{ch} \psi(\lambda) = \{i \in \{1, \dots, m\} \mid g_i(x, \lambda) = 0 \ \forall x \in \psi(\lambda)\}, \qquad \lambda \in \Lambda. \tag{4.3.18}$$

By ψ^I we denote the point-to-set mapping defined by

$$\psi^I(\lambda) =_{\mathrm{Df}} \{x \in E_n \mid f(x, \lambda) \leqq \varphi(\lambda), g_i(x, \lambda) \leqq 0, i \in I\}, \qquad \lambda \in \Lambda, \tag{4.3.19}$$

where $I \subset \{1, \dots, m\}$.

Lemma 4.3.1. *Consider a parametric optimization problem of the form* (4.3.17) *and let $\lambda^0 \in \mathfrak{A}$ and $\delta > 0$ be given such that*

(1) *$f(\cdot, \lambda^0)$ and $g_i(\cdot, \lambda^0)$ are weakly analytic functions for all $i \in I =_{\mathrm{Df}} \operatorname{ch} \psi(\lambda^0)$, and*

(2) $\dim L^+\psi^I(\lambda) = \dim L^+\psi^I(\lambda^0)$.[1]

If $M|_{\mathfrak{A}}$ is l.s.c.-B at λ^0 then $\psi|_{\mathfrak{A}}$ is also l.s.c.-B at λ^0.

Proof. Let $x^0 \in \psi(\lambda^0)$ and $\{\lambda^k\} \subset \mathfrak{A}$ where $\lambda^k \to \lambda^0$ be chosen arbitrarily. We define

$$M_1\lambda =_{\mathrm{Df}} \{x \in M(\lambda) \mid \|x - x^0\| \leqq 1\}, \qquad \lambda \in \mathfrak{A},$$

$$\varphi_1(\lambda) =_{\mathrm{Df}} \inf \{f(x, \lambda) \mid x \in M_1\lambda\}, \qquad \lambda \in \mathfrak{A},$$

$$\psi_1(\lambda) =_{\mathrm{Df}} \{x \in M_1\lambda \mid f(x, \lambda) \leqq \varphi_1(\lambda)\}, \qquad \lambda \in \mathfrak{A},$$

$$\psi_1^I(\lambda) =_{\mathrm{Df}} \{x \in E_n \mid f(x, \lambda) \leqq \varphi_1(\lambda), g_i(x, \lambda) \leqq 0, i \in I\}, \qquad \lambda \in \mathfrak{A}.$$

Since $M|_{\mathfrak{A}}$ is l.s.c.-B at $\lambda^0 \in \mathfrak{A} \subset \mathfrak{B}$ it follows from Lemma 2.2.4 that M_1 is l.s.c.-B at λ^0 and in particular there exists a $\delta_1 > 0$ such that $M_1\lambda \neq \emptyset \ \forall \lambda \in V_{\delta_1}\{\lambda^0\} \cap \mathfrak{A}$. Because $M_1\lambda$ is even bounded and closed for each $\lambda \in V_{\delta_1}\{\lambda^0\} \cap \mathfrak{A}$ we then find that the sets $\psi_1(\lambda)$ are non-empty and compact for all $\lambda \in V_{\delta_1}\{\lambda^0\} \cap \mathfrak{A}$. The continuity properties of f and g_i, $i = 1, \dots, m$, ensure that M_1 is closed at λ^0. If the convexity conditions imposed on f and g_i are taken into account Theorem 4.3.3 may be applied, thus the function φ_1 is continuous at λ^0.

Because of $I = \operatorname{ch} \psi(\lambda^0)$ one finds

$$\psi^I(\lambda^0) = \{x \in E_n \mid f(x, \lambda^0) = \varphi(\lambda^0), g_i(x, \lambda) = 0, i \in I\}$$

(see (3.2.10)). As obviously $\varphi_1(\lambda^0) = \varphi(\lambda^0) = f(x^0, \lambda^0)$ we have

$$\psi^I(\lambda^0) = \psi_1^I(\lambda^0), \qquad I = \operatorname{ch} \psi(\lambda^0). \tag{4.3.20}$$

Because of (3.2.12) and (3.2.13) the vertex sets of $\psi^I(\lambda)$ and $\psi_1^I(\lambda)$, $\lambda \in \mathfrak{A}$, are equal: $L^+\psi^I(\lambda) = L^+\psi_1^I(\lambda)$ and it follows from hypothesis (2) that

$$\dim L^+\psi_1^I(\lambda) = \dim L^+\psi_1^I(\lambda^0), \qquad \lambda \in V_{\delta_1}\{\lambda^0\} \cap \mathfrak{A}. \tag{4.3.21}$$

---

[1]) Where the symbol L^+ is defined by (3.2.6).

There further exists a $y^0 \in \operatorname{ri} \psi(\lambda^0)$ such that $||y^0 - x^0|| < 1$ and because of $I = \operatorname{ch} \psi(\lambda^0)$ the inequality $g_i(y^0, \lambda^0) < 0$ holds for all $i \in I$. Hence $I = \operatorname{ch} \psi_1(\lambda^0)$. If in Theorem 3.2.2 we set $\Lambda = V_{\delta_1}\{\lambda^0\} \cap \mathfrak{A}$ and $M = \psi_1$ then noting Corollary 3.2.2.1 and the continuity of φ_1 at λ^0 we find that all conditions of Theorem 3.2.2 are fulfilled, it follows that ψ_1 is l.s.c.-B at λ^0. In particular $x^0 \in \psi_1(\lambda^0)$, there thus exists a sequence $\{y^k\} \subset E_n$ satisfying $y^k \in \psi_1(\lambda^k)$ and $||y^k - x^0|| < 1$ for sufficiently large k. The points y^k are, for sufficiently large k, local optimal points of $\min \{f(x, \lambda^k) \mid x \in M(\lambda^k)\}$ and because of the convexity conditions imposed on f and g_i, $i = 1, \dots, m$, they are also global optimal points, thus $y^k \in \psi(\lambda^k)$ if k is sufficiently large. Since the sequence $\{\lambda^k\}$, $\lambda^k \to \lambda^0$; and the point $x^0 \in \psi(\lambda^0)$ were assumed to be arbitrary we have shown that ψ is l.s.c.-B at λ^0. □

Let the problem (4.3.17) be given. If $I \subset \{1, \dots, m\}$ and $0 \leqq d \leqq n$ then the set

$$\mathfrak{S}(I, d) =_{\mathrm{Df}} \{\lambda \in \mathfrak{A} \mid \operatorname{ch} \psi(\lambda) = I, \dim \psi(\lambda) = d\} \tag{4.3.22}$$

is called *local stability set* with respect to (I, d) if $\mathfrak{S}(I, d) \neq \emptyset$. Concrete cases of such local stability sets for linear and quadratic optimization problems will be examined in Chapter 5.

Theorem 4.3.6. *Let the parametric optimization problem* (P_λ) *satisfy the conditions of* (4.3.17). *Further, let* $I \subset \{1, \dots, m\}$ *and* $d \in \{0, 1, \dots, n\}$ *be given such that the set* $\mathfrak{S}(I, d)$ *is non-empty.*

If the functions $f(\cdot, \lambda)$ *and* $g_i(\cdot, \lambda)$ *are weakly analytic for all* $i \in I$ *and* $\lambda \in \mathfrak{S}(I, d)$ *and if the mapping* $M|_G$ *is l.s.c.-B at* $\lambda^0 \in G$, *where* $G \subset \mathfrak{S}(I, d)$, *then* $\psi|_G$ *is also l.s.c.-B at* λ^0.

Proof. We use Lemma 4.3.1 and identify Λ with G (this is possible since G is a subset of a metric space and thus also a metric space with the induced metric). By definition of $\mathfrak{S}(I, d)$ we have $I = \operatorname{ch} \psi(\lambda^0)$ and thus condition (1) of Lemma 4.3.1 is fulfilled. Because $I = \operatorname{ch} \varphi(\lambda)$ for all $\lambda \in G$ we find from (3.2.9), if there M is replaced by ψ

$$\operatorname{ri} \psi(\lambda) = \{x \in E_n \mid g_i(x, \lambda) \leqq 0, i \in I, f(x, \lambda) \leqq \varphi(\lambda), g_j(x, \lambda) < 0, j \in \bar{I}\}\,, \quad \lambda \in G\,,$$

and

$$\dim \psi(\lambda) = \dim \psi^I(\lambda)$$

follows for all $\lambda \in G$.

In a fashion similar to that of the proof of Corollary 3.2.2.1 we may show that $\psi^I(\lambda)$ is an affine subspace of E_n for all $\lambda \in G$, therefore $L^+\psi^I(\lambda) = \psi^I(\lambda)$ and consequently

$$\dim L^+\psi(\lambda) = \dim \psi^I(\lambda) = \dim \psi(\lambda) = d\,, \qquad \forall \lambda \in G\,,$$

condition (2) is thus also satisfied. Since $M|_G$ is l.s.c.-B at λ^0 the lemma then implies that $\psi|_G$ is l.s.c.-B at λ^0. □

If in the preceding theorem the condition that $M|_G$ is l.s.c.-B at λ^0 is omitted the lower semicontinuity (B) of $\psi|_G$ at λ^0 is no longer ensured, as the following example shows.

Example 4.3.6. Consider the problem

$$\min \{x \mid x^2 + y^2 \leqq 1, y \geqq 0, y = tx - |t|\}\,, \qquad t \in \mathbb{R} \quad \text{variable}\,.$$

We have

$$\psi(t) = \begin{cases} \{(1, 0)\} & \text{if} \quad t > 0\,, \\ \{(-1, 0)\} & \text{if} \quad t \leqq 0\,. \end{cases}$$

Although ch $\psi(t) = \{1, 2, 3\}$ $\forall t \in \mathbb{R}$ and dim $\psi(t) = 0$ $\forall t \in \mathbb{R}$ hold, ψ is neither closed nor l.s.c.-B at $t = 0$.

Corollary 4.3.6.1. *Under the conditions of Theorem 4.3.6 and the condition that $M|_G$ is l.s.c.-B on G the following is true:*

(1) *$\varphi|_G$ is continuous on G.*

(2) *$\psi|_G$ is closed on G.*

(3) *There exists a vector function $x: G \to \mathbb{E}_n$ where $x(\lambda) \in \psi(\lambda)$ for all $\lambda \in G$, which is continuous on G.*

(4) *If $\psi(\lambda)$ is bounded for all $\lambda \in G$ then $\psi|_G$ is H-continuous on G.*

Proof. The first two propositions follow from Theorems 4.2.4 (2) and 4.2.1 (3) respectively, the last is a consequence of Theorem 4.3.3 and Lemma 2.2.3. In order to prove (3) set $x(\lambda)$ equal to the element of smallest norm in $\psi(\lambda)$ and observe Theorem 4.3.3. □

In the case that, in addition to the conditions of (4.3.17), (P_λ) fulfills the further condition that $f(\cdot, \lambda)$ and $g_i(\cdot, \lambda)$, $i = 1, \dots, m$, are quadratic functions, i.e. if they are given by

$$f(x, \lambda) = x^T C^0(\lambda)\, x + p^0(\lambda)^T x\,, \qquad g_i(x, \lambda) = x^T C^i(\lambda)\, x + p^i(\lambda)^T x + b_i(\lambda)$$

where $C^j(\lambda)$, $j \in \{0, 1, \dots, m\}$, $\lambda \in \Lambda$, are symmetric positive-definite (n, n)-matrices and $p^j(\lambda) \in \mathbb{E}_n$, $j \in \{0, 1, \dots, m\}$, $\lambda \in \Lambda$, then each local stability set $\mathfrak{S}(I, d)$ may be represented as

$$\mathfrak{S}(I, d) = \{\lambda \in \mathfrak{S} \mid \text{ch}\, \psi(\lambda) = I, \text{rg}\, D^I(\lambda) = n - d\} \tag{4.3.23}$$

where

$$D^I(\lambda) =_{\text{Df}} \left(C^{\alpha_1}(\lambda) \dots C^{\alpha_r}(\lambda)\; p^{\alpha_1}(\lambda) \dots p^{\alpha_r}(\lambda)\right), \quad \lambda \in \Lambda, \quad I \cup \{0\} = \{\alpha_1, \dots, \alpha_r\}\,. \tag{4.3.24}$$

This follows from considerations similar to those employed in Corollary 3.2.3.1.

We now give an example to indicate that in Theorem 4.3.6 $f(\cdot, \lambda)$ and $g_i(\cdot, \lambda)$ $\forall i \in \text{ch}\, \psi(\lambda^0)$ must not simply be convex but rather also weakly analytic.

Example 4.3.7. Consider the parametric problem

$$\min \{\mu z - y \mid f_1(x, y, z, \lambda) \leqq 0\}\,, \qquad (\mu, \lambda) \in \mathbb{R} \times \mathbb{R}^+ \text{ variable}\,,$$

$$\mathbb{R}^+ = \{\lambda \in \mathbb{R} \mid \lambda \geqq 0\}\,, \qquad f_1(x, y, z, \lambda) = g_1(x, \lambda) + g_2(z) + y - 1\,,$$

where

$$g_1(x, \lambda) = \begin{cases} (x - \lambda)^2 & \text{if} \quad x \geqq \lambda\,, \\ (x + \lambda)^2 & \text{if} \quad x \leqq -\lambda\,, \\ 0 & \text{if} \quad -\lambda < x < \lambda \end{cases}$$

and

$$g_2(z) = \begin{cases} (z + 4)^2 & \text{if} \quad z \leqq -4\,, \\ z^2 & \text{if} \quad z \geqq 0\,, \\ 0 & \text{if} \quad -4 < z < 0 \end{cases}$$

(cf. fig 4.1). The functions g_1 and g_2 are obviously continuous on their domains, f_1 as the sum of continuous functions is itself continuous, $g_1(\cdot, \lambda)$, $\lambda \geqq 0$, and g_2 are convex with respect to x and z respectively, $f_1(\cdot, \cdot, \cdot, \lambda)$ is therefore also convex on $\mathbb{E}_3$: the conditions of (4.3.17) are thus satisfied.

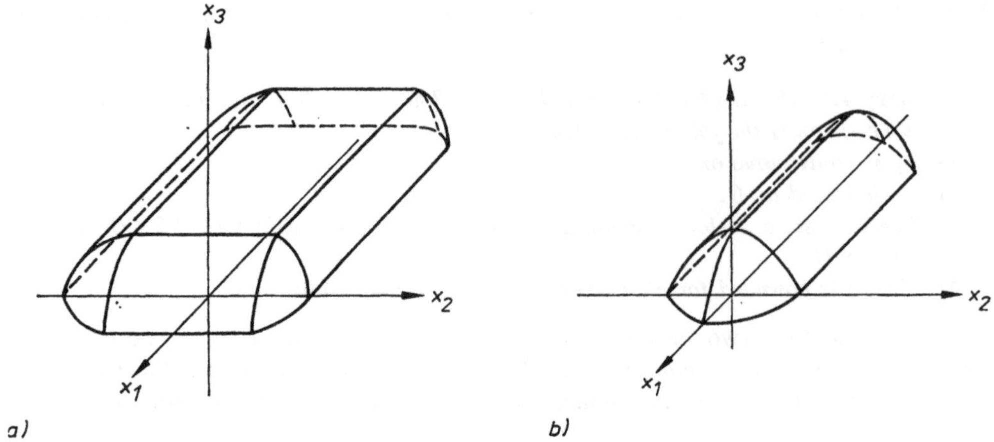

Fig. 4.1. Constraint set a) for the case $\lambda > 0$, b) for the case $\lambda = 0$

The constraint set mapping is l.s.c.-B at $\lambda = 0$ since $f_1(0, 0, 0, 0) < 0$ and Theorem 4.3.6 enables one to apply Theorem 3.1.6. With some patience one may confirm that (see also fig. 4.1)

$$\psi(\mu, \lambda) = \begin{cases} \{(x, y, z) \in E_3 \mid -\lambda \leqq x \leqq \lambda, y = 1 - \frac{1}{4}\mu^2, z = -\frac{1}{2}\mu) & \text{if } \mu < 0, \lambda \geqq 0, \\ \{(x, y, z) \in E_3 \mid -\lambda \leqq x \leqq \lambda, y = 1 - \frac{1}{4}\mu^2, z = -4 - \frac{1}{2}\mu\} & \text{if } \mu > 0, \lambda \geqq 0, \\ \{(x, y, z) \in E_3 \mid -\lambda \leqq x \leqq \lambda, y = 1, -4 \leqq z \leqq 0\} & \text{if } \mu = 0, \lambda \geqq 0. \end{cases}$$

It follows that $G =_{\mathrm{Df}} \mathfrak{S}(\{1\}, 1) = \{(0, 0)\} \cup \{(\mu, \lambda) \in R^+ \times R \mid \lambda > 0, \mu \neq 0\}$, $\psi|_G$ is however not l.s.c.-B at $(\mu, \lambda) = (0, 0)$.

In this section we have up to now demonstrated the individual continuity properties directly, in what follows we however consider propositions relating to the continuity of φ which may be implied from the convexity of this function itself. We find that the individual problems (P_λ) needs not necessarily be "convex" if φ is to be convex or concave. First we present a useful auxiliary theorem.

Lemma 4.3.2. *Let P be a convex polyhedron in E_n and $f\colon P \to R$ be a function which is convex on P.*
Then f is upper semicontinuous on P.

Proof. See R. T. Rockafellar [2]. □

We now consider the problems

$$\inf \{f_0(x, \lambda) \mid x \in X, f(x, \lambda) \leqq 0 \; \forall i \in I\}, \quad \lambda \in \Lambda, \tag{4.3.25}$$

subject to the conditions

I is an arbitrary index set and X is a linear space; (4.3.26)

Λ is a convex polyhedron contained in E_m; (4.3.27)

f_0, f_i, $\forall i \in I$, are real-valued convex functions on $X \times \Lambda$; (4.3.28)

the extreme values $\varphi(\lambda)$ are finite $\forall \lambda \in \Lambda$. (4.3.29)

One easily confirms that φ is convex on Λ and that the following theorem follows immediately.

Theorem 4.3.7. *The extreme value function φ corresponding to problem* (4.3.25) *is upper semicontinuous on Λ if the conditions* (4.3.26) *to* (4.3.29) *are satisfied.*

Another possibility of applying Lemma 4.3.2 presents itself if the objective function exhibits linear or concave dependence on the parameter and the constraint set is fixed. To be more exact let the problems

$$\inf \{f(x, \lambda) \mid x \in M\}, \qquad \lambda \in \Lambda, \tag{4.3.30}$$

satisfy the following conditions

M is an arbitrary non-empty set, Λ is a convex polyhedron contained in $\mathbb{E}_m$; (4.3.31)

$f: M \times \Lambda \to \mathbb{R}$ is concave at λ for each fixed x; (4.3.32)

the extreme values $\varphi(\lambda)$ are finite $\forall \lambda \in \Lambda$. (4.3.33)

Theorem 4.3.8. *The extreme value function φ corresponding to problem* (4.3.30) *is lower semicontinuous on Λ if the conditions* (4.3.31) *to* (4.3.33) *are satisfied.*

Proof. For arbitrary $\lambda^1, \lambda^2 \in \Lambda$ and $\alpha \in (0, 1)$ we have

$$\begin{aligned}\varphi(\alpha\lambda^1 + (1-\alpha)\lambda^2) &\geqq \inf \{\alpha f(x, \lambda^1) + (1-\alpha) f(x, \lambda^2) \mid x \in M\} \\ &\geqq \alpha\varphi(\lambda^1) + (1-\alpha)\varphi(\lambda^2).\end{aligned}$$

Hence φ is concave on M and by Lemma 4.3.2 lower semicontinuous. □

We employ this last theorem in studies on parametric quadratic problems presented in Section 4.5.

4.4. Duality and Non-Linear Optimization

In the last fifteen years R. T. Rockafellar [1], P.-J. Laurent [1], I. Ekeland and R. Teman [1], R. J.-B. Wets [2] and others have established a theory on the relations between stability and duality of optimization problems, this theory is, at least in the convex case almost complete. Of particular interest is the fact that the validity of various duality theorems is equivalent to stating that the extreme value function associated with the parametrization of the initial problem has certain properties, as we will now show in a more precise manner.

Consider the problem

(P) $\quad \inf \{f(x) \mid x \in X, g(x) \in K\}$

where X is a normed space, K is a closed convex cone contained in a normed space Y int $K \neq \emptyset$, $f: X \to \mathbb{R}$ is a convex continuous function and $g: X \to Y$ is a continuous mapping with the property

$$\alpha g(x) + (1-\alpha) g(y) - g(\alpha x + (1-\alpha) y) \in -K$$
$$\forall \alpha \in (0, 1), \qquad x, y \in X.$$

The following problem is associated with (P)

(D) $\quad \sup \{h(u) \mid u \in K^p\}$

where

$$h(u) =_{\mathrm{Df}} \inf [f(x) + u(g(x))],$$

i.e. we consider the Lagrangian dual problem corresponding to (P).

We denote the extreme values of each problem by φ_P and φ_D respectively. Because of

$$\varphi_P = \inf_{x \in X} \sup_{u \in K^P} [f(x) + u(g(x))] \geqq \sup_{u \in K^P} \inf_{x \in X} [f(x) + u(g(x))] = \varphi_D$$

we have $\varphi_P \geqq \varphi_D$.

Of particular interest are the conditions which establish the equality of both extreme values (in this case we speak of weak duality) or which in addition ensure the existence of a solution of the dual problem (strong duality).

We now introduce the parameter-dependent problem

(P_y) $\quad \inf \{f(x) \mid x \in X, g(x) \in y + K\}$

where $y \in Y$ and $y + K =_{\mathrm{Df}} \{y + k \mid k \in K\}$ and consider its extreme value function $\varphi = \varphi(y)$.

Theorem 4.4.1. *Let φ_P be finite. Then*

(1) $\varphi_P = \varphi_D \Leftrightarrow \varphi$ *is lower semicontinuous at* $y = 0$;

(2) $\varphi_P = \varphi_D$ *and* (D) *possesses a solution* $\Leftrightarrow \varphi$ *is subdifferentiable at* $y = 0$.[1])

Proof. See for instance I. Ekeland and R. Teman [1]. □

It is thus obvious that statements relating to the lower semicontinuity of the function $\varphi = \varphi(y)$ may be employed to demonstrate weak duality. Using the results of the previous section and the conditions formulated for problem (P) we in particular obtain the following corollary.

Corollary 4.4.1.1. *If φ_P is finite and the mapping M defined by*

$$M(y) = \{x \in X \mid g(x) \in y + K\}$$

is u.s.c.-H at $y = 0$, then $\varphi_P = \varphi_D$.

Proof. This follows from (1) and Theorem 4.3.2. □

We now consider the finite-dimensional case

$$X = \mathbb{E}_n, \qquad Y = \mathbb{E}_m, \qquad g(x) = (g_1(x), \ldots, g_m(x))^{\mathrm{T}},$$

$$K = \{y \in \mathbb{E}_m \mid y_i \leqq 0 \ \forall i\},$$

and the corresponding problems

$(\overline{P})$ $\quad \inf \{f(x) \mid g_i(x) \leqq 0, i = 1, 2, \ldots, m\}$,

$(\overline{D})$ $\quad \sup \{h(u) \mid u \in \mathbb{E}_m, u \geqq 0\}$

where $h(u) = \inf_{x \in \mathbb{E}_n} [f(x) + u^{\mathrm{T}} g(x)]$ and $f, g_i: \mathbb{E}_n \to \mathbb{R}$ are convex functions.

Corollary 4.4.1.2. *If $\varphi_{\overline{P}}$ is finite and a point x exists such that for all i we have either $g_i(x) \leqq 0$ or $g_i(x) < 0$ according to whether g_i is weakly analytic or not, then $\varphi_{\overline{P}} = \varphi_{\overline{D}}$.*

Proof. See (1) and Theorem 4.3.5. □

We would like to remark that convex and weakly analytic functions play the same role with respect to weak duality as affine-linear functions in regard to strong duality.

---

1) i.e. there exists a vector $v \in Y^*$ satisfying

$$\varphi(y) - \varphi(0) \geqq v(y) \qquad \forall y \in Y.$$

For comparison we note the well-known duality theorem for finite-dimensional convex problems:

If $\varphi_{\bar{P}}$ is finite and a point x exists such that for all i we have either $g_i(x) \leqq 0$ or $g_i(x) < 0$ according to whether g_i is affine-linear or not, then $\varphi_{\bar{P}} = \varphi_{\bar{D}}$ and $(\bar{D})$ has a solution.

Continuing with the finite-dimensional convex problem $(\bar{P})$ we refer to the fact that the existence of a Slater point x, $g_i(x) < 0 \ \forall i$, and the finiteness of $\varphi_{\bar{P}}$ not only imply strong duality but also the continuity of φ at the point $y = 0$ (the convex function φ is in this case finite on a neighbourhood of the zero point and thus also continuous). This last statement may be given in a more general form.

Corollary 4.4.1.3. *Let $\varphi_{\bar{P}}$ be finite. Consider the following statements:*

(1) *$\psi_{\bar{P}}$ is non-empty and bounded.*

(2) *There exists an $\varepsilon > 0$ such that all infima*

$$a(c) = \inf \{f(x) + c^{\mathsf{T}}x \mid g_i(x) \leqq 0 \ \forall i\}$$

are finite for $||c|| < \varepsilon$.

(3) *The weak duality holds for $(\bar{P})$ and $(\bar{D})$.*

(1′) *$\psi_{\bar{D}}$ is non-empty and bounded.*

(2′) *There exists an $\varepsilon > 0$ such that all infima*

$$\varphi(y) = \inf \{f(x) \mid g_i(x) \leqq y_i \ \forall i\}$$

are finite for $||y|| < \varepsilon$.

(3′) *There exists a Slater point for* (P).

(4′) *The function φ is continuous at $y = 0$.*

Then (1) *and* (2) *are equivalent and the same holds for* (1′), (2′), (3′) *and* (4′). *Further,* (1) *implies* (3).

Proof. (1) ⇒ (2), (3). By Theorem 4.3.3 the functions a and φ are l.s.c. at $c = 0 \in E_n$ and $y = 0 \in E_m$ respectively. Thus (2) is true. From Theorem 4.4.1 (3) also follows.

(2) ⇒ (1). If $\psi_{\bar{P}}$ is empty or unbounded then there exists a $u \in O^+M$ (M is the constraint set of $(\bar{P})$) such that $u \neq 0$ and $f(x + \lambda u) \leqq f(x) \ \forall x \in E_n \ \forall \lambda \geqq 0$. For arbitrary $c = -\alpha u$, $\alpha > 0$, it immediately follows that $h(c) = -\infty$.

(1′) ⇒ (2′). Let $\bar{u} \in \psi_{\bar{D}}$. Because of

$$\varphi(y) \geqq \inf_{x \in E_n} [f(x) + \bar{u}^{\mathsf{T}}(g(x) - y)] = \varphi_{\bar{D}} - \bar{u}^{\mathsf{T}}y$$

the only possibility is $\varphi(y) > -\infty$.

In order to show that $0 \in \operatorname{int} \operatorname{dom} \varphi$, which implies that φ is finite for all y having sufficiently small norms, we assume that the contrary is true. Since $\operatorname{int} \operatorname{dom} \varphi$ is non-empty there exist a $v \in E_m$ satisfying $v \neq 0$ and $0 = v^{\mathsf{T}}0 \leqq v^{\mathsf{T}}y \ \forall y \in \operatorname{dom} \varphi$. In particular we then have for an arbitrary $\lambda \geqq 0$

$$h(\bar{u} + \lambda v) = \inf_{x \in E_n} \left(f(x) + \bar{u}^{\mathsf{T}}g(x) + \lambda v^{\mathsf{T}}g(x)\right) \geqq h(\bar{u})$$

because $g(x) \in \operatorname{dom} \varphi \ \forall x \in E_n$, thus all points $\bar{u} + \lambda v$ are contained in $\psi_{\bar{D}}$ and this set is unbounded, in contradiction to the above.

(2′) ⇒ (3′). This is obvious.

(3′) ⇒ (4′). Follows directly from Theorem 4.3.5.

(4′) ⇒ (1′). The implication (4′) ⇒ (3′) is easy to see. Now, let x^0 be a Slater point, i.e. $g_i(x^0) < 0 \ \forall i$. Then one obtains

$$\underline{\lim}\, h(u^t) \leqq \underline{\lim} \left[f(x^0) + \sum_{i=1}^{m} u_i^t g_i(x^0) \right] = -\infty$$

for each sequence $\{u^t\}$ satisfying $\|u^t\| \to \infty$ and $u^t \geqq 0$. The non-empty set $\psi_{\bar{\mathrm{D}}}$ is thus bounded. □

We further remark that strong duality may hold for ($\bar{\mathrm{P}}$) and ($\bar{\mathrm{D}}$) even if the extreme value function φ is not continuous at $y = 0$, as is illustrated by the following example.

Example 4.4.1. Let g be defined as in (3.3.6) and Example 3.3.1. Then the problem

$$\inf \{x_1 \mid x \in E_2, g(x) \leqq 0, -x_2 \leqq 0\}$$

is strongly dualizable with the dual solution set

$$\psi_{\bar{\mathrm{D}}} = \{u \in E_2 \mid u_1 \geqq 1, u_2 = 0\}$$

(where u_1 is associated to g) and with the extreme values $\varphi_{\bar{\mathrm{P}}} = \varphi_{\bar{\mathrm{D}}} = 0$. The function φ however is not u.s.c. at $(0, 0)^{\mathrm{T}}$ as the discussion of Example 3.3.1 shows. In fact we have for small $\varepsilon > 0$:

$$\inf \{x_1 \mid g(x) \leqq \varepsilon\, \mathrm{e}^{-1/\varepsilon}, -x_2 \leqq -\varepsilon\} = 1\,.$$

Finally, in certain forms of parametrization of the original problem continuity properties of the dual optimal set mapping may be inferred from certain properties of the primal optimal set mapping, the primal constraint set must however be a fixed convex polyhedron. Or more precisely:

Let the primal problem be given by

$$(\mathrm{P}_\lambda) \qquad \min \{f(x, \lambda) \mid Ax \leqq b\}$$

where $x \in E_n$, $b \in E_m$, A is a (m, n)-matrix, $f\colon E_n \times \Lambda \to R$, and $f(\cdot, \lambda)$ is a convex function for all $\lambda \in \Lambda$. Further, let f be differentiable with respect to x and f and $\nabla_x f(\cdot, \cdot)$ be continuous on $E_n \times \{\lambda^0\}$. Assume also that the problems (P_λ) are soluble for all $\lambda \in \Lambda$ (Λ is a metric space). The problem class (P_λ) is said to be *dual-stable* if for each sequence $\{\lambda^t\} \subset \Lambda$ with $\lambda^t \to \lambda^0$ the following holds: There exist an infinite subsequence $\{\lambda^{t_k}\}$ and solutions x^k of the problems ($\mathrm{P}_{\lambda^{t_k}}$) such that

$$\nabla_x f(x^k, \lambda^{t_k}) \to \nabla_x f(x^0, \lambda^0)$$

and

$$I(x^k) \subset I(x^0) \qquad \forall k$$

where x^0 is a suitable solution of the problem (P_{λ^0}) and $I(x)$ is the index set of the constraints active at the point x: $I(x) =_{\mathrm{Df}} \{i \mid A_i x = b_i\}$. By Theorem 4.3.3 the problem class (P_λ) is dual-stable if the solution set of (P_{λ^0}) is non-empty and bounded.

Corollary 4.4.1.4. *If the problem class* (P_λ) *is dual-stable then the optimal set mapping* ψ_{D} *corresponding to the dual problems*

$$(\mathrm{D}_\lambda) \qquad \sup \{h(u, \lambda) \mid u \geqq 0, u \in E_m\}$$

where $h(u, \lambda) = \inf\, [f(x, \lambda) + u^{\mathrm{T}}(Ax - b)]$ *is H-continuous at* $\lambda = \lambda^0$.

Proof. We refer the reader to B. Kummer [6] where in addition examples are given to show that the condition of dual stability is essential to the corollary and that the corollary is not valid if non-linear constraints are involved. □

4.5. Mixed-Integer Quadratic Problems

Additional difficulties arise in the study of parametric mixed-integer optimization problems as compared to the "continuous" case, since generally no practicable necessary optimality conditions are available and theorems on the continuity of the constraint set mapping without compactness requirements presented in Chapter 3 of this volume are only of use in this respect for the case of linear constraints. First extensive investigations were therefore limited to the special case that the constraint sets are always compact (M. Radke [1], continuity properties of φ, ψ and M) and to linear problems with constraint sets as in Section 3.4 (B. Bank [1], continuity of φ, ψ and the constraint set mapping M).

A notable improvement in the situation is due to R. Hansel [1], who for the first time studied mixed-integer quadratic problems involving parameter-dependent linear parts of the quadratic objective function without assuming convexity; the constraint sets are defined as in Section 3.4. The principal results of the section are derived from the above paper.

We consider problems of the form

$$(\mathrm{P}_s(p, \lambda)) \quad \inf \{x^{\mathrm{T}}Dx + p^{\mathrm{T}}x \mid Ax \leqq \lambda, x \in E_n, x_1, \dots, x_s \in \mathbb{Z}\} ,$$

where $s \in [0, n]$ is integer and fixed ($s = 0$ is taken to mean that no integer condition is imposed), D is a symmetric (n, n)-matrix, A a (m, n)-matrix with only rational elements, and $(p, \lambda) \in E_{n+m}$.

The rationality of the matrix A as well as further rationality conditions imposed on D below, are only necessary in the following theorems if $s \geqq 1$, this means that "continuous" quadratic and mixed-integer linear problems are also included as special cases. In the following we use the notations of Section 3.4:

$$f(x, p) =_{\mathrm{Df}} x^{\mathrm{T}}Dx + p^{\mathrm{T}}x ,$$

$$G(\lambda) = \{x \in E_n \mid Ax \leqq \lambda, x_1, \dots, x_s \in \mathbb{Z}\} ,^{1)}$$

$$U_G = O^+G = \{x \in E_n \mid Ax \leqq 0, x_1, \dots, x_s \in \mathbb{Z}\} ,$$

$$= G(0) .$$

The extreme value and the solution set of the problem $(\mathrm{P}_s(p, \lambda))$ are as usual denoted by $\varphi_s(p, \lambda)$ and $\psi_s(p, \lambda)$ respectively.

The remainder of the section is subdivided into two parts, the first of which contains subsidiary theorems which, although they are essential for the proof of the level set theorem (Theorem 4.5.3) below, are also individually significant in the case $s = 0$ (see for instance Lemma 4.5.2 and Theorem 4.5.1).

Preliminary Theorems

Lemma 4.5.1. *Let D be a symmetric (n, n)-matrix, A be a (m, n)-matrix and U be the cone*

$$U = \{x \in E_n \mid Ax \leqq 0\} .$$

Further, let

$$x^{\mathrm{T}}Dx \geqq 0 \qquad \forall x \in U \tag{4.5.1}$$

[1]) If $M(\lambda)$ is the polyhedron defined by (3.4.2) then $G(\lambda) \subset M(\lambda)$ holds for all $s \geqq 0$ and in particular $G(\lambda) = M(\lambda)$ for $s = 0$.

hold. The set $K^0 = \{x \in U \mid x^T D x = 0\}$ *may then be represented as a union of a finite number of convex polyhedral cones which possess integer generating vectors if A and D are rational.*

Proof. The proposition is trivial if $\dim U = 0$. Assume therefore that the lemma is true for $\dim U < k$ and then set $\dim U = k$. The cone U may be represented as

$$U = \operatorname{ri} U \cup \bigcup_{\nu} S_\nu\,, \qquad \dim S_\nu < k$$

where each S_ν is one of the finite number of closed faces of the cone, we then obviously have

$$K_0 = \{x \in \operatorname{ri} U \mid x^T D x = 0\} \cup \bigcup_{\nu} \{x \in S_\nu \mid x^T D x = 0\} \tag{4.5.2}$$

and, taking (4.5.1) into account,

$$\{x \in \operatorname{ri} U \mid x^T D x = 0\} = \{x \in \operatorname{ri} U \mid x^T D v = 0\ \forall v \in \operatorname{lin} U\}\,. \tag{4.5.3}$$

From

$$x^T D v = 0 \qquad \forall v \in \operatorname{lin} U\,, \qquad x \in U\,,$$

it follows that $x \in K_0$. (4.5.2) and (4.5.3) then imply

$$K_0 = \{x \in U \mid x^T D v = 0\ \forall v \in \operatorname{lin} U\} \cup \bigcup_{\nu} \{x \in S_\nu \mid x^T D x = 0\}\,.$$

The first set of this representation is the intersection of a subspace of E_n and U and is thus a convex polyhedral cone, since A and D are rational it follows that this cone has integer generators. By the original assumption of the induction this also applies to the remaining sets, thus completing the proof. □

Lemma 4.5.2. *Let A be rational and*

$$\mathfrak{C}_s = \{p \in E_n \mid \varphi_s(p, 0) > -\infty\}\,.$$

Then $\mathfrak{C}_s = \mathfrak{C}_0$, *and* $\mathfrak{C}_0$ *is a convex polyhedral cone.*

Proof. We first consider $\mathfrak{C}_0$. If the quadratic form $x^T D x$ is negative for an x contained in $M(0)$ (3.4.2) $(M(0) = U)$ then $\mathfrak{C}_0 = \emptyset$ and nothing remains to be said. If this is not the case then, using the notation of Lemma 4.5.1

$$p \in \mathfrak{C}_0 \Leftrightarrow p^T x \geqq 0 \qquad \forall x \in K_0 \tag{4.5.4}$$

(B. C. Eaves [3]). Since K_0 may be written as a union of a finite number of polyhedral cones a finite number of vectors $u^1, \dots, u^N$, the generators of the cones, exist such that

$$p^T x \geqq 0 \qquad \forall x \in K_0 \Leftrightarrow p^T u^\nu \geqq 0 \qquad \forall \nu \in \{1, \dots, N\}$$

holds. Hence $\mathfrak{C}_0$ must be a convex polyhedral cone.

In order to prove $\mathfrak{C}_s = \mathfrak{C}_0$ one need only demonstrate that $\mathfrak{C}_s \subset \mathfrak{C}_0$ holds, the reverse inclusion is obvious. We again employ proof by induction with respect to the dimension of the cone $U = M(0)$. Assume therefore that $\mathfrak{C}_s = \mathfrak{C}_0$ is true for $\dim U < k$ and then set $\dim U = k$ (we note that the case $\dim U = 0$ is trivial). Further, assume $p \in \mathfrak{C}_s \setminus \mathfrak{C}_0$. If $x^T D x < 0$ for a certain $x \in U$ then, since A is by hypothesis rational, there also exists a vector $y \in U$ with only rational elements and which satisfies $y^T D y < 0$. One can then find a $\mu > 0$ such that the points $t\mu y$ are contained in $G(0)$ if t are arbitrary natural numbers, because of

$$f(t\mu y, p) \to -\infty\,, \qquad t \to \infty\,,$$

it then follows that $p \notin \mathfrak{G}_s$ which contradicts the above assumption. One concludes that $x^\mathsf{T}Dx \geqq 0$, $x \in U$, the existence of a point $u \in K_0$ satisfying

$$u^\mathsf{T}Du = 0 , \qquad p^\mathsf{T}u < 0 \tag{4.5.5}$$

follows from (4.5.4). The representation (4.5.2) with (4.5.3) also apply to K_0.

If u is contained in one of the sets

$$\{x \in S_\nu \mid x^\mathsf{T}Dx = 0\}$$

we find that this contradicts the above assumption concerning the dimension of U, since $\dim S_\nu < k$, the cone S_ν has a rational representation of the form

$$S_\nu = \{x \in E_n \mid A^{(\nu)}x \leqq 0\}$$

and

$$\inf\,\{f(x, p) \mid x \in S_\nu\} = -\infty ,$$

$$\inf\,\{f(x, p) \mid x \in S_\nu, x_1, \ldots , x_s \in \mathbb{Z}\} > -\infty$$

would then hold. It follows that u is contained in the set

$$\{x \in \operatorname{ri} U \mid x^\mathsf{T}Dv = 0\ \forall v \in \operatorname{lin} U\} .$$

By (4.5.5) the objective function $f(\cdot, p)$ decreases and is not lower bounded on the ray $S = \{\alpha u \mid \alpha \geqq 0\}$. Since $p \in \mathfrak{G}_s$ the only integer point contained in S is the origin. By KRONECKER's Theorem (cf. E. G. BELOUSOV [1], p. 138) there thus exist integer points z^t, $t = 1, 2, \ldots$, such that $\|z^t\| \to \infty$ and $d(z^t, S) \to 0$ hold for $t \to \infty$. Because of the rationality of all elements of the matrix A and since $d(z^t, \operatorname{lin} U) \to 0$ it follows that for large t $z^t \in \operatorname{lin} U$ (note also E. G. BELOUSOV [1], Chapter III).

Choosing $u^t \in S$ such that

$$w^t = z^t - u^t , \qquad \|w^t\| \to 0 , \qquad t \to 0 ,$$

and taking $u \in \operatorname{ri} U$ and $z^t \in \operatorname{lin} U$ into account it follows that

$$z^t \in U , \qquad t \geqq t_0 ,$$

also holds. Finally, we obtain $z^t \in G(0)$ and

$$\begin{aligned} f(z^t, p) &= f(w^t + u^t, p) = f(u^t, p) + 2u^{t\mathsf{T}}Dw^t + p^\mathsf{T}w^t , \\ &= f(u^t, p) + 2\alpha_t u^\mathsf{T}Dw^t + p^\mathsf{T}w^t , \\ &= f(u^t, p) + 0 + \varepsilon_t \end{aligned}$$

where $\varepsilon_t \to 0$ and $f(u^t, p) = \alpha_t p^\mathsf{T}u \to -\infty$. In contradiction to the choice of p it follows that

$$p \notin \mathfrak{G}_s$$

which verifies $\mathfrak{G}_s \setminus \mathfrak{G}_0 = \emptyset$ and thus completes the proof. □

We now pass from the special case that the "continuous" feasible region is a cone to the general case.

Lemma 4.5.3. *If A is rational then the following is true:*

$$\inf_{x \in G(\lambda)} f(x, p) > -\infty \Leftrightarrow \inf_{x \in \operatorname{co} G(\lambda)} f(x, p) > -\infty .$$

Proof. Because A is rational it follows that conv $G(\lambda)$ is a convex polyhedron and thus may be represented in the form

$$\operatorname{conv} G(\lambda) = C + U \tag{4.5.6}$$

where C is a compact polyhedron and U is the cone $\{v \in \mathbb{E}_n \mid Av \leqq 0\}$.

Assume $\inf\limits_{x \in \operatorname{conv} G(\lambda)} f(x, p) = -\infty$, there then exist points $x \in C$ and $u \in U$ such that

$$\inf_{\alpha \geqq 0} f(x + \alpha u, p) = -\infty \tag{4.5.7}$$

(cf. B. C. Eaves [3]). From $u^T D u < 0$ one immediately obtains $\inf\limits_{x \in G(\lambda)} f(x, p) = -\infty$ if one chooses an arbitrary rational point $\bar{u} \in U$ satisfying $\bar{u}^T D \bar{u} < 0$ and an arbitrary $\bar{x} \in G(\lambda)$ and then considers the behaviour of $f(\cdot, p)$ on the ray $\{\bar{x} + \alpha\bar{u} \mid \alpha \geqq 0\}$. Assume therefore $u^T D u = 0$. One may write x in the form

$$x = \sum_i \alpha_i x^i$$

where $\alpha_i \geqq 0$, $\sum\limits_i \alpha_i = 1$, and x^i are a finite number of elements of $G(\lambda)$. From (4.5.7) it then follows that

$$0 > 2x^T D u + p^T u = \sum_i \alpha_i (2x^{iT} D u + p^T u)$$

and there thus exists an $x^i \in G(\lambda)$ satisfying

$$2x^{iT} D u + p^T u < 0\,,$$

from which we obtain

$$\inf_{v \in U} f(x^i + v, p) = -\infty\,.$$

Applying Lemma 4.5.2 we then find

$$\inf_{v \in U_G} f(x^i + v, p) = -\infty$$

where $U_G = \{v \in U \mid v_1, \ldots, v_s \in \mathbb{Z}\}$.

Finally, since $x^i + U_G \subset G(\lambda)$ it follows that $\inf\limits_{x \in G(\lambda)} f(x, p) = -\infty$ and the proof is completed. □

The behaviour of the objective function under perturbations of the linear part of the same may be characterized by the preceding lemmas alone.

Theorem 4.5.1. *The following statements hold for the problem*

$(\mathrm{P}_s(p))$ $\quad \inf \{f(x, p) \mid x \in G(\lambda)\}$ [1])

providing the constraint matrix A is rational:

(1) *The set $\mathfrak{E}_s(\lambda) = \{p \in \mathbb{E}_n \mid \varphi_s(p) > -\infty\}$ is a convex polyhedron.*

(2) *The restriction $\varphi_s|_{\mathfrak{E}_s(\lambda)}$ of the function φ_s to $\mathfrak{E}_s(\lambda)$ is continuous.*

Proof. (1) By Lemma 4.5.3 we have

$$\mathfrak{E}_s(\lambda) = \left\{p \in \mathbb{E}_n \,\Big|\, \inf_{x \in \operatorname{co} G(\lambda)} f(x, p) < -\infty\right\}.$$

Employing (4.5.6) we find from the results of B. C. Eaves [3]

$$\begin{aligned} p \in \mathfrak{E}_s(\lambda) &\Leftrightarrow \inf_{\alpha \geqq 0} f(x + \alpha u, p) > -\infty \qquad \forall x \in C\,, \qquad \forall u \in U\,,\\ &\Leftrightarrow \inf_{u \in U} f(x + u, p) > -\infty \qquad \forall x \in C\,. \end{aligned}$$

[1]) $\lambda \in \mathbb{E}_m$ is a fixed right-hand side.

Because of

$$f(x + u, p) = f(x, p) + u^{\mathsf{T}} D u + (2Dx + p)^{\mathsf{T}} u$$

it follows that

$$p \in \mathfrak{C}_s(\lambda) \Leftrightarrow 2Dx + p \in \mathfrak{C}_s \qquad \forall x \in C$$

where $\mathfrak{C}_s = \left\{q \in E_n \mid \inf_{u \in U} f(u, q) > -\infty\right\}$. By Lemma 4.5.2 $\mathfrak{C}_s$ is a polyhedral convex cone. Thus

$$\mathfrak{C}_s(\lambda) = \{p \in E_n \mid 2Dcx + p \in \mathfrak{C}_s \ \forall x \in C\}$$

must be a convex polyhedron.

Proposition (2) then follows immediately from Theorems 4.3.8 and 4.2.2 (1). □

The next lemma provides a means of describing the behaviour of the level sets

$$N_s(p, \lambda, a) =_{\mathrm{Df}} \{x \in E_n \mid x \in G(\lambda), f(x, p) \leqq a\} \tag{4.5.8}$$

Lemma 4.5.4. *Let the matrices D and A of the problem $(\mathrm{P}_s(p, \lambda))$ be both rational, let further $(p^t, \lambda^t, a_t) \to (p^0, \lambda^0, a_0)$, $t \to \infty$, and for $t = 1, 2, \ldots$ $\varphi_s(p^t, \lambda^t) > -\infty$ hold.*

If $\{x^t\}$ is a sequence satisfying

$$x^t \in N_s(p^t, \lambda^t, a_t)\,, \qquad ||x^t|| \to \infty$$

then there exist an integer vector $u \in E_n$ and an infinite subsequence $\{x^{t_\nu}\}$ with the properties

(1) *for each positive integer k there exists an index $\nu(k)$ such that*

$$x^{t_\nu} - ku \in N_s(p^{t_\nu}, \lambda^{t_\nu}, a_{t_\nu})\,, \qquad \nu > \nu(k);$$

(2) $||x^{t_\nu} - u|| < ||x^{t_\nu}|| \quad \forall \nu$;

(3) $u^{\mathsf{T}} D u = 0$;

(4) $Au \leqq 0$.

Proof. Assume that without loss of generality *

$$v = \lim_{t \to \infty} \frac{x^t}{||x^t||}\,, \qquad \alpha_t = ||x^t||\,, \qquad v^t = \frac{x^t}{||x^t||}$$

then $Av \leqq 0$ and because of

$$f(x^t, p^t) = \alpha_t^2 v^{t\mathsf{T}} D v^t + \alpha^t p^{t\mathsf{T}} v^t \leqq a_t$$

we also have

$$v^{\mathsf{T}} D v = \lim_{t \to \infty} v^{t\mathsf{T}} D v^t \leqq 0\,.$$

Because of $\varphi_s(p^t, \lambda^t) > -\infty$ Lemma 4.5.3 then gives

$$Au \leqq 0 \Rightarrow u^{\mathsf{T}} D u \geqq 0.$$

Let A_i denote the i-th row of the matrix A and λ_i the corresponding component of λ. Further $I = \{i \mid A_i v = 0\}$. The set

$$U_I = \{u \in E_n \mid Au \leqq 0, A_i u = 0, i \in I, u^{\mathsf{T}} D u = 0\}$$

contains v. By Lemma 4.5.1 U_I may be represented as the union of a finite number of convex polyhedral cones possessing rational generators. It follows that a rational vector $\bar{v} \in U_I$ exists such that $\bar{v}^{\mathsf{T}} v > \frac{1}{2}$ (i.e. the distance between v and the rational vector $\bar{v} \in U_I$ is sufficiently small). With a suitably chosen $\mu \geqq 1$ we obtain an integer vector

$$u = \mu \bar{v} \in U_I.$$

For large t

$$v^{t\mathsf{T}}\bar{v} > \tfrac{1}{2}\,, \qquad v^{t\mathsf{T}}u > \tfrac{1}{2} \quad \text{and} \quad x^{t\mathsf{T}}u \to +\infty$$

then hold, and we obtain, again for large t

(a') $\quad ||x^t - u||^2 = ||x^t||^2 - 2x^{t\mathsf{T}}u + ||u||^2 < ||x^t||^2\,.$

Furthermore, $u \in U_I$ gives

$$f(x^t - ku, p^t) = f(x^t, p^t) - k(2Dx^t + p^t)^\mathsf{T} u\,. \tag{4.5.9}$$

For negative integer k we have $x^t - ku \in G(\lambda^t)$, and one concludes

$$(2Dx^t + p^t)^\mathsf{T} u \geqq 0\,. \tag{4.5.10}$$

Now assume $k \geqq 0$. If $i \in I$ then

$$A_i(x^t - ku) = A_i x^t \leqq \lambda_i^t$$

follows, if this is not the case ($i \notin I$) we have $A_i v < 0$ and thus

$$A_i x^t \to -\infty\,, \qquad t \to \infty\,.$$

Consequently

$$A_i(x^t - ku) \leqq \lambda_i^t \qquad \forall t > t(i, k)\,,$$

i.e. for $t > t(k)$ we always have

$$A(x^t - ku) \leqq \lambda^t$$

and this and (4.5.9), (4.5.10) yield:

(b') For each positive integer k there exists an index $t(k)$ such that $x^t - ku \in N_s(p^t, \lambda^t, a_t)$, $t > t(k)$, holds.

The lemma then follows from (a'), (b') and the manner of construction of u. □

Stability and Consequences. As before we consider problems of the type

$(\mathrm{P}_s(p, \lambda))\quad \inf\,\{f(x, p) \mid x \in G(\lambda)\}$

where f and G are defined as above. By $\mathfrak{C}_s$ we now denote the parameter set

$$\mathfrak{C}_s = \{(p, \lambda) \in E_{n+m} \mid \varphi_s(p, \lambda) > -\infty\}\,.$$

Theorem 4.5.2. *If A has only rational elements then the restriction of φ_s to $\mathfrak{C}_s$ is lower semicontinuous.*

Proof. By Theorem 3.4.2 we can employ the representation

$$G(\lambda) = K_G(\lambda) + U_G$$

where K_G is u.s.c.-H on E_m and $K_G(\lambda)$ is compact. We have

$$\begin{aligned}\varphi_s(p, \lambda) &= \inf_{y \in K_G(\lambda)} \inf_{u \in U_G} f(y + u, p)\\ &= \inf_{y \in K_G(\lambda)} \left[f(y, p) + \inf_{u \in U_G} \left(u^\mathsf{T} D u + (2Dy + p)^\mathsf{T} u\right)\right].\end{aligned}$$

The function in square brackets is continuous with respect to (y, p) for $y \in K_G(\lambda)$, $(p, \lambda) \in \mathfrak{C}_s$, since by Theorem 4.5.1 this holds for the extreme value function involved. As K_G is u.s.c.-H the lower semicontinuity of φ_s follows in the same manner as in the proof of Theorem 4.2.2 (2). □

The following theorem could well be called the main theorem of mixed-integer quadratic optimization, as may be seen below it constitutes an extension of the existence theorem of quadratic optimization to mixed-integer problems and establishes the behaviour of

(i) the optimal and ε-optimal sets if the linear parts of the objective function and the constraints are parameter-dependent and

(ii) the level sets of the objective function.

Theorem 4.5.3. *Assume* $(p^t, \lambda^t, a_t) \in E_{n+m+1}$, $t = 0, 1, 2, \ldots$, *and let*

$$(p^t, \lambda^t, a_t) \to (p^0, \lambda^0, a_0,) \qquad t \to \infty,$$
$$\varphi_s(p^t, \lambda^t) > -\infty, \qquad t = 0, 1, 2, \ldots,$$

hold. Then the following holds for the level sets

$$N_t =_{\mathrm{Df}} \{x \in G(\lambda^t) \mid f(x, p^t) \leqq a_t\}.$$

For each $\varepsilon > 0$ *there exists an index* $t(\varepsilon)$ *such that*

$$N_t \subset U_\varepsilon N_0 \qquad \forall t \geqq t(\varepsilon).$$

Proof. Assume that the theorem is not valid. Of all possible counter-examples we can find one for which the dimension of the convex hull conv $G(\lambda^0)$ is minimal (we set dim $\emptyset = -1$) and consider it in the following. Without loss of generality * assume

$$\Delta_t =_{\mathrm{Df}} N_t \setminus U_\varepsilon N_0 \neq \emptyset, \qquad t = 1, 2, \ldots, \qquad \varepsilon > 0 \quad \text{fixed}.$$

Choose from the closed sets Δ_t elements x^t of smallest norm, these obviously do not possess an accumulation point, since this would be contained in N_0, and $x^t \in U_\varepsilon N_0$ would follow for an infinite number of x^t. Thus $||x^t|| \to \infty$ $t \to \infty$ and Lemma 4.5.4 may be applied: there exists an integer vector $u \in E_n$ such that without loss of generality * the following holds:

(a) For each positive integer k there exists an index $t(k)$ such that $x^t - ku \in N_t$ $\forall t > t(k)$ holds;

(b) $||x^t - u|| < ||x^t||$;

(c) $u^{\mathsf{T}} D u = 0$;

(d) $Au \leqq 0$;

Defining $y^t = x^t - u$ we find $y^t \in U_\varepsilon N_0$ (for large t) because of (a), (b) and the choice of x^t. There then exist points $z^t \in N_0$ such that

$$||y^t - z^t|| < \varepsilon, \qquad t > t_0.$$

Since $x^t \notin U_\varepsilon N^0$ and $||x^t - (z^t + u)|| < \varepsilon$ it follows that $z^t + u \notin N_0$. Taking (c) and (d) into account we then have

$$a_0 < f(z^t + u, p^0) = f(z^t, p^0) + (2Dz^t + p^0)^{\mathsf{T}} u$$

and thus

$$(2Dz^t + p^0)^{\mathsf{T}} u > 0. \tag{4.5.11}$$

It further follows from

$$f(x^t - ku, p^t) = f(x^t, p^t) - k(2Dx^t + p^t)^{\mathsf{T}} u$$

and $\varphi_s(p^t, \lambda^t) > -\infty$ that

$$(2Dx^t + p^t)^{\mathsf{T}} u \geqq 0$$

holds. We also show that

$$\lim_{t\to\infty} (2Dx^t + p^t)^\mathsf{T} u = 0 . \tag{4.5.12}$$

This is the case since otherwise without loss of generality *

$$(2Dx^t + p^t)^\mathsf{T} u \geqq \alpha > 0 , \qquad t \geqq 1 ,$$

and from (a) and (4.5.9) it would then follow that the following is true

$$\varliminf_{t\to\infty} \varphi_s(p^t, \lambda^t) \leqq \varliminf_{t\to\infty} f(x^t - ku, p^t) \leqq a_0 - k\alpha$$

i.e. $\varliminf \varphi_s(p^t, \lambda^t) = -\infty$. By Theorem 4.5.2 we would then have $\varphi_s(p^0, \lambda^0) = -\infty$ which contradicts the above hypothesis.

We now use (4.5.11) and (4.5.12) to construct a further counter-example with constraint sets

$$\widetilde{G}(\lambda^t, \beta_t) \subset G(\lambda^t)'$$

for which

$$\dim \operatorname{conv} \widetilde{G}(\lambda^0, \beta_0) < \dim \operatorname{conv} G(\lambda^0)$$

holds, in contradiction to the assumed minimal dimension of conv $G(\lambda^0)$. In order to demonstrate this we define

$$\widetilde{G}(\lambda^t, \beta_t) = \{x \in G(\lambda^t) \mid 2u^\mathsf{T} Dx = \beta_t\}$$

where $\beta_t = 2u^\mathsf{T} Dx^t$. Then $\beta_t \to \beta_0 = -p^0 u$ follows from (4.5.12), since u is an integer vector $\widetilde{G}$ may be described by a rational matrix $\widetilde{A}$. The manner of construction of the constraint sets further implies

$$\inf \{f(x, p^t) \mid x \in \widetilde{G}(\lambda^t, \beta_t)\} > -\infty .$$

If $\widetilde{N}_t =_{\mathrm{Df}} \{x \in \widetilde{G}(\lambda^t, \beta_t) \mid f(x, p^t) \leqq a_t\}$ we then find

$$x^t \in \widetilde{N}_t \setminus U_\varepsilon \widetilde{N}_0 .$$

Finally, since the points $z^t \in G(\lambda^0)$ are by (4.5.11) not contained in the hyperplane

$$H = \{x \in E_n \mid (2Du)^\mathsf{T} x = \beta_0\}$$

which however obviously contains $\widetilde{G}(\lambda^0, \beta_0)$, we also have dim conv $\widetilde{G}(\lambda^0, \beta_0) <$ dim conv $G(\lambda^0)$ thus completing the proof. □

An immediate consequence of the preceding theorem is the following existence theorem for mixed-integer quadratic problems.

Theorem 4.5.4. *If A and D are rational matrices then a feasible point at which the infimum is attained exists if*

$$\inf \{x^\mathsf{T} Dx + p^\mathsf{T} x \mid Ax \leqq \lambda, x \in E_n, x_1, \dots, x_s \in \mathbb{Z}\}$$

is finite.

Note. This theorem has been proved by R. Mandel [1] under the additional condition that p and λ are rational.

Proof. If the infimum is denoted by a_0 and likewise

$$p^t = p , \quad \lambda^t = \lambda , \quad t = 0, 1, 2, \dots ,$$

and

$$\alpha_t = a_0 + \frac{1}{t} , \qquad t = 1, 2, \dots ,$$

then we have with the notation of Theorem 4.5.3

$$N_t \neq \emptyset\,, \qquad t = 1, 2, \ldots .$$

Since $N_t \subset U_\varepsilon N_0\ \forall t > t(\varepsilon)$ it follows that N_0 is non-empty. □

The next theorem is a proposition on the upper semicontinuity (H) of certain ε-optimal sets.

Theorem 4.5.5. *Let A and D be rational and let the following hold:*

$$\varphi_s(p^t, \lambda^t) > -\infty\,, \qquad t = 0, 1, 2, \ldots,$$

$$(p^t, \lambda^t) \to (p^0, \lambda^0)\,, \qquad t \to \infty\,,$$

$$\varepsilon_t \geqq 0\,, \qquad t = 1, 2, \ldots, \qquad \varepsilon_t \to 0\,, \qquad t \to \infty\,.$$

If then $P_{x_1,\ldots,x_s}(\lambda^t) = P_{x_1,\ldots,x_s}(\lambda^0)$[1] *and*

$$\psi_s(p, \lambda, \varepsilon) =_{\mathrm{Df}} \{x \in G(\lambda) \mid f(x, p) \leqq \varphi_s(p, \lambda) + \varepsilon\}$$

then a $t(\delta)$ exists for each $\delta > 0$ such that

$$\psi_s(p^t, \lambda^t, \varepsilon_t) \subset U_\delta \psi_s(p^0, \lambda^0, 0) \qquad \forall t > t(\delta)\,.$$

Proof. By Theorem 4.5.2

$$\varliminf_{t\to\infty} \varphi_s(p^t, \lambda^t) \geqq \varphi_s(p^0, \lambda^0)$$

and according to Theorem 3.4.3 the mapping G is lower semicontinuous on the set $\Lambda = \{\lambda^t \mid t = 0, 1, 2, \ldots\}$ from which

$$\varlimsup \varphi_s(p^t, \lambda^t) \leqq \varphi_s(p^0, \lambda^0)$$

follows because of Theorem 4.2.2. We set

$$a_t = \varphi_s(p^t, \lambda^t) + \varepsilon_t\,, \qquad t \geqq 1\,,$$

and then Theorem 4.5.3 may be applied, which immediately proves the proposition. □

Corollary 4.5.5.1. *If $\Lambda \subset E_{n+m}$ is a subset such that for $(p, \lambda) \in \Lambda$ $\varphi_s(p, \lambda)$ is always finite and the equality*

$$P_{x_1,\ldots,x_s} G(\lambda) = P_{x_1,\ldots,x_s} G(\lambda')$$

follows from $(p, \lambda), (p, \lambda') \in \Lambda$ then the rationality of A implies that $\varphi_s\colon \Lambda \to R$ is continuous.

If D is also rational then the mapping

$$\psi_s\colon \Lambda \to 2^{E_n}$$

is u.s.c.-H and $\psi_s(p, \lambda) \neq \emptyset$ holds for all $(p, \lambda) \in \Lambda$.

Proof. The continuity of $\varphi_s\colon \Lambda \to R$ again follows from Theorems 3.4.3, 4.2.2 and 4.5.2 whereas the upper semicontinuity (H) of ψ_s follows from Theorem 4.5.5. □

As a last remark we note that the rationality conditions hypothesized in the theorems of this section are for $s > 0$ all essential (examples illustrating this may be

[1]) cf. Theorem 3.4.3.

found in R. HANSEL [1]). The condition that the corresponding infima are finite may also not be relaxed as the reader easily confirms. The level set theorem (Theorem 4.5.3) may be sharpened in the convex case (D is positive semidefinite) by demonstrating that the level sets N_t considered in that theorem may be represented in the form

$$N_t = C_t + U_G^t$$

where C_t is compact and

$$U_G^t = \{x \in E_n \mid Dx = 0, p^{tT}x = 0, Ax \leqq 0, x_1, \dots, x_s \in \mathbb{Z}\} .$$

5. Properties of Characteristic Parameter Sets for Special Classes of Optimization Problems

5.1. Introduction

In this chapter we restrict our considerations to parametric optimization problems in finite-dimensional spaces, some results may however be extended to linear normed spaces. We are concerned with characteristic parameter sets for various classes of parametric optimization problems, such sets are the feasible parameter set $\mathfrak{B}$ (the set of parameters for which the constraint set is non-empty) and the solubility set $\mathfrak{A}$, i.e. the set of parameters for which the optimal set is non-empty. Moreover the characterization of the convexity set $\mathfrak{C}_f$ (the set of parameters for which the objective function is convex) is of interest with respect to convex parametric problems. This chapter however centres around local stability studies in which we consider certain subsets of the solubility set on which the optimal solution and extreme values of the corresponding optimization problems are, in a certain sense, stable.

It appears reasonable to study the set $\mathfrak{K}(x^0)$ of all parameters for which a given feasible x^0 is an optimal point. This concept is suitable for parametric optimization problems with fixed constraint sets, which are examined in Section 5.2.

Because of their properties in quite general parametric optimization problems with fixed constraint set M_0 the local stability sets $\mathfrak{K}(x^0)$ can be calculated for many applications of interest. The concept however has the disadvantage that one may only obtain a finite decomposition or partitioning of the solubility set into local stability sets in the case of linear parametric optimization problems. For the convenience of the reader we now give definitions of the concepts decomposition, covering and partitioning as they are used here.

A system of sets

$$\{\mathfrak{A}^j \subset E_n \mid j \in I\}$$

where I is an arbitrary subset of R, is called a *decomposition* of $\mathfrak{A} \subset E_n$, $\mathfrak{A} \neq \emptyset$, if

(i) $\bigcup_{j \in I} \mathfrak{A}^j = \mathfrak{A}$,

(ii) $\mathfrak{A}^{j_1} \cap \mathfrak{A}^{j_2} = \emptyset$, $\quad j_1, j_2 \in I$, $\quad j_1 \neq j_2$.

hold. As in the usual sense of the term a *covering* is a system of sets which only has the property (i). If $\mathfrak{A}, \mathfrak{A}^1, \dots, \mathfrak{A}^N$ are non-empty convex subsets of E_n then the system $\{\mathfrak{A}^1, \dots, \mathfrak{A}^N\}$ is called a *partitioning* of $\mathfrak{A}$ if the conditions

(i) $\bigcup_{j=1}^{N} \mathfrak{A}^j = \mathfrak{A}$,

(ii) $\operatorname{ri} \mathfrak{A}^{j_1} \cap \operatorname{ri} \mathfrak{A}^{j_2} = \emptyset$, $\quad j_1, j_2 \in \{1, \dots, N\}$, $\quad j_1 \neq j_2$,

(iii) $\dim \mathfrak{A} = \dim \mathfrak{A}^j$, $\quad j = 1, \dots, N$,

are satisfied.

For problems of greater generality than linear problems we obtain an infinite decomposition or merely a partitioning with local stability sets of the form $\mathfrak{R}(x^0)$. As far as establishing solution procedures with the object of determining a solution of the corresponding optimization problem for each parameter value (if such a solution exists) is concerned, one is however interested in those local stability sets which lead to a finite decomposition or partitioning. Investigations in this direction on various classes of parametric optimization problems constitute in our opinion an essential theoretical foundation for finding solution procedures.

The results of this chapter indicate for what classes a decomposition or partitioning of the solubility set $\mathfrak{A}$ and the corresponding optimal set mapping ψ or at least a selection function $\hat{x}(\lambda) \in \psi(\lambda)$ and finally the extreme value function φ may be determined using our approach. The analysis of linear optimization problems with variable coefficients in the objective function and the constraint vector (see Corollaries 5.5.2.1 and 5.5.2.2) shows that good prospects exist for calculating a partitioning $\mathfrak{A}^1, \dots, \mathfrak{A}^N$ (which sets are here polyhedral cones or convex polyhedra) with an associated selection function

$$\hat{x}(\lambda) = \begin{cases} \hat{x}^1(\lambda) & \text{if } \lambda \in \mathfrak{A}^1, \\ \vdots & \\ \hat{x}^N(\lambda) & \text{if } \lambda \in \mathfrak{A}^N \end{cases}$$

where the $\hat{x}^j(\lambda)$ are affine-linear in λ for $j = 1, \dots, N$) using the simplex technique see also W. Dinkelbach [1], T. Gal [4], [6], F. Nožička [2], and F. Nožička et al. [2]). The investigations of Sections 5.4 and 5.5 demonstrate further that we also have a similar situation in the case of convex quadratic optimization problems with parameters in the linear part of the objective function and in the constraint vector, as well as in the case of parameter-dependent linear complementarity problems with parameters in the right-hand sides and with positive semidefinite constraint matrix K. However even for the class of convex quadratic optimization problems with variable coefficients in the quadratic as well as the linear part of the objective function, we find that the corresponding local stability sets are only arcwise connected and in general not convex. One may then conjecture that the parametric linear complementarity problems mentioned above together with certain parametric linear fractional problems (cf. E.-Ch. Tammer [1], [3]) constitute the largest actually known class for which one may calculate a finite partitioning in the sense considered without undue effort (note also Section 6.1). A corresponding algorithm is proposed in Chapter 6. We have thus indicated that local stability sets should also be defined with respect to the property of constituting a finite decomposition or partitioning. The starting point for such considerations is to characterize those subsets of the solubility set on which the mappings M, ψ and φ have suitable properties. Upper semicontinuity of ψ at an arbitrary parameter point of the solubility set may be demonstrated for quite general classes of parametric optimization problems (cf. Theorems 4.2.2 and 4.3.3). But lower semicontinuity, for example, does not in general hold globally even in linear problems, and can only be shown under very restrictive conditions in Theorem 4.3.5. The question then arises of how one can find a subset of the solubility set such that the optimal set mapping is lower semicontinuous on that subset. In particular the existence of a continuous selection function is then ensured according to Section 2.3. In the definition of a local stability set the position of the optimal set in the constraint set is taken into account and the characteristic index set of constraints "active" at the optimal points is taken to be invariant (this is based on F. Nožička et al. [2]).

The concept of a local stability set of course depends on the class of problems considered (see (5.3.9) and (5.5.30)).

The results presented in this chapter on lower semicontinuity of the optimal set mapping ψ and the related existence of continuous selection functions $\hat{x}(\lambda) \in \psi(\lambda)$ also have immediate significance for the investigations of Section 6.3.

We would like to make some remarks on the historical development of local stability studies. Previously these investigations have only been undertaken for special classes of optimization problems, in particular linear and related classes (quadratic, linear fractional and linear complementarity problems). Up to the end of the Sixties the first phase of the development was characterized by the fact that the simplex technique was employed in a suitably modified form to solve such parametric problems. Here we only note work by A. S. Manne [1], S. I. Gass and T. L. Saaty [1], Ph. Wolfe [1], and C. E. Lemke [1].

Theoretical studies could only be undertaken in a restricted form in this manner. It became apparent that the partitioning determined for the parameter set considered is not only dependent on the method employed but is also influenced by certain method-specific aspects (such as degeneracy in the simplex technique). The geometrical approach of F. Nožička for a theory of linear optimization made a qualitative analysis of linear optimization problems with variable coefficients in the objective function and the constraint vector possible (see F. Nožička [1], [2]). Here already the concept of a local stability set based on the characteristic index set of the optimal set proved to be fundamental. The results presented in this chapter are due to students of F. Nožička in that the present authors see themselves in this capacity. We would here also like to place emphasis on work by B. C. Eaves [1], [2] which establishes important theorems on the solubility set of quadratic optimization problems and linear complementarity problems.

As a large number of different classes of parametric optimization problems is examined in this chapter a summary of the most important results would appear to be convenient.

In Section 5.2 we examine parametric optimization problems with fixed constraint sets and for which the objective function $f(x, \lambda)$ is affine-linear in the parameter vector λ. The local stability set $\mathfrak{R}(x^0)$ is then convex and closed for feasible x^0 and the optimal set mapping ψ proves to be constant on the relative interior of $\mathfrak{R}(x^0)$. If the objective function has the special form

$$f(x, \lambda) = g(x) + \sum_{i=1}^{s} \lambda_i h_i(x) \tag{5.1.1}$$

where this function is continuously differentiable at x on E_n and

$$M_0 = \{x \in E_n \mid f_j(x) \leqq 0, j = 1, \dots, m\} \tag{5.1.2}$$

holds, then the restriction of the local stability set $\mathfrak{R}(x^0)$ to the convexity set $\mathfrak{C}_f$ may be represented as

$$\mathfrak{R}^+(x^0) := \mathfrak{R}(x^0) \cap \mathfrak{C}_f = \{\lambda \in \mathfrak{C}_f \mid -\nabla_x f(x^0, \lambda) \in K^p(x^0, M_0)\} .$$

Here $K^p(x^0, M_0)$ is the polar cone of the cone $K(x^0, M_0)$ of feasible directions at x^0 with respect to M_0.

For an implicit representation of the cone $K^p(x^0, M_0)$ and thus also of the cone $\{\lambda \in E_s \mid -\nabla_x f(x^0, \lambda) \in K^p(x^0, M_0)\}$ it suffices to find the extremal vectors of $K(x^0, M_0)$ which may be done using a method of T. S. Motzkin and E. Burger (see for example

S. N. Černikov [1]) if the cone cl $K(x^0, M_0)$ is a polyhedral one. A modification of this method adapted to linear parametric optimization problems may be found in F. Nožička et al. [2, Chapter 11]. However as shown by K. Wendler [2] technical limits become apparent since the number of rows in the calculation tableau increases with each iteration step.

It follows from Theorem 5.2.1 that the convexity set $\mathfrak{C}_f$ is convex and closed.

If we set $f(x, \lambda) = g(x) + \lambda^T x$, which is of a more specialized form than in (5.1.1), then the set $\mathfrak{K}(x^0) + \{\nabla g(x^0)\}$ is itself a convex polyhedral cone and has the representation

$$\mathfrak{K}(x^0) + \{\nabla g(x^0)\} = -K^p(x^0, M_0) .$$

We also examine quadratic parametric optimization problems

$$\min \{x^T C x + p^T x \mid x \in M_0\} , \qquad (C, p) \text{ variable} ,$$

where M_0 is given in (5.1.2). Here the convexity set is a closed convex cone of full dimension, $\mathfrak{K}^+(x^0) = \mathfrak{K}(x^0) \cap \mathfrak{C}_f$ is a non-empty, convex and closed cone with a vertex at 0 if one chooses $x^0 \in M_0$.

Properties of the optimal set mapping and the extreme value function may be found in Theorems 5.2.2 to 5.2.5, and 5.2.7, and in Corollary 5.2.2.2.

In Section 5.3 we analyse the optimization problem $\min \{x^T C x + p^T x \mid x \in M(b)\}$, (C, p, b) variable, where

$$M(b) = \{x \in E_n \mid a^{iT} x \leqq b_i, i = 1, \dots, m\} .$$

The feasible parameter set $\mathfrak{B} = \{b \in E_m \mid M(b) \neq \emptyset\}$ is a polyhedral convex cone (Theorem 5.3.1). The solubility set $\mathfrak{A} = \{(C, p, b) \mid \psi(C, p, b) \neq \emptyset\}$ is a convex cone (Theorem 5.3.1) which in general is not necessarily closed. Continuity properties of the constraint set mapping M, the optimal set mapping ψ and the extreme value function φ on $\mathfrak{B}$ and $\mathfrak{A}$ are given in Theorem 5.3.2. Using the characteristic index set introduced in (4.3.18) with respect to $\psi(C, p, b)$ we define the local stability set

$$\mathfrak{S}(I, d) = \{(C, p, b) \in \mathfrak{A} \mid \operatorname{ch} \psi(C, p, b) = I, \dim \psi(C, p, b) = d\} . \qquad (5.1.3\text{a})$$

If one considers the set

$$\Sigma^I(b) = \{x \in E_n \mid a^{iT} x = b_i, i \in I, a^{iT} x < b_i, i \in \bar{I}\}$$

which, if not empty, is either a face of $M(b)$ or ri $M(b)$, then one obtains another description for $\mathfrak{S}(I, d)$:

$$\mathfrak{S}(I, d) = \{(C, p, b) \in \mathfrak{A} \mid \operatorname{ri} \psi(C, p, b) \subset \Sigma^k(b), \dim \psi(C, p, b) = d\} . \ (5.1.3\text{b})$$

There then exists a finite decomposition of the solubility set into local stability sets of the form $\mathfrak{S}(I, d)$. The central theorem of Section 5.3 demonstrates that the set $\mathfrak{S}(I, d)$ is arcwise connected (Theorem 5.3.3).

Theorem 5.3.4 provides continuity properties of the optimal set mapping and the extreme value function on $\mathfrak{S}(I, d)$ as an application of results given in Chapter 4 The closure of a local stability set of the form

$$\mathfrak{S}(I) = \{(C, p) \in \mathfrak{A} \mid \operatorname{ri} \psi(C, p) \subset \Sigma^I\} \qquad (5.1.4)$$

is identical to $\mathfrak{K}^+(x^0)$ for the fixed constraint set $M_0 = \{x \in E_n \mid a^{iT} x \leqq b_i, i = 1, \dots, m\}$ providing $\Sigma^I = \{x^0\}$ is a vertex of M_0 (Theorem 5.3.5).

Section 5.4 is concerned with parametric linear complementarity problems of the following form:

Determine for each $\lambda \in E_n$ the set

$$\psi(\lambda) = \{(x, y) \in M(\lambda) \mid x^{\mathsf{T}}y = 0\} \tag{5.1.5}$$

where $M(\lambda) = \{(x, y) \in E_{2n} \mid -Kx + y = \lambda, x \geqq 0, y \geqq 0\}$.
Here K is a member of the class of positive semidefinite matrices or of the class of square matrices with positive main minoıs.

The solubility set $\mathfrak{A}$ is a convex polyhedral cone and we have the representation

$$\mathfrak{A} = -(\{q \in E_n \mid K^{\mathsf{T}}q \leqq 0, q \geqq 0\})^p$$

(Theorems 5.4.1 and 5.4.2) the calculation of which has already been remarked upon. The optimal set mapping $\psi|_{\mathfrak{A}}$ is u.s.c.-H on $\mathfrak{A}$. Since the optimal set $\psi(\lambda)$ is a closed face of $M(\lambda)$ under the conditions given, the concept of a characteristic index set of the optimal set is also suitable for defining a local stability set $\mathfrak{A}^r$ (see (5.4.14)). We obtain a finite decomposition of the solubility set $\mathfrak{A}$ and, using the closures of the local stability sets of maximal dimension likewise a partitioning of $\mathfrak{A}$. We emphasize that cl $\mathfrak{A}^r$ is a convex polyhedral cone and that $\psi|_{\mathfrak{A}^r}$ is H-continuous on $\mathfrak{A}^r$.

As is the case in other sections of this chapter we obtain results corresponding to a parameter-dependence of the special form $q(t) = q^0 + \sum_{i=1}^{d} q^i t_i$ from those of the fully parametric problem (5.1.4). Since the single-parametric problem ($d = 1$) offers the best possibilities for an efficient numerical treatment this case is considered in some detail. These results constitute the theoretical basis for an algorithm presented in Section 6.2, the parameter t is then taken to be the $(2n + 1)$-th variable. The sets

$$N = \{(x, y, t) \in E_{2n+1} \mid -Kx + y - q^1 t = q^0, x \geqq 0, y \geqq 0\}\,.$$

$$G_1 = \{(x, y, t) \in N \mid x^{\mathsf{T}}y = 0\}$$

are the graphs of the mappings M and ψ respectively. The set

$$N^r = \{(x, y, t) \in N \mid x_i = 0, i \in I^r, y_j = 0, j \in J^r\}$$

is a closed face of N or is equal to N where the index set pair (I^r, J^r) defines the local stability set $\mathfrak{A}^r$.

In Section 5.5 we examine convex quadratic optimization problems

$$\min \{\tfrac{1}{2}\, x^{\mathsf{T}}Cx + p^{\mathsf{T}}x \mid Ax \leqq b, x \geqq 0\}\,, \quad (p, b) \text{ variable}\,,$$

which constitute a special case of those studied in Section 5.3 on the one hand and Section 5.4 on the other. Since here the matrix C is constant we of course obtain the sharper results of the theorems demonstrated in the latter. Using the approach employed in Section 5.4 we obtain as the main result the existence of a uniquely determined partitioning of the solubility set $\mathfrak{A}$ into polyhedral cones on whose relative interiors the restriction of the optimal set mapping is H-continuous, the optimal set has constant dimension and the extreme value function is quadratic (Theorem 5.5.2 and Note 5.5.5). It then in particular follows that we may employ the local stability set $\mathfrak{S}(I)$ given in (5.1.4). Here the local stability set is characterized by the position of the optimal set in the constraint polyhedron, whereas the concept employed for linear complementarity problems is based on the corresponding Kuhn-Tucker polyhedron.

Another type of "local stability set" is the subject of Section 5.6. In mixed-integer parametric optimization problems with constraint sets of the form

$$G(\lambda) = \{x \in E_n \mid Ax \leqq \lambda, x_1, \dots, x_s \in \mathbb{Z}\}$$

the feasible parameter set

$$\mathfrak{B} = \{\lambda \in E_m \mid G(\lambda) \neq \emptyset\}$$

may be entirely decomposed into subsets on which the restriction of the constraint set mapping

$$G\colon \mathfrak{B} \to 2^{E_n}$$

is l.s.c.-B. This last section is concerned with characterizing these local stability sets and the constraint mapping.

5.2. Parametric Optimization Problems with Fixed Constraint Sets

In the analysis of linear parametric optimization problems where the parameter-dependence is confined to the objective function, as for example in the book of F. Nožička et al. [2, Chapters 2 and 6] a number of results do not fully exploit the particular structure of the problem, thus in the class of parametric problems with fixed constraint sets characteristics of the extreme value function, the solubility set and certain local stability sets are derived from the affine-linear parameter-dependence of the objective function alone.

The basic problem considered in this section has the form

$$\min \{f(x, \lambda) \mid x \in M_0\}, \qquad \lambda \in \Lambda \subset E_s, \tag{5.2.1}$$

where we assume that

Λ[1]) is a non-empty convex subset of E_s, (5.2.1a)

M_0 is a non-empty subset of E_n, and (5.2.1b)

f is a real-valued function defined on $E_n \times E_s$ and $f(x, \cdot)$ is for each $x \in E_n$ an affine-linear function on E_s.[2]) (5.2.1c)

In order to achieve a concise exposition we restrict ourselves to problems in finite-dimensional spaces. Certain results of this section remain valid if Λ (or M_0) is a subset of a linear normed space.

Besides problem (5.2.1) we also consider certain specialized parametric problems which however may all be seen as special cases of the former, for instance they may be represented in a concrete fashion by convex functions, or the class of objective functions is restricted. We always denote the solubility set of the corresponding parametric problem by $\mathfrak{A}$, likewise the extreme value function by φ and the optimal set mapping by ψ.[3]) φ and ψ are only defined on Λ, the former is therefore a function with values in $\mathbb{R} \cup \{-\infty\}$. We also define

$$\operatorname{fin} \varphi =_{\mathrm{Df}} \operatorname{dom}(-\varphi) = \{\lambda \in \Lambda \mid \varphi(\lambda) > -\infty\}. \tag{5.2.2}$$

If $X \subset E_n$ is a non-empty convex set then the parameter set

$$\mathfrak{C}_{f,X} =_{\mathrm{Df}} \{\lambda \in E_s \mid f(\cdot, \lambda) \text{ convex on } X\} \tag{5.2.3}$$

---

[1]) Λ is taken to be a metric subspace of E_s (with the induced metric).

[2]) A function $g\colon E_s \to \mathbb{R}$ is said to be *affine-linear* if an $a \in E_s$ and a $b \in \mathbb{R}$ exist such that
$$g(u) = a^{\mathrm{T}} u + b\,.$$

[3]) We occasionally refer to the number of the problem involved to avoid misunderstandings, i.e. "ψ (5.2.1)", "φ (5.2.8)" etc.

is called the *convexity set* of the function f with respect to X. We also write

$$\mathfrak{C}_f = \mathfrak{C}_{f, E_n} \,. \tag{5.2.4}$$

Theorem 5.2.1. *Let a function f satisfying* (5.2.1c) *and a non-empty convex set $X \subset E_n$ be given. Then*

(1) *the convexity set $\mathfrak{C}_{f,X}$ is a convex closed subset of E_s;*

(2) *if $\mathfrak{C}_{f,X}$ is non-empty then the functions $f(\cdot, \lambda)$ are strictly convex on X either for all $\lambda \in \operatorname{ri} \mathfrak{C}_{f,X}$ or for no $\lambda \in \mathfrak{C}_{f,X}$.*

Proof. (1) We define the function

$$F_{x,y,\alpha}(\lambda) =_{\text{Df}} f(\alpha x + (1-\alpha)\, y, \lambda) - \alpha f(x, \lambda) - (1-\alpha)\, f(y, \lambda)$$

on E_s for all $x, y \in X$ and all $\alpha \in [0, 1]$. By condition (5.2.1c) $F_{x,y,\alpha}$ is affine-linear for all $x, y \in X$ and $\alpha \in [0, 1]$ and therefore the set

$$\Lambda_{x,y,\alpha} =_{\text{Df}} \{\lambda \in E_s \mid F_{x,y,\alpha}(\lambda) \leqq 0\}$$

is convex and closed, which then also applies to $\mathfrak{C}_{f,X} = \bigcap_{\substack{0 \leqq \alpha \leqq 1 \\ x,y \in X}} \Lambda_{x,y,\alpha}$.

(2) Because of $\emptyset \neq \mathfrak{C}_{f,X} \subset E_s$ (and thus $\operatorname{ri} \mathfrak{C}_{f,X} \neq \emptyset$) it suffices to show that if a $\lambda^0 \in \mathfrak{C}_{f,X}$ exists such that $f(\cdot, \lambda^0)$ is strictly convex on X then $f(\cdot, \lambda)$ is also strictly convex for all $\lambda \in \operatorname{ri} \mathfrak{C}_{f,X}$. Assume there exist a $\mu \in \operatorname{ri} \mathfrak{C}_{f,X}$, points $x, y \in X$ and a $\gamma \in (0, 1)$ such that we have

$$f(z, \mu) = \gamma f(x, \mu) + (1-\gamma)\, f(y, \mu) \tag{5.2.5}$$

where $z = \gamma x + (1-\gamma)\, y \in X$, then there also exists a $\beta > 0$ such that

$$\mu_\beta = \mu + \beta(\mu - \lambda^0) \in \operatorname{ri} \mathfrak{C}_{f,X} \tag{5.2.6}$$

holds. Since $f(\xi, \cdot)$ is affine-linear on E_s for each $\xi \in E_n$ and $f(\cdot, \lambda^0)$ is strictly convex on E_n then, taking (5.2.5) into account we conclude

$$\begin{aligned} f(z, \mu_\beta) &= (1+\beta)\, f(z, \mu) - \beta f(z, \lambda^0) \\ &> (1+\beta)\, \big(\gamma f(x, \mu) + (1-\gamma)\, f(y, \mu)\big) - \beta\big(\gamma f(x, \lambda^0) + (1-\gamma)\, f(y, \lambda^0)\big) \\ &= \gamma f(x, y_\beta) + (1-\gamma)\, f(y, \mu_\beta)\,. \end{aligned}$$

This contradicts (5.2.6), the above assumption is false and thus the proposition is valid. □

Theorem 5.2.1 is still valid if X and Λ are assumed to be general linear normed spaces (as long as $\operatorname{ri} \mathfrak{C}_{f,X} \neq \emptyset$). In the following theorem properties of the extreme value function φ and the optimal set mapping ψ of problem (5.2.1) are established.

Theorem 5.2.2

(1) *The extreme value function φ of problem* (5.2.1) *is concave and* fin φ *is a convex subset of Λ.*

(2) *Let f be continuous on $M_0 \times \Lambda$. Then φ is upper semicontinuous on Λ and the mapping ψ is closed on Λ.*

Proof. The first proposition follows immediately from the fact that $f(x, \cdot)$ is concave on E_s for each $x \in E_n$. The second is a consequence of Theorems 4.2.2 (1) and 4.2.1 (3) since the constraint set mapping of problem (5.2.1), being a constant mapping is in particular also l.s.c.-B. □

The last theorem also does not use the condition that X and Λ are finite-dimensional spaces.

Since the domain Λ of φ is a convex subset of E_s and a finite-valued concave function is continuous on the relative interior of its domain we have the following corollary of Theorem 5.2.2 (1).

Corollary 5.2.2.1. *The function φ is continuous on* ri (fin φ).

Proof. See the above remarks. □

The next corollary demonstrates that certain particular conclusions regarding the continuity of the mappings φ and ψ with respect to the relative interior of the convexity set $\mathfrak{C}_{f,M_0}$ may be drawn from the conditions (5.2.1a) to (5.2.1c) of problem (5.2.1).

Corollary 5.2.2.2. *Let M_0 be a non-empty convex and closed subset of E_n and f be a function which is continuous on $M_0 \times \Lambda$. Further, let the convexity set $\mathfrak{C}_{f,M_0}$ be non-empty and $f(\cdot, \lambda)$ be strictly convex on M_0 for at least one $\lambda \in \mathfrak{C}_{f,M_0}$.*

If $\mathfrak{Y} = \mathfrak{A} \cap \mathfrak{C}_{f,M_0}$ then the extreme value function $\varphi_{\mathfrak{Y}}$ of problem (5.2.1) *restricted to $\mathfrak{Y}$ is continuous on $\mathfrak{A} \cap$ ri $\mathfrak{C}_{f,M_0}$ and the corresponding optimal set mapping $\psi_{\mathfrak{Y}}$ restricted to $\mathfrak{Y}$ is H-continuous on $\mathfrak{A} \cap$ ri $\mathfrak{C}_{f,M_0}$.*

Proof. By Theorem 5.2.1 (2) $f(\cdot, \lambda)$ is strictly convex on M_0 for all $\lambda \in$ ri $\mathfrak{C}_{f,M_0}$, $\psi(\lambda)$ thus contains only one element for all $\lambda \in \mathfrak{A} \cap$ ri $\mathfrak{C}_{f,M_0}$. It follows that $\psi_{\mathfrak{Y}}$ is H-continuous on $\mathfrak{A} \cap$ ri $\mathfrak{C}_{f,M_0}$. One also immediately confirms that the hypotheses of the corollary then imply that conditions (4.3.5) to (4.3.10) are fulfilled: the continuity of $\varphi_{\mathfrak{Y}}$ and the upper semicontinuity (B) of $\psi_{\mathfrak{Y}}$ on $\mathfrak{A} \cap$ ri $\mathfrak{C}_{f,M_0}$ follow from Theorem 4.3.3. □

We refer the reader to Sections 4.2 and 4.3 for further statements concerning the continuity of the extreme value function φ and the optimal set mapping ψ of problem (5.2.1). In particular φ and ψ are continuous and u.s.c.-B respectively at a point $\lambda^0 \in \Lambda$ if f is continuous on $M_0 \times \{\lambda^0\}$ and a $\delta > 0$ and a bounded closed set $K \subset E_n$ exist such that $\psi(\lambda)$ is contained in K for all $\lambda \in V_\delta\{\lambda^0\}$ (this follows from Theorem 4.2.1). As a special case this is of course true if the set M_0 is bounded and closed.

That Theorem 5.2.2. (2) does not in general imply the continuity of φ on Λ or the upper semicontinuity (H) of ψ is illustrated by the following two examples relating to the rather specialized class of parametric convex quadratic problems with fixed polyhedral constraint sets.

Example 5.2.1 See Example 1.1 where the objective function is modified as $\lambda_1 x^2 + \lambda_2 x$. In Example 1.1 the optimal set mapping ψ is u.s.c.-H on R, the extreme value function φ however is not lower semicontinuous at $\lambda = 0$.

Example 5.2.2. In this example ψ is not u.s.c.-H. Consider the problem

$$\min \{\lambda_1 x_1^2 + \lambda_2 x_2^2 + 2\lambda_3 x_1 x_2 \mid (x_1, x_2)^\mathsf{T} \in E_2\}$$

where $\lambda = (\lambda_1, \lambda_2, \lambda_3)^\mathsf{T} \in \Lambda = E_3$ is variable. For $\lambda^0 = (1, 1, 1)^\mathsf{T}$ and $\lambda^k = \left(1, (1 + 1/k)^2, 1 + 1/k\right)^\mathsf{T}$, $k = 1, 2, \ldots$, we find

$$\psi(\lambda^0) = \{(x_1, x_2)^\mathsf{T} \in E_2 \mid x_1 = -x_2\},$$

$$\psi(\lambda^k) = \left\{(x_1, x_2)^\mathsf{T} \in E_2 \;\middle|\; x_1 = -\left(1 + \frac{1}{k}\right) x_2\right\}$$

which means, as indicated above, that ψ is not u.s.c.-H at λ^0.

In the following we extend in a natural fashion the notion of local stability employed in linear optimization (F. Nožička et al. [2]) to the parametric problem (5.2.1); we find that essential features of the local stability sets considered in the former context also appear in the more general case dealt with here. Let $\mathfrak{A}$ be the solubility set of problem (5.2.1). We then define for each $x^0 \in M_0$ the set

$$\mathfrak{R}(x^0) =_{\mathrm{Df}} \{\lambda \in \mathfrak{A} \mid x^0 \in \psi(\lambda)\} \tag{5.2.7}$$

and call it a *local stability set with respect to x^0*. The system of sets $\{\mathfrak{R}(x^0) \mid x^0 \in M_0\}$ constitutes a covering of $\mathfrak{A}$.

Theorem 5.2.3. *Consider the problem* (5.2.1) *and assume that $x^0 \in M_0$ is a point satisfying the condition that the parameter set $\mathfrak{R}(x^0)$ is non-empty. The following then holds:*

(1) *$\mathfrak{R}(x^0)$ is a convex closed subset of E_s;*

(2) *if $\mathfrak{Y} \subset \mathfrak{R}(x^0)$ is a non-empty convex and closed set then*

$$\text{(a)} \quad \psi(\lambda) = \psi(\mu) \quad \textit{for all} \quad \lambda, \mu \in \operatorname{ri} \mathfrak{Y},$$

$$\text{(b)} \quad \psi(\lambda) \subset \psi(\mu) \quad \textit{for all} \quad \lambda \in \operatorname{ri} \mathfrak{Y}, \qquad \mu \in \mathfrak{Y};$$

(3) *the extreme value function φ is affine-linear on $\mathfrak{R}(x^0)$*

$$\varphi(\lambda) = f(x^0, \lambda), \qquad \lambda \in \mathfrak{R}(x^0).$$

Proof. (1) Let λ and μ be arbitrary parameter points contained in $\mathfrak{R}(x^0)$. It then follows from

$$f(x^0, \lambda) \leqq f(x, \lambda) \qquad \forall x \in M_0,$$

$$f(x^0, \mu) \leqq f(x, \mu) \qquad \forall x \in M_0$$

and

$$f(x, \alpha\lambda + (1 - \alpha)\mu) = \alpha f(x, \lambda) + (1 - \alpha) f(x, \mu) \qquad \forall x \in M_0, \quad \forall \alpha \in R$$

that $f(x^0, \alpha\lambda + (1-\alpha)\mu) \leqq f(x, \alpha\lambda + (1 - \alpha)\mu)$ holds for all $x \in M_0$ and all $\alpha \in (0, 1)$. Hence $\mathfrak{R}(x^0)$ is convex. If $\{\lambda^k\} \subset \mathfrak{R}(x^0)$ is a sequence with $\lambda^k \to \lambda$ then $f(x^0, \lambda^k) \leqq f(x, \lambda^k)$ for all k. Since the function $f(x, \cdot)$ is continuous on E_s for each $x \in M_0$ we have $f(x^0, \lambda) = \lim_{k\to\infty} f(x^0, \lambda^k) \leqq \lim_{k\to\infty} f(x, \lambda^k) = f(x, \lambda)$ and $\mathfrak{R}(x^0)$ is thus a closed subset of E_s.

(2) In order to prove (a) assume that parameters $\lambda, \mu \in \operatorname{ri} \mathfrak{Y}$ and a $z \in \psi(\lambda)$ exist such that $z \notin \psi(\mu)$. Bearing $z \in \psi(\lambda) \subset M_0$ and $\mathfrak{R}(x^0) \subset \mathfrak{A}$ in mind one then finds

$$f(z, \lambda) = f(x^0, \lambda) = \min \{f(x, \lambda) \mid x \in M_0\},$$

$$f(z, \mu) > f(x^0, \mu) = \min \{f(x, \mu) \mid x \in M_0\}.$$

Setting $\lambda_\alpha = \lambda + \alpha(\lambda - \mu)$ for all $\alpha > 0$ one then obtains, noting (5.2.1c),

$$f(z, \lambda_\alpha) = (1 + \alpha)\,(z, \lambda) - \alpha f(z, \mu)$$

$$< (1 + \alpha) f(x^0, \lambda) - \alpha f(x^0, \mu) = f(x^0, \lambda_\alpha)$$

for all $\alpha > 0$. For sufficiently small $\alpha > 0$ we however have $\lambda_\alpha \in \operatorname{ri} \mathfrak{Y} \subset \mathfrak{R}(x^0)$ in contradiction to the above inequality $f(z, \lambda_\alpha) < f(x^0, \lambda_\alpha)$, thus (a) is valid.

Now assume $\lambda \in \operatorname{ri} \mathfrak{Y}$ and $\mu \in \operatorname{bd} \mathfrak{Y}$, because of $\operatorname{cl} \mathfrak{Y} = \operatorname{cl} (\operatorname{ri} \mathfrak{Y})$ there then exists a sequence $\mu^k \in \operatorname{ri} \mathfrak{Y}$ with $\mu^k \to \mu$. By (a) the sets $\psi(\mu^k)$ are identical to $\psi(\lambda)$ for all k. The continuity of the λ-component of the objective function gives

$$f(y, \mu) = \lim_{k\to\infty} f(y, \mu^k) \leqq \lim_{k\to\infty} f(x, \mu^k) = f(x, \mu)$$

for all $y \in \psi(\lambda)$ and all $x \in M_0$. Thus $\psi(\lambda) \subset \psi(\mu)$ holds.

Proposition (3) follows immediately from (5.2.1c) and the definition of $\mathfrak{R}(x^0)$. □

In order to establish the relation between local stability sets of type $\mathfrak{R}(x^0)$ and the local stability sets introduced in Chapter 4 (4.3.22) (on which the characteristic index set and the dimension of the optimal sets are invariant) we now specialize problem (5.2.1) and obtain information on the position of the optimal sets.

Consider the parametric optimization problem

$$\min \{f(x, \lambda) \mid x \in M_0\}, \qquad \lambda \in \Lambda = E_s, \tag{5.2.8}$$

subject to the conditions

$f: E_n \times E_s \to R$ is defined by

$$f(x, \lambda) =_{\mathrm{Df}} g(x) + \sum_{i=1}^{s} \lambda_i h_i(x) \tag{5.2.8a}$$

where $g: E_n \to R$, $\quad h_i: E_n \to R$, $\quad i = 1, \ldots, s$,
are functions which are continuously differentiable on E_n;

the convexity set $\mathfrak{C}_f$ is non-empty; (5.2.8b)

the functions $f_j : E_n \to R, j = 1, \ldots, m$, are given convex continuously differentiable functions on E_n and
$M_0 = \{x \in E_n \mid f_j(x) \leqq 0, j = 1, \ldots, m\}$ is non-empty; (5.2.8c)

for each $x \in M_0$ the cone
$K(x, M_0) = \{z \in E_n \mid \exists \bar{\alpha} > 0 : x^0 + \alpha z \in M_0, 0 \leqq \alpha \leqq \bar{\alpha}\}$
has the representation (5.2.8d)
$\operatorname{cl} K(x, M_0) = \{z \in E_n \mid z^{\mathrm{T}} \nabla f_j(x) \leqq 0, j \in I(x)\}$
where $I(x) = \{j \in \{1, \ldots, m\} \mid f_j(x) = 0\}$.

These conditions make it possible to apply the classic Kuhn-Tucker theory of convex optimization to the corresponding fixed optimization problem for each $\lambda \in \mathfrak{C}_f \cap \Lambda$. The constraint qualification (5.2.8d) may be replaced by a stronger condition (for instance by a constraint qualification of Slater type). The cone $K(x, M_0)$ is termed the *cone of feasible directions at* x with respect to M_0. Its polar cone $K^p(x, M_0)$ may by (5.2.8d) be represented as

$$K^p(x, M_0) = \left\{z \in E_n \;\middle|\; z = \sum_{j \in I(x)} u_j \nabla f_j(x), u_j \geqq 0, j \in I(x)\right\}. \tag{5.2.9}$$

If $\lambda \in \Lambda$ is fixed then the relation

$$-\nabla_x f(x^0, \lambda) \in K^p(x^0, M_0) \tag{5.2.10}$$

follows from (5.2.8a) to (5.2.8d) and is a necessary condition for the function $f(\cdot, \lambda)$ to attain its minimum with respect to M_0 at the point $x^0 \in M_0$. If $\lambda \in \mathfrak{C}_f \cap \Lambda$ then (5.2.10) is also a sufficient condition for the optimality of x^0.

In the following we characterize the parameter set for which a point $x^0 \in M_0$ satisfies (5.2.10).

Lemma 5.2.1. *Let $x_0 \in M_0$ be a point for which the set*

$$\mathfrak{Z}(x^0) =_{\mathrm{Df}} \{\lambda \in E_s \mid -\nabla_x f(x, \lambda) \in K^p(x^0, M_0)\}$$

is non-empty. Then $\mathfrak{Z}(x^0)$ is a convex polyhedral cone in E_s.

Proof. Defining the (n, s)-matrix

$$Z(x^0) = (\nabla h_1(x^0) \ldots \nabla h_s(x^0))$$

and the index set

$$I(x^0) = \{j \in \{1, \ldots, m\} \mid f_j(x^0) = 0\}$$

we find that for each $\lambda \in \mathfrak{Z}(x^0)$ there exists a $u \in E_m$ with the property $u_1, \ldots, u_m \geqq 0$ such that

$$-\nabla g(x^0) - Z(x^0)\,\lambda = \sum_{j \in I(x_0)} u_j \nabla f_j(x^0)$$

is satisfied. The set

$$\widetilde{X} = \left\{z \in E_n \,\middle|\, z =_{\mathrm{Df}} -\nabla g(x^0) - \sum_{j \in I(x_0)} u_j \nabla f_j(x^0),\ u_j \geqq 0, j \in I(x^0)\right\}$$

is a convex polyhedral cone. We have

$$\mathfrak{Z}(x^0) = Z(x^0)^{-1}(\widetilde{X}) =_{\mathrm{Df}} \{\lambda \in E_s \mid Z(x^0)\,\lambda \in \widetilde{X}\}\,,$$

i.e. $\mathfrak{Z}(x^0)$ is the full inverse image of a convex polyhedral cone under the linear mapping $Z(x^0)$. It follows from well-known theorems of the theory of convex polyhedrons (cf. for instance R. T. ROCKAFELLAR [2, Theorem 19.3]) that $\mathfrak{Z}(x^0)$ is also a convex polyhedral cone. □

We define

$$\mathfrak{R}^+(x^0) =_{\mathrm{Df}} \mathfrak{R}(x^0) \cap \mathfrak{C}_f \tag{5.2.11}$$

where $x^0 \in M_0$. By condition (5.2.8b), Theorem 5.2.1 and Theorem 5.2.3 (1) $\mathfrak{R}^+(x^0)$ is convex and closed. Since (5.2.10) is in the case $\lambda \in \mathfrak{A} \cap \mathfrak{C}_f$ a necessary and sufficient condition for $x^0 \in \psi(\lambda)$ we have

$$\mathfrak{R}^+(x^0) = \{\lambda \in \mathfrak{C}_f \mid -\nabla_x f(x^0, \lambda) \in K^p(x^0, M_0)\}. \tag{5.2.12}$$

Note 5.2.1. The representation (5.2.12) may in principle be used to calculate the sets $\mathfrak{R}^+(x^0)$ for fixed $x^0 \in M_0$. In particular an implicit representation of the cone $K^p(x^0, M_0)$, and thus also of the cone $\mathfrak{Z}(x^0)$ may be found with the help of any one of the known procedures for determining a fundamental system of edge vectors of a polyhedral cone (cf. for example S. N. ČERNIKOV [1, Chapter III]). Computing experiments with this method (K. WENDLER [2]) however also indicate the limits of its applicability.

In the following studies on the position of the optimal sets $\psi(\lambda)$ of problem (5.2.8) we always restrict our considerations to the convexity set $\mathfrak{C}_f$ of the (parametric) objective function f. Assume $x^0 \in M_0$ and $\mathfrak{R}^+(x^0) \neq \emptyset$. By Theorem 5.2.3 (2) there then exists a set $M_{\mathrm{opt}}(x^0)$ corresponding uniquely to the point x^0 such that $\psi(\lambda) = M_{\mathrm{opt}}(x^0)$ holds for all $\lambda \in \operatorname{ri} \mathfrak{R}^+(x^0)$. From

$$I(x^0) = \{i \in \{1, \ldots, m\} \mid f_i(x^0) = 0\} \tag{5.2.13}$$

it immediately follows that

$$\operatorname{ch} \psi(\lambda) \subset I(x^0) \qquad \forall \lambda \in \operatorname{ri} \mathfrak{R}^+(x^0) \tag{5.2.14}$$

where the characteristic index set $\operatorname{ch} \psi(\lambda)$ is for each $\lambda \in \mathfrak{R}^+(x^0)$ defined by (3.2.8), i.e. $\operatorname{ch} \psi(\lambda) = \{i \in \{1, \ldots, m\} \mid f_i(x) = 0\ \forall x \in \psi(\lambda)\}$. In general (5.2.14) does not imply equality (see Example 5.2.3).

Recalling Lemma 3.2.1 we find that $\operatorname{ch} \psi(\lambda) = I(x^0)$ may be interpreted in another way: for parametric convex problems we have

$$\operatorname{ri} \psi(\lambda) \subset \Sigma^{\operatorname{ch} \psi(\lambda)} \qquad \forall \lambda \in \mathfrak{A} \tag{5.2.15}$$

where the set

$$\Sigma^I =_{\mathrm{Df}} \{x \in E_n \mid f_i(x) = 0, i \in I, f_j(x) < 0, j \in \{1, \dots, m\} \setminus I\} \tag{5.2.16}$$

defined for $I \subset \{1, \dots, m\}$ is called a *pseudo-face* of M_0; it then follows that

$$\operatorname{ch} \psi(\lambda) = I(x^0) \Leftrightarrow \operatorname{ri} \psi(\lambda) \subset \Sigma^{I(x_0)} \tag{5.2.17}$$
$$\forall \lambda \in \operatorname{ri} \mathfrak{R}^+(x^0) \qquad \forall \lambda \in \operatorname{ri} \mathfrak{R}^+(x^0)$$

holds.

The following lemma provides a criterion for determining whether equality holds in (5.2.14) and thus whether the relative interior of the optimal set $\psi(\lambda)$ (5.2.8) is contained in the same pseudo-face as the (given) point x^0 for each $\lambda \in \operatorname{ri} \mathfrak{R}^+(x^0)$.

Lemma 5.2.2. *Consider problem* (5.2.8) *under the conditions* (5.2.8a) *to* (5.2.8d). *Assume* $x^0 \in M_0$ *and that the set* $I(x^0)$ *is defined according to* (5.2.13).

If a $\lambda^0 \in \mathfrak{C}_f$ *exists such that*

$$-\nabla_x f(x^0, \lambda^0) \in \operatorname{ri} K^p(x^0, M_0) \tag{5.2.18}$$

is satisfied then $\operatorname{ch} \psi(\lambda) = I(x^0)$ *for all* $\lambda \in \operatorname{ri} \mathfrak{R}^+(x^0)$.

Proof. It is sufficient to show that

$$(x - x^0)^{\mathrm{T}} \nabla f_i(x^0) = 0 \qquad \forall x \in \psi(\lambda^0) \qquad \forall i \in I(x^0) \tag{5.2.19}$$

holds. Because of $f_i(x^0) = 0$, $i \in I(x^0)$, the convexity of the functions f_i then gives for each $x \in \psi(\lambda^0)$

$$0 \geqq f_i(x) \geqq (x - x^0)^{\mathrm{T}} \nabla f_i(x^0) = 0 \qquad \forall i \in I(x^0)\,. \tag{5.2.20}$$

By (5.2.18) and (5.2.12) we have $\lambda^0 \in \mathfrak{R}^+(x^0)$ and Theorem 5.2.3 (2) then yields the relation $\psi(\lambda) \subset \psi(\lambda^0)$ for all $\lambda \in \operatorname{ri} \mathfrak{R}(x^0)$. From (5.2.20) we then have

$$I(x^0) \subset \operatorname{ch} \psi(\lambda) \qquad \forall \lambda \in \operatorname{ri} \mathfrak{R}^+(x^0)\,,$$

and the lemma follows from this and the inclusion (5.2.14).

We now demonstrate the validity of (5.2.19). By (5.2.9) we have

$$K^p(x^0, M_0) = \Big\{h \in E_n \,\Big|\, h = \sum_{i \in I(x_0)} u_i \nabla f_i(x^0),\ u_i \geqq 0,\ i \in I(x^0)\Big\}\,.$$

If $L(x^0)$ is the vertex set of $K^p(x^0, M_0)$ and

$$I' = \{j \in I(x^0) \mid \nabla f_j(x^0) \in L(x^0)\}$$

then there exists an index set $I'' \subset I(x^0) \setminus I'$ such that $\{\nabla f_i(x^0) \mid i \in I''\}$ is a fundamental system (of edge vectors) of the cone $K^p(x^0, M_0)$. It follows from this, (5.2.19) and Theorem A.1.1 that a $\bar{z} \in L(x^0)$ and numbers $v_i > 0$, $i \in I''$, exist such that

$$-\nabla_x f(x^0, \lambda^0) = \bar{z} + \sum_{i \in I''} v_i \nabla f_i(x^0) \tag{5.2.21}$$

holds. Now assume $x \in \psi(\lambda^0)$. Since $\lambda^0 \in \mathfrak{C}_f$ the optimization problem is convex for λ^0 (and fixed) and

$$(x - x^0)^{\mathrm{T}} \nabla f(x^0, \lambda^0) = 0 \tag{5.2.22}$$

follows, which in particular means that $(x - x^0) \in \big(K^p(x^0, M_0)\big)^p$. Thus

$$(x - x^0)^{\mathrm{T}} z = 0 \qquad \forall z \in L(x^0) \tag{5.2.23}$$

from which

$$(x - x^0)^{\mathrm{T}} \nabla f_j(x^0) = 0 \qquad \forall j \in I'$$

immediately follows. From (5.2.21), (5.2.22), and (5.2.23) we have

$$\sum_{i \in I''} v_i(x - x^0)^\mathsf{T} \nabla f_i(x^0) = 0 .$$

The convexity of f_i, $i \in I''$, yields $(x - x^0)^\mathsf{T} \nabla f_i(x^0) \leqq 0$ and thus, because of $v_i > 0$, $i \in I''$,

$$(x - x^0)^\mathsf{T} \nabla f_i(x^0) = 0 \qquad \forall i \in I(x^0) \setminus I'$$

if one takes the fact into account that $\{\nabla f_i(x^0) \mid i \in I''\}$ is a fundamental system. As $x \in \psi(\lambda^0)$ is arbitrary (5.2.19) follows. □

Corollary 5.2.2.1. *The following holds under the conditions of Lemma 5.2.2.:* $\lambda^0 \in \mathfrak{R}^+(x^0)$ *and* $\operatorname{ch} \psi(\lambda^0) = I(x^0)$.

Proof. See the proof of Lemma 5.2.2. □

The next theorem summarizes the results contained in Theorems 5.2.1 and 5.2.3 and Lemmas 5.2.1 and 5.2.2 for the case of parametric convex problems.

Theorem 5.2.4. *Consider problem* (5.2.8) *subject to the conditions* (5.2.8a) *to* (5.2.8d). *For each* $x^0 \in M_0$ *the local stability set* (5.2.11) *restricted to* $\mathfrak{C}_f$ *is either empty or it possesses the following properties:*

(1) $\mathfrak{R}^+(x^0) = \{\lambda \in \mathfrak{C}_f \mid -\nabla_x f(x^0, \lambda) \in K^p(x^0, M_0)\}$;

(2) $\mathfrak{R}^+(x^0)$ *is a convex closed subset of* $\mathbb{E}_s$;

(3) *there exists a non-empty convex closed set* $M_{\text{opt}}(x^0) \subset M_0$ *such that*

$$\psi(\lambda) = M_{\text{opt}}(x^0) \qquad \forall \lambda \in \operatorname{ri} \mathfrak{R}^+(x^0) ,$$

$$\psi(\mu) \supset M_{\text{opt}}(x^0) \qquad \forall \mu \in \mathfrak{R}^+(x^0);$$

(4) *the extreme value function* φ *is affine-linear on* $\mathfrak{R}^+(x^0)$;

(5) *if* $\operatorname{ch} \psi(\lambda)$ *is the characteristic index set corresponding to* $\psi(\lambda)$, $\lambda \in \mathfrak{A}$, *and if* $I(x^0)$ *is defined by* (5.2.13) *then*

$$\operatorname{ch} \psi(\lambda) \subset I(x^0) \qquad \forall \lambda \in \operatorname{ri} \mathfrak{R}^+(x^0);$$

furthermore, if a $\lambda^0 \in \Lambda$ *exists such that* $-\nabla_x f(x^0, \lambda^0) \in \operatorname{ri} K^p(x^0, M_0)$ *then*

$$\operatorname{ch} \psi(\lambda) = I(x^0) \qquad \forall \lambda \in \operatorname{ri} \mathfrak{R}^+(x^0)$$

holds.

Proof. See the preceding remarks. □

The following special case may be dealt with in a fashion which for the main part is similar to that of parametric linear problems.

Theorem 5.2.5. *Consider the parametric convex problem*

$$\min \{f(x, \lambda) \mid x \in M_0\} , \qquad \lambda \in \mathbb{E}_n , \tag{5.2.24}$$

where

$$f(x, \lambda) = g(x) + \lambda^\mathsf{T} x , \qquad M_0 = \{x \in \mathbb{E}_n \mid a^{j\mathsf{T}} x \leqq b_j, j = 1, \dots, m\} ,$$

$a^j \in \mathbb{E}_n$, $b_j \in \mathbb{R}$, $j = 1, \dots, m$, *and* g *is a function which is convex and continuously differentiable on* $\mathbb{E}_n$.

For each $x^0 \in M_0$ *there then exists a corresponding non-empty local stability set* $\mathfrak{R}(x^0)$ *such that the following statements are true if* $I(x^0) = \{i \in \{1, \dots, m\} \mid a^{i\mathsf{T}} x^0 = b_i\}$:

(1) $\mathfrak{R}(x^0) + \{\nabla g(x^0)\}$ *is a convex polyhedral cone and the following representation holds*

$$\mathfrak{R}(x^0) = \mathfrak{R}^+(x^0) = \{-\nabla g(x^0)\} - K^p(x^0, M_0)$$

where

$$K(x_0, M_0) = \{z \in E_n \mid a^{j\mathrm{T}} z \leqq 0, j \in I(x^0)\};$$

(2) $\dim \mathfrak{R}(x^0) = n - \dim \Sigma^{I(x^0)}$;

(3) *a non-empty convex and closed subset* $M_{\text{opt}}(x^0) \subset M_0$ *exists such that*

$$\psi(\lambda) = M_{\text{opt}}(x^0) \qquad \forall \lambda \in \operatorname{ri} \mathfrak{R}(x^0),$$

$$\psi(\mu) \subset M_{\text{opt}}(x^0) \qquad \forall \mu \in \mathfrak{R}(x^0)$$

hold;

(4) *the extreme value function* φ *of problem* (5.2.24) *is affine-linear on* $\mathfrak{R}(x^0)$;

(5) $\operatorname{ri} \psi(\lambda) \subset \Sigma^{I(x_0)}$ *holds for all* $\lambda \in \operatorname{ri} \mathfrak{R}(x^0)$ (i.e. $I(x^0) = \operatorname{ch} \psi(\lambda)$);

(6) $\mathfrak{R}(x) = \mathfrak{R}(x^0)$ *for all* $x \in M_{\text{opt}}(x^0)$.

Proof. The problem (5.2.24) obviously fulfils all conditions of problem (5.2.8), Theorem 5.2.4 may be invoked and (1), (3), and (4) then follow. By Theorem A1.3 we have $\dim K^p(x^0, M_0) = n - \dim L(x^0)$ where $L(x^0)$ is the vertex set of $K(x^0, M_0)$. This relation and the representation of $\mathfrak{R}(x^0)$ in (1) verify proposition (2). For $\lambda^0 = -\nabla g(x^0)$ the point x^0 is an optimal solution of the problem $\min \{f(x, \lambda) \mid x \in E_n\}$ for each $x^0 \in M_0$, it follows that $\mathfrak{R}(x^0)$ is non-empty. If $\lambda \in \operatorname{ri} \mathfrak{R}(x^0)$ then

$$-\nabla_x f(x^0, \lambda) = -(\nabla g(x^0) + \lambda) \in \operatorname{ri} K^p(x^0, M_0)$$

and (5) follows from (5.2.17) and Theorem 5.2.4 (5). Finally, (6) is a consequence of the fact that for all $x \in M_{\text{opt}}(x^0)$ the gradients $\nabla g(x)$ are identical and $I(x) = I(x^0)$ holds because of (5). □

As special cases Theorem 5.2.5 contains the main results of studies on local stability of parametric linear problems with parameter-dependent objective functions (cf. F. Nožička et al. [2, Chapters 2 and 6]) and parametric quadratic problems with parameters in the linear part of the objective function (cf. J. Guddat [3]).

The following example demonstrates that in a general convex parametric problem the properties (5) and (6) of the preceding theorem relating to the local stability sets $\mathfrak{R}^+(x^0)$ are not always given.

Example 5.2.3

(1) In general, in a problem of type (5.2.8)

$$\operatorname{ch} \psi(\lambda) = I(x^0) =_{\text{Df}} \{i \in \{1, \dots, m\} \mid f_i(x^0) = 0\}$$

does not hold for all $\lambda \in \operatorname{ri} \mathfrak{R}^+(x^0)$. Consider the problem

$$\min \{f(x, \mu) \mid x \in M_0\}, \qquad \mu \in R \text{ variable},$$

where $f(x, \mu) = -x_1 - x_2 - x_3 + \mu x_3^2$ and

$$M_0 = \{x \in E_3 \mid f_1(x) = -x_1 \leqq 0, \quad f_2(x) = -x_2 \leqq 0, \quad f_3(x) = -x_3 \leqq 0, \quad f_4(x) = x_1 + x_2 + x_3 - 1 \leqq 0\}.$$

The convexity set of f is given by $\mathfrak{C}_f = \{\mu \in R \mid \mu \geqq 0\}$. Conditions (5.2.8a) to (5.2.8d) are satisfied and we set $\Lambda = \mathfrak{C}_f$. Obviously

$$\psi(\mu) = \begin{cases} \{x \in M_0 \mid x_1 + x_2 + x_3 = 1, x_3 = 0\} & \text{if } \mu > 0, \\ \{x \in M_0 \mid x_1 + x_2 + x_3 = 1\} & \text{if } \mu = 0. \end{cases}$$

We consider $x^0 = (1, 0, 0)$, then $I(x^0) = \{2, 3, 4\}$ and $\mathfrak{R}^+(x^0) = \Lambda$. However

$$\operatorname{ch} \psi(\mu) = \begin{cases} \{3, 4\} & \text{if } \mu > 0, \\ \{4\} & \text{if } \mu = 0. \end{cases}$$

Thus (see (5.2.14)) $\operatorname{ch}\psi(\mu) \subset I(x^0)$, however no $\mu \in \mathfrak{R}^+(x^0)$ exists such that $\operatorname{ch}\psi(\mu) = I(x^0)$ holds.

(2) An extremely simple example showing that $\mathfrak{R}^+(x)$ is not identical to $\mathfrak{R}^+(x^0)$ for each x contained in $M_{\text{opt}}(x^0)$ is given by the following problem involving the set of real numbers:

$$\min \{\mu x^2 \mid x \in \mathbb{R}\}, \qquad \mu \in \mathbb{R} \quad \text{variable}.$$

We find

$$\psi(\mu) = \begin{cases} \{0\} & \text{if } \mu > 0, \\ \mathbb{R} & \text{if } \mu = 0, \\ \emptyset & \text{otherwise}. \end{cases}$$

On the other hand $M_{\text{opt}}(1) = \mathbb{R}$. Thus $0 \in M_{\text{opt}}(1)$, however

$$\mathfrak{R}^+(0) = \{\mu \in \mathbb{R} \mid \mu \geqq 0\} \neq \{0\} = \mathfrak{R}^+(1).$$

Quadratic functions in which all (or certain) coefficients are taken to be parameters constitute an important class of real-valued functions with affine-linear parameter-dependence. The remainder of the present section is devoted to quadratic parametric optimization problems with fixed constraint sets.

Consider the parametric problem

$$\min \{f(x, C, p) \mid x \in M_0\}, \qquad (C, p) \in \Lambda =_{\text{Df}} \mathbb{H}_{s(n)} \times \mathbb{E}_n \quad \text{variable} \tag{5.2.25}$$

where

$\mathbb{H}_{s(n)}$ is the *linear normed space of all symmetric* (n, n)*-matrices* and has the Euclidean norm of the space $\mathbb{E}_{n^2}$; (5.2.25a)

$f(x, C, p) =_{\text{Df}} x^{\mathsf{T}}Cx + p^{\mathsf{T}}x \ \forall (x, C, p) \in \mathbb{E}_n \times \mathbb{H}_{s(n)} \times \mathbb{E}_n$; (5.2.25b)

M_0 is given according to (5.2.8c) and (5.2.8d). (5.2.25c)

The number $s(n)$ is the dimension of the vector space $\mathbb{H}_{s(n)}$, i.e.

$$s(n) = \tfrac{1}{2}(n^2 + n). \tag{5.2.26}$$

Note 5.2.2. In the following and in Section 5.3 convex quadratic objective functions are frequently considered, and since we often use the fact that a quadratic function is convex (strictly convex) if and only if the coefficient matrix of its quadratic form is positive semidefinite (positive definite) it makes sense to base our investigations on the matrix space $\mathbb{H}_{s(n)}$. One however could just as well assume $\Lambda = \mathbb{E}_{s(n)} \times \mathbb{E}_n$ and represent the coefficients of the quadratic form as vectors

$$\hat{c} = (\hat{c}_{11}, \hat{c}_{12}, \dots, \hat{c}_{1n}, \hat{c}_{22}, \hat{c}_{23}, \dots, \hat{c}_{2n}, \dots, \hat{c}_{n-1,n-1}, \hat{c}_{n-1,n}, \hat{c}_{nn})^{\mathsf{T}} \in \mathbb{E}_{s(n)}.$$

If $c_{ij}\ \forall i, j \in \{1, \dots, m\}$ are the elements of a matrix $C \in \mathbb{H}_{s(n)}$ then $\hat{c}_{ij}$ are given by

$$\hat{c}_{ij} = \begin{cases} c_{ij} & \text{if } i = j, \\ 2c_{ij} & \text{if } i < j, \end{cases} \qquad i, j \in \{1, \dots, n\}, \tag{5.2.27}$$

and $\mathbb{H}_{s(n)}$ is isomorph to $\mathbb{E}_{s(n)}$. All conclusions made in the remainder of this and in the next section refering to some subset $\Lambda_0 \subset \mathbb{H}_{s(n)} \times \mathbb{E}_n$ are true if and only if they are true for the subset of $\mathbb{E}_{s(n)+n}$ assigned to Λ_0 by (5.2.27) (e.g. statements such as Λ_0 is a relatively open set, a convex set, a convex closed cone, a convex polyhedron or is an arcwise connected set; furthermore dimensional considerations, convergence of sequences $(C^k, p^k) \to (C, p)$ for $k \to \infty$, linearity of a given function on Λ_0, etc.).

Theorem 5.2.6. *Assume* $\Lambda = \mathbb{H}_{s(n)} \times \mathbb{E}_n$ *and let f be a function given by* (5.2.25b). *Then the convexity set*

$$\mathfrak{C}_f = \{(C, p) \in \mathbb{H}_{s(n)} \times \mathbb{E}_n \mid C \text{ positive semidefinite}\}$$

is a convex closed cone in $\mathbb{H}_{s(n)} \times \mathbb{E}_n$ *with full dimension, i.e.*

$$\dim \mathfrak{C}_f = s(n) + n .$$

Furthermore, $\operatorname{int} \mathfrak{C}_f = \{C \in \mathbb{H}_{s(n)} \mid C \text{ positive definite}\} \times \mathbb{E}_n$.

Proof. The zero element $0 \in \Lambda$ is contained in $\mathfrak{C}_f$. Using the notation of (5.2.27) we set $f(x, \hat{C}, p) = \sum_{i=1}^{n} \sum_{j=i}^{n} \hat{c}_{ij} x_i x_j$, by Theorem 5.2.1 (1) the set

$$\{(\hat{C}, p) \in \mathbb{E}_{s(n)+n} \mid f(\cdot, \hat{C}, p) \text{ convex on } \mathbb{E}_n\}$$

is a closed convex set. Since (5.2.27) defines an isomorphism $\mathfrak{C}_f$ is also closed and convex in Λ. Thus from $x^{\mathrm{T}}Cx \geqq 0$, $x \in \mathbb{E}_n$, $C \in \mathbb{H}_{s(n)}$, we have $x^{\mathrm{T}}(\alpha C)\, x \geqq 0$ $\forall \alpha \geqq 0$ and $\mathfrak{C}_f$ is a cone.

One easily confirms that the coefficient matrices $C^{ij} \in \mathbb{H}_{s(n)}$ corresponding to the positive semidefinite quadratic forms

$$g_{ij}(x) = x_i^2 + 2x_i x_j + x_j^2 \qquad \forall i: 1 \leqq i \leqq n \qquad \forall j: i \leqq j \leqq n , \tag{5.2.28}$$

are linearly independent. The pairs

$$(0, e^i) \in \mathfrak{C}_f \qquad \forall j \in \{1, \dots, n\} ,$$
$$(C^{ij}, 0) \in \mathfrak{C}_f \qquad \forall i \in \{1, \dots, n\} \qquad \forall j \in \{i, \dots, n\}$$

then form a system of $s(n) + n$ linearly independent vectors of the vector space $\mathbb{H}_{s(n)} \times \mathbb{E}_n$; e^j denotes as usual the j-th unit vector in $\mathbb{E}_n$. Finally, bearing Note 5.2.2 in mind the above representation of $\operatorname{int} \mathfrak{C}_f$ follows immediately from Theorem 5.2.1 (2) since obviously

$$(\tilde{C}, \tilde{p}) = \sum_{j=1}^{n} (0, e^j) + \sum_{i=1}^{n} \sum_{j=i}^{n} (C^{ij}, 0)$$

is contained in $\operatorname{int} \mathfrak{C}_f$ and $f(\cdot, \tilde{C}, \tilde{p})$ is strictly convex on $\mathbb{E}_n$. □

If $n \geqq 2$, $\mathfrak{C}_f$ is not a polyhedral cone one easily confirms that the set

$$\{(c_{11}, c_{12}, c_{22}) \in \mathbb{E}_3 \mid c_{11}c_{22} \geqq c_{12}^2, c_{11} \geqq 0, c_{22} \geqq 0\}$$

is not polyhedral.

The extreme value function φ of problem (5.2.25) is by Theorem 5.2.2 concave on Λ. Continuity properties of φ and the optimal set mapping ψ may be deduced from Theorem 5.2.2, Corollaries 5.2.2.1, and 5.2.2.2, and the remarks following the latter; some of the conditions imposed in these propositions are automatically fulfilled for problem (5.2.25): f is continuous on Λ, M_0 is convex, and closed and $\mathfrak{C}_f$ is non-empty.

In the following theorem on problem (5.2.25) properties of the local stability sets $\mathfrak{R}(x^0) = \{(C, p) \in \mathfrak{A} \mid x^0 \in \psi(C, p)\}$ or $\mathfrak{R}^+(x^0) = \mathfrak{R}(x^0) \cap \mathfrak{C}_f$ are established.

Theorem 5.2.7. *Given the parametric optimization problem* (5.2.25) *the following hold for each* $x^0 \in M_0$:

(1) $\mathfrak{R}(x^0)$ *is a non-empty convex and closed subset of* Λ;

(2) $\psi(C, p) = \{x^0\}$ *for all* $(C, p) \in \operatorname{ri} \mathfrak{R}(x^0)$;

(3) *the extreme value function is linear on* $\mathfrak{R}(x^0)$ *and the relation* $\varphi(C, p) = x^{0\mathsf{T}}Cx^0 + p^\mathsf{T}x^0$ *holds for all* $(C, p) \in \mathfrak{R}(x^0)$;

(4) $\mathfrak{R}^+(x^0) = \mathfrak{R}(x^0) \cap \mathfrak{C}_f$ *is a non-empty convex and closed cone with*

$$\dim \mathfrak{R}^+(x^0) \geqq s(n) + \dim K^p(x^0, M_0)\,.$$

Proof. (1) Let $E \in \mathbb{H}_{s(n)}$ be the unit matrix. The point x^0 obviously constitutes a minimum of the problem

$$\min \{x^\mathsf{T}x - 2x^{0\mathsf{T}}x \mid x \in \mathbb{E}_n\} \tag{5.2.29}$$

and thus $(E, -2x^0) \in \mathfrak{R}(x^0)$. By Theorem 5.2.3 (1) and Note 5.2.2 we find that $\mathfrak{R}(x^0)$ is closed and convex.

(2) The point x^0 is the only minimal point of problem (5.2.29) and we therefore have from Theorem 5.2.3 (2) and Note 5.2.2

$$\psi(C, p) \subset \psi(E, -2x^0) = \{x^0\} \quad \text{for all} \quad (C, p) \in \operatorname{ri} \mathfrak{R}(x^0)\,.$$

(3) This follows trivially from the definition of $\mathfrak{R}(x^0)$.

(4) It follows from $(E, -2x^0) \in \mathfrak{C}_f$, Lemma 5.2.1, the relation (5.2.12), and Theorem 5.2.6 that $\mathfrak{R}^+(x^0)$ is a non-empty convex closed cone. Let $\{C^\nu \mid \nu = 1, \dots, s(n)\}$ be a system of linearly independent positive semidefinite matrices, e.g. the system $\{C^{ij}\}$ of coefficient matrices given by (5.2.28). Then the unconstrained minimum (global minimum) of the convex function g^ν, $g^\nu(x) =_{\mathrm{Df}} x^\mathsf{T}C^\nu x - (2C^\nu x^0)^\mathsf{T} x$ is attained at x^0 for each $\nu \in \{1, \dots, s(n)\}$, consequently we have $(C^\nu, -2C^\nu x^0) \in \mathfrak{R}^+(x^0)$ $\forall \nu \in \{1, \dots, s(n)\}$. Furthermore, let $h^1, \dots, h^r$ be linearly independent vectors contained in $K^p(x^0, M_0)$ and $r = \dim K^p(x^0, M_0)$. Since the zero matrix $0 \in \mathbb{H}_{s(n)}$ may not be represented as a linear combination of $C^1, \dots, C^{s(n)}$ the pairs

$$(C^\nu, -2C^\nu x^0)\,, \qquad \nu = 1, \dots, s(n)\,, \qquad (0, h^i)\,, \qquad i = 1, \dots, r\,,$$

are linearly independent and obviously $(0, h^i) \in \mathfrak{R}^+(x^0)$ $\forall i \in \{1, \dots, r\}$. The dimension of the cone $\mathfrak{R}^+(x^0)$ is thus equal to or greater than $s(n) + \dim K^p(x^0, M_0)$. □

An interesting special case of problem (5.2.25) follows if M_0 is described by a system of linear inequalities. Such parametric problems are considered in the next section in the context of parametric quadratic problems with variable polyhedral constraint sets.

5.3. Convex Quadratic Optimization Problems with Parameters in the Objective Function and the Right-Hand Sides of the Linear Constraints

This section is concerned with the analysis of parametric convex quadratic optimization problems in n-dimensional Euclidean space; perturbations of the coefficients of the objective function and of the right-hand sides of the linear constraints are allowed.

Besides characterizing the feasible parameter set and the solubility set we also examine the decomposition of the solubility set into a finite number of local stability sets of type $\mathfrak{S}(I, d)$ as defined in Section 4.3. The extreme value function φ and the optimal set mapping ψ (definitions are given in Section 4.3) are continuous on these sets and the existence of a continuous selection function $\hat{x}(\lambda) \in \psi(\lambda)$ is also ensured.

Such local stability investigations are motivated by the fact that in general the extreme value function is not continuous on the solubility set $\mathfrak{A}$ (see Example 1.1) and likewise an optimal vector function which is continuous on $\mathfrak{A}$ does not exist for the problem class considered here.

We examine the parametric convex quadratic problem

$$\min \{f(x, C, p) \mid x \in M(b)\}\,, \qquad (C, p, b) \in \Lambda \text{ variable} \tag{5.3.1}$$

under the conditions

$$M: \mathbb{E}_m \to 2^{\mathbb{E}_n}\,,$$

$$M(b) =_{\mathrm{Df}} \{x \in \mathbb{E}_n \mid a^{i\mathrm{T}} x \leqq b_i,\, i = 1, \dots, m\} \tag{5.3.1a}$$

where $a^i \in \mathbb{E}_n\ \forall i \in \{1, \dots, m\}$ are given and fixed and $b = (b_1, \dots, b_m)^{\mathrm{T}}$;

$$f: \mathbb{E}_n \times \mathbb{H}_{s(n)} \times \mathbb{E}_n \to \mathbb{R} \quad \text{and} \quad f(x, C, p) =_{\mathrm{Df}} x^{\mathrm{T}} C x + p^{\mathrm{T}} x; \tag{5.3.1b}$$

$$\Lambda =_{\mathrm{Df}} \mathfrak{C}_f \times \mathbb{E}_m \tag{5.3.1c}$$

where $\mathfrak{C}_f$ is given by

$$\mathfrak{C}_f = \{(C, p) \in \mathbb{E}_{s(n)} \times \mathbb{E}_n \mid C \text{ positive semidefinite}\}$$

and $\mathbb{H}_{s(n)}$ by (5.2.25).

A denotes the matrix

$$A = (a^1 \dots a^m)\,. \tag{5.3.2}$$

As usual φ is the extreme value function, ψ the optimal set mapping and $\mathfrak{A}$ the solubility set of problem (5.3.1). By the *feasible parameter set of problem (5.3.1)* we mean the set

$$\mathfrak{B} =_{\mathrm{Df}} \{b \in \mathbb{E}_m \mid M(b) \neq \emptyset\}\,. \tag{5.3.3}$$

We remark here that in (5.3.1a) the representation of the constraint polyhedra in the form of inequalities is not a restrictive condition. The considerations that follow may be carried out in completely analogous fashion if different descriptions are employed, this is specially indicated in some of the notes of this section. Restricting the parameter space to the convexity set is part of our objective of specializing the local stability studies of Section 4.3 to the present problem. Note 5.3.5 gives some idea of the non-convex case.

In what follows we use the notation (n and r are positive integers)

$$\begin{aligned} \mathbb{H}^+(n) &=_{\mathrm{Df}} \{C \in \mathbb{H}_{s(n)} \mid C \text{ positive semidefinite}\}\,,\\ \mathbb{H}^{++}(n) &=_{\mathrm{Df}} \{C \in \mathbb{H}_{s(n)} \mid C \text{ positive definite}\}\,,\\ \mathbb{H}^+(n, r) &=_{\mathrm{Df}} \{C \in \mathbb{H}^+(n) \mid \operatorname{rank} C = r\}\,. \end{aligned} \tag{5.3.4}$$

Theorem 5.3.1

(1) *The feasible parameter set* $\mathfrak{B}$ *of problem* (5.3.1) *is an m-dimensional convex polyhedral cone with the vertex set* $\mathfrak{B}_0 = \{Az \mid z \in \mathbb{E}_n\}$.

(2) *The solubility set* $\mathfrak{A}$ *of* (5.3.1) *is a convex cone of dimension* $s(n) + n + m$ *and we have*

$$\operatorname{int}_{\operatorname{aff}\Lambda} \mathfrak{A} = \mathbb{H}^{++}(n) \times \mathbb{E}_n \times \operatorname{int}_{\mathbb{E}_m} \mathfrak{B}\,.$$

Proof. (1) This follows from condition (5.3.1a) and the following representation of $\mathfrak{B}$:

$$\mathfrak{B} = \{Az + E^{(m)} y \mid z \in \mathbb{E}_n,\, 0 \leqq y \in \mathbb{E}_m\}\,^{1)}\,. \tag{5.3.5}$$

[1]) $E^{(m)}$ is the unit (m, m)-matrix.

(2) By (1) and Theorem 5.2.6 $\mathfrak{B}$ and $\mathfrak{C}_f$ are convex sets, thus ri $(\mathfrak{C}_f \times \mathfrak{B}) = \text{ri}\,\mathfrak{C}_f \times \text{ri}\,\mathfrak{B}$ follows. Again by Theorem 5.2.6 we have ri $\mathfrak{C}_f = \mathbb{H}^{++}(n) \times \mathbb{E}_n$. Since each problem $\min \{x^T C^0 x + p^{0T} x \mid x \in M(b^0)\}$ where $C^0 \in \mathbb{H}^{++}(n)$, $p^0 \in \mathbb{E}_n$, and (in particular) $b^0 \in \text{ri}\,\mathfrak{B}$, is soluble it follows that ri $\mathfrak{C}_f \times \text{ri}\,\mathfrak{B} \subset \mathfrak{A}$. A classic theorem of convex analysis (see for instance R. T. Rockafellar [2, Corollary 6.3.1]) and $\mathfrak{A} \subset \mathfrak{C}_f \times \mathfrak{B}$ then imply ri $\mathfrak{A} = \text{ri}\,\mathfrak{C}_f \times \text{ri}\,\mathfrak{B}$.

Since $\dim \mathbb{H}^{++}(n) = s(n)$ (Theorem 5.2.6) and $\dim \mathfrak{B} = m$ (1) we have ri $\mathfrak{A} = \text{int}_{\text{aff}\,A}\,\mathfrak{A}$ and ri $\mathfrak{B} = \text{int}_{\mathbb{E}_m}\,\mathfrak{B}$. In particular $\dim \mathfrak{A} = s(n) + n + m$ then holds.

One need then only show that $\mathfrak{A}$ is a convex cone. The zero element 0 of $\mathbb{H}_{s(n)} \times \times \mathbb{E}_n \times \mathbb{E}_m$ is obviously contained in $\mathfrak{A}$. Assume then $(\tilde{C}, \tilde{p}, \tilde{b}) \in \mathfrak{A}$ and $(C, p, b) \in \mathfrak{A}$. Suppose that $\mathfrak{A}$ is not a convex cone, one may then find numbers $t > 0$ and $s > 0$ such that

$$\inf \{x^T(tC + s\tilde{C})\, x + (tp + s\tilde{p})^T x \mid Ax \leqq tb + s\tilde{b}\} = -\infty \tag{5.3.6}$$

holds if one applies the existence theorem of quadratic optimization and notes that $(tb + s\tilde{b}) \in \mathfrak{B}$ by (1) of the present theorem. Since a quadratic objective function $g(x) = x^T Q x + q^T x$ is not lower bounded on a non-empty convex polyhedron P if and only if a $u \in O^+P$ (the recession cone of P) and an $\bar{x} \in P$ exist such that $\gamma(\beta) = g(\bar{x} + \beta u)$ is not lower bounded on $\mathbb{R}^+ = \{\beta \mid \beta \geqq 0\}$ (see for instance B. C. Eaves [3, § 3]) it follows from (5.3.6) that a $u^0 \in O^+M(tb + s\tilde{b})$ and an $x^0 \in M(tb + s\tilde{b})$ exist such that the following holds:

$$\begin{aligned} &f(x^0 + \alpha u^0, tC + s\tilde{C}, tp + s\tilde{p}) = f(x^0, tC + s\tilde{C}, tp + s\tilde{p}) \\ &\quad + \alpha[\alpha u^{0T}(tC + s\tilde{C})\, u^0 + 2u^{0T}(tC + s\tilde{C})\, x^0 + (tp + s\tilde{p})^T u^0] \xrightarrow[\alpha \to +\infty]{} -\infty\,. \end{aligned}$$

The matrix $tC + s\tilde{C}$ is positive semidefinite; if $u^{0T}(tC + s\tilde{C})\, u^0 > 0$ then the expression in square brackets would attain positive values for large α, which means that the objective functions does not exhibit the property of being a decreasing and not lower bounded function for $\alpha \to +\infty$. Consequently $tu^{0T}Cu^0 + su^{0T}\tilde{C}u^0 = 0$, we thus obtain from $C \in \mathbb{H}^+(n)$ and $\tilde{C} \in \mathbb{H}^+(n)$ the relations $Cu^0 = 0$ and $\tilde{C}u^0 = 0$ and finally $p^T u^0 < 0$ or $\tilde{p}^T u^0 < 0$.

Assume then, without loss of generality, $p^T u^0 < 0$, it follows that

$$\begin{aligned} &f(x + \alpha u^0, C, p) = f(x, C, p) + \alpha^2 u^{0T} C u^0 + 2\alpha u^{0T} C x + \alpha p^T u^0 \xrightarrow[\alpha \to +\infty]{} -\infty\,, \\ &x \in M(b)\,. \end{aligned}$$

However, $u^0 \in O^+M(tb + s\tilde{b}) = O^+M(b)$ and thus $x + \alpha u^0 \in M(b)$ $\forall \alpha \geqq 0$ then hold, in contradiction to $(C, p, b) \in \mathfrak{A}$. It follows that $\mathfrak{A}$ is a convex cone. □

The solubility set $\mathfrak{A}$ of problem (5.3.1) is in general not closed in $\mathbb{H}_{s(n)} \times \mathbb{E}_n \times \mathbb{E}_m$ since for instance all parameter points $(C = 0, p, b)$ corresponding to unsolvable linear optimization problems with non-empty constraint sets but having unbounded decreasing objective functions, are contained in cl $\mathfrak{A} \setminus \mathfrak{A}$.

If $M(b)$ is compact for all $b \in \mathfrak{B}$ we obviously have $\mathfrak{A} = \mathfrak{C}_f \times \mathfrak{B}$ which implies the following corollary.

Corollaray 5.3.1.1. If $\{h \in \mathbb{E}_n \mid Ah \leqq 0\} = \{0\}$ then $\mathfrak{A} = \mathfrak{C}_f \times \mathfrak{B}$.

Proof. See the preceding remarks. □

Theorem 5.3.2

(1) *The constraint set mapping* $M|_{\mathfrak{B}}$ (5.3.1a) *restricted to* $\mathfrak{B}$ *is H-continuous on* $\mathfrak{B}$.

(2) *If* $\mathfrak{D} = \mathfrak{C}_f \times \mathfrak{B}$ *then* $\varphi|_{\mathfrak{D}}$ *is upper semicontinuous on* $\mathfrak{A}$ *and* $\psi|_{\mathfrak{D}}$ *is closed on* $\mathfrak{A}$.

(3) $\varphi|_{\mathfrak{D}}$ *is continuous on the set* $\mathfrak{A}_0 =_{\mathrm{Df}} \{(C, p, b) \in \mathfrak{A} \mid \psi(C, p, b)$ *bounded*$\}$ *and* $\psi|_{\mathfrak{D}}$ *is u.s.c.-B on* $\mathfrak{A}_0$.

(4) *The optimal set mapping* $\psi|_{\mathfrak{A}_{00}}$ *restricted to* $\mathfrak{A}_{00} =_{\mathrm{Df}} \mathbb{H}^{++}(n) \times \mathbb{E}_n \times \mathfrak{B}$ *is a vector function which is continuous on* $\mathfrak{A}_{00}$.

Proof. (1) See Corollary 3.4.1.1.

(2) Since $M|_{\mathfrak{B}}$ is l.s.c.-B on $\mathfrak{B}$ it follows from Theorem 4.2.2 (1) that $\varphi|_{\mathfrak{D}}$ is upper semicontinuous on $\mathfrak{A}$ and by Theorem 4.2.1 (3) $\psi|_{\mathfrak{D}}$ is closed on $\mathfrak{A}$.

(3) Because of (1) and the conditions (5.3.1a) to (5.3.1c) all hypotheses of Theorem 4.3.3 are fulfilled and thus (3) follows if $\lambda^0 = (C, p, b) \in \mathfrak{A}_0$.

(4) For each $(C, p, b) \in \mathfrak{A}_{00}$ the optimal set contains only one element. Since $\mathfrak{A}_{00} \subset \mathfrak{A}_0$ (4) then follows immediately from (3). □

We note here that the continuous optimal vector function considered in proposition (4) of the preceding theorem is piece-wise differentiable any number of times (cf. Note 5.3.3).

Note 5.3.1. If instead of (5.3.1) we consider the problem

$$\min \{x^{\mathsf{T}}Cx + p^{\mathsf{T}}x \mid x \in \tilde{M}(\tilde{b})\}\,, \qquad (C, p, \tilde{b}) \text{ variable}\,, \tag{5.3.7}$$

under the conditions (5.3.1b) and (5.3.1c) but also subject to

$$\tilde{M}(\tilde{b}) = \{x \in \mathbb{E}_n \mid Ax \leqq b^0 + B\tilde{b}\}\,, \qquad \tilde{b} \in \mathbb{E}_r \text{ variable}\,,$$

$$b^0 \in \mathbb{E}_m \text{ fixed}\,, \qquad B \text{ is a fixed } (m, r)\text{-matrix}\,, \tag{5.3.7a}$$

then we find that

$$\mathfrak{B} - b^0 = B \cdot \tilde{\mathfrak{B}}$$

holds for the feasible parameter set $\tilde{\mathfrak{B}} =_{\mathrm{Df}} \{\tilde{b} \in \mathbb{E}_r \mid \tilde{M}(\tilde{b}) \neq \emptyset\}$ where $\mathfrak{B} - b^0 =_{\mathrm{Df}} \{b - b^0 \mid b \in \mathfrak{B}\}$ and $B \cdot \tilde{\mathfrak{B}} =_{\mathrm{Df}} \{Bb \mid b \in \tilde{\mathfrak{B}}\}$. It follows that since $\tilde{\mathfrak{B}}$ is the inverse image of the convex polyhedral cone $\mathfrak{B} - b^0$ under the linear mapping B it is itself a convex polyhedral cone.

In particular (5.3.7a) also includes the description of the constraint polyhedron in the form of equations subject to the condition $x \geqq 0$.

Finally, we remark that the continuity propositions (1), (2), and (3) of the last theorem may be extended without difficulty to the constraint set mapping M', the extreme value function φ' and the optimal set mapping ψ' of the problem

$$\min \{x^{\mathsf{T}}C(\lambda)\, x + p(\lambda)^{\mathsf{T}}\, x \mid Ax \leqq b(\lambda)\}\,, \qquad \lambda \in \Lambda'\,, \tag{5.3.8}$$

$$C\colon \Lambda' \to \mathbb{H}^{+}(n), \qquad p\colon \Lambda' \to \mathbb{E}_n \quad \text{and} \quad b\colon \Lambda' \to \mathbb{E}_m \text{ are continuous on } \Lambda' \tag{5.3.8a}$$

and

$$\Lambda' \text{ is a metric space, } A \text{ a fixed } (m, n)\text{-matrix}\,, \tag{5.38.b}$$

are assumed.

We now specialize the concept of a local stability set introduced in Chapter 4 to the parametric convex quadratic problem. Setting $\lambda = (C, p, b)$ we then denote the characteristic index set with respect to $\psi(C, p, b)$ (containing the indices of the constraints active for all $x \in \psi(C, p, b)$ assuming $(C, p, b) \in \mathfrak{A}$) introduced in (4.3.18) by ch $\psi(C, p, b)$, likewise $\mathfrak{S}(I, d)$ is the set defined according to (4.3.22)

$$\mathfrak{S}(I, d) = \{(C, p, b) \in \mathfrak{A} \mid \operatorname{ch} \psi(C, p, b) = I, \dim \psi(C, p, b) = d\}\,. \tag{5.3.9}$$

The special form of problem (5.3.1) guarantees that the set $\mathfrak{S}(I, d)$ is non-empty and is thus a local stability set in the sense of (4.3.22) for each $(I, d) \in \mathfrak{T}$, where

$$\mathfrak{T} = 2^{\{1,\dots,m\}} \times \{0, 1, \dots, n\} . \tag{5.3.10}$$

Given $(I, d) \in \mathfrak{T}$ one may construct a $(\hat{C}, \hat{p}, \hat{b}) \in \mathfrak{S}(I, d)$ in the following manner: We set $\hat{b}_i = a^{i\mathsf{T}}\bar{e}$, $i \in I$, $\hat{b}_j = a^{j\mathsf{T}}\bar{e} + 1$, $j \in \bar{I}$, where $\bar{e} = (1, \dots, 1)^{\mathsf{T}}$ and $C = 0$ if the rank of the matrix A^I formed from the vectors a_i, $i \in I$, is equal to $n - d$, otherwise (if rank $A^I = s < n - d$)

$$\hat{C} = (\hat{c}_{ij}), \qquad \hat{c}_{ij} = \begin{cases} 1 & \text{if } \ 1 \leqq i = j \leqq n - d - s , \\ 0 & \text{otherwise} , \end{cases}$$

and finally $\hat{p} = -2\hat{C}\bar{e}$.

From the preceding and the definition of the sets $\mathfrak{S}(I, d)$ it follows that the system

$$\{\mathfrak{S}(I, d) \mid (I, d) \in \mathfrak{T}\} \tag{5.3.11}$$

is a finite decomposition of the solubility set $\mathfrak{A}$ of problem (5.3.1) into local stability sets. If

$$\Sigma^I(b) = \{x \in E_n \mid a^{i\mathsf{T}}x = b_i, i \in I, a^{j\mathsf{T}}x < b_j, j \in \{1, \dots, m\}\backslash I\,,\ b \in \mathfrak{B}\,, \tag{5.3.12}$$

then the following relation holds (cf. (5.2.17)):

$$\operatorname{ch} \psi(C, p, b) = I \Leftrightarrow \operatorname{ri} \psi(C, p, b) \subset \Sigma^I(b) \tag{5.3.13}$$

where $(C, p, b) \in \mathfrak{A}$ and $I \subset \{1, \dots, m\}$.

Lemma 5.3.1. *Let $\mathfrak{A}$ be the solubility set of problem* (5.3.1) *and $y\colon \mathfrak{A} \to E_n$ be an arbitrary vector function satisfying $y(C, p, b) \in \psi(C, p, b)$ $((C, p, b) \in \mathfrak{A})$. Then the following holds for all $(C, p, b) \in \mathfrak{A}$:*

$$\operatorname{ri} \psi(C, p, b) = \{x \in \Sigma^{\operatorname{ch}\psi(C,p,b)}(b) \mid Cx = Cy(C, p, b)\} . \tag{5.3.14}$$

Proof. Follows from Theorem A.4.3. □

The following lemma gives a sufficient criterion for establishing whether a parameter triple $(C, p, b) \in \mathfrak{C}_f \times \mathfrak{B}$ is contained in the set

$$\mathfrak{S}(I) = \{(C, p, b) \in \mathfrak{A} \mid \operatorname{ch} \psi(C, p, b) = I\} .$$

Lemma 5.3.2. *Given $(C, p, b) \in \mathfrak{C}_f \times \mathfrak{B}$ and an index set $I \subset \{1, \dots, m\}$ then $\psi(C, p, b) \neq \emptyset$ and $\operatorname{ch} \psi(C, p, b) = I$ if a vector $y \in M(b)$* (5.3.1a) *and multipliers $\mu_i \in R$, $i \in I$, exist such that*

$$a^{i\mathsf{T}}y = b_i\,, \qquad i \in I\,, \qquad a^{j\mathsf{T}}y < b_j\,, \qquad j \in \{1, \dots, m\}\backslash I\,,$$
$$2Cy + p = -\sum_{i\in I} \mu_i a^i\,, \qquad \mu_i > 0\,, \qquad i \in I\,,$$

hold. In particular we then have $y \in \operatorname{ri} \psi(C, p, b)$.

Proof. The local Kuhn-Tucker conditions immediately give $y \in \psi(C, p, b)$. If we set $M_0 = M(b)$ then by (5.2.9) the polar cone corresponding to $K(y, M_0)$ (= the direction cone at y with respect to M_0) is of the form

$$K^p(y, M_0) = \Big\{z \in E_n \mid z = \sum_{i\in I} u_j a^j,\ u_j \geqq 0, j \in I(y)\Big\} .$$

By hypotheses $\nabla_x f(y, C, p) = 2Cy + p \in -\operatorname{ri} K^p(y, M_0)$. Consider the parametric problem

$$\min \{x^{\mathsf{T}}Dx + q^{\mathsf{T}}x \mid x \in M_0\}\,, \qquad (D, q) \in \mathfrak{C}_f . \tag{5.3.15}$$

Let $\hat{\psi}$ denote the optimal set mapping of (5.3.15); $\mathfrak{C}_f$ is given according to (5.3.1c). Bearing Note 5.2.2 in mind we find that problem (5.3.15) is obviously a special case of the parametric problem (5.2.8) in which the parameter set is restricted to the convexity set. Consequently the conditions of Lemma 5.2.2 are satisfied, Corollary 5.2.2.1 gives $\operatorname{ch} \hat{\psi}(C, p) = \operatorname{ch} \psi(C, p, b)$. By Lemma 5.3.1 we also have $y \in \operatorname{ri} \psi(C, p, b)$. □

The two preceding lemmas as well as the next following are needed to prove Theorem 5.3.3, in which it is shown that the local stability sets $\mathfrak{S}(I, d)$, $(I, d) \in \mathfrak{T}$, are arcwise connected.

The following simple example is intended to show that the sets $\mathfrak{S}(I, d)$ are in general not convex.

Example 5.3.1. If λ^1 and λ^2 denote collectively the coefficient vectors of the objective function and the right-hand sides of the constraint polyhedra of the problems

$$\min \{y^2 - 8y \mid y - x = 4, x \geqq 0, y \geqq 0\} \tag{5.3.16}$$

and

$$\min \{3x - 2y \mid y - x = 2, x \geqq 0, y \geqq 0\} \tag{5.3.17}$$

respectively, then

$$\min \{\tfrac{3}{2}\, x + \tfrac{1}{2}\, y^2 - 5y \mid y - x = 3, x \geqq 0, y \geqq 0\} \tag{5.3.18}$$

corresponds to $\frac{1}{2}(\lambda^1 + \lambda^2)$. All three problems have unique solutions: (0, 4), (0, 2), and $(\frac{1}{2}, \frac{7}{2})$ in the above order. The characteristic index sets are $\{1\}$ in the first two cases and the empty set in the last.

We now present definitions, notations, and various results needed in the following with regard to arcwise connected sets (cf. H. Schubert [1]). A topological space X is said to be *arcwise connected* if for any two points $x \in X$ and $y \in X$ there exists a continuous mapping $\beta: [0, 1] \to X$ such that $\beta(0) = x$ and $\beta(1) = y$ hold. Such a mapping is called an *arc with initial point x and endpoint y*, we write simply $x\beta y$. If Y is a subset of X then Y is arcwise connected if the topological subspace Y is arcwise connected. An arcwise connected space X is connected, i.e. the only subsets which are both open and closed are the empty set and X itself.

By *coupling* two arcs $x\beta^1 y$ and $y\beta^2 z$ according to

$$\beta^1\beta^2(t) = \begin{cases} \beta^1(2t) & \text{if } 0 \leqq t \leqq \frac{1}{2}\,, \\ \beta^2(-1 + 2t) & \text{if } \frac{1}{2} \leqq t \leqq 1 \end{cases} \tag{5.3.19}$$

we obtain the arc $x\beta^1\beta^2 z$. The arc $y\beta^- x$ defined by

$$\beta^-(t) = \beta(1 - t)\,, \qquad 0 \leqq t \leqq 1\,, \tag{5.3.20}$$

is termed the arc which is inverse to $x\beta y$. Before we come to Theorem 5.3.3 we need an important property of the sets (l and r are positive integers)

$$\mathbb{H}^+(l, r) = \{C \in \mathbb{H}^+(l) \mid \operatorname{rank} C = r\}\,, \qquad 1 \leqq r \leqq l\,. \tag{5.3.21}$$

By $Q^{l,r} = (q_{ij}^{l,r})$ we denote the diagonal (l, l)-matrix for which

$$q_{ij}^{l,r} = \begin{cases} 1 & \text{if } 1 \leqq i = j \leqq r\,, \\ 0 & \text{otherwise}\,, \end{cases} \qquad i, j = 1, \dots, l\,, \qquad r \leqq l\,, \tag{5.3.22}$$

holds. Further, let $E^{(l)} = (e_{ij}^{(l)})$ be the unit (l, l)-matrix and $\hat{E}^{(l)} = (\hat{e}_{ij}^{(l)})$ be the matrix with elements

$$\hat{e}_{ij}^{(l)} = \begin{cases} -1 & \text{if } i = j = 1\,, \\ e_{ij}^{(l)} & \text{otherwise}\,. \end{cases} \tag{5.3.23}$$

Lemma 5.3.3. *If r and l are positive integers and $1 \leqq r \leqq l$ then the set $\mathbb{H}^+(l, r)$ is arcwise connected.*

Proof. If $l = r$ then $\mathbb{H}^+(l, r) = \mathbb{H}^+(l)$ is by Theorem 5.2.6 in fact convex, assume therefore $r < l$. Further, assume that the matrices $C^1, C^2 \in \mathbb{H}^+(l, r)$ are given and arbitrary. Since C^1 and C^2 are positive semidefinite there exist, as is well-known from inear algebra, regular (l, l)-matrices P^1 and P^2 such that

$$C^i = P^{i\mathsf{T}} Q^{l,r} P^i , \qquad i = 1, 2 ,$$

holds. One easily confirms that for each $i \in \{1, 2\}$ there exists a continuous mapping $B^i\colon [0, 1] \to \mathbb{H}^{l,l}$ with the properties

$$B^i(0) = P^i , \qquad B^i(1) = \begin{cases} E^{(l)} & \text{if} \quad \det P^i > 0 , \\ \hat{E}^{(l)} & \text{if} \quad \det P^i < 0 , \end{cases}$$

and $\det B^i(\alpha) \neq 0 \ \forall \alpha \in [0, 1]$ where $\det P$ denotes the determinant of a matrix P (see for instance E. ZEIDLER[1, Part I, p. 144f.]). By

$$D^i(\alpha) = B^i(\alpha)^\mathsf{T}\, Q^{l,r} B^i(\alpha) , \qquad 0 \leqq \alpha \leqq 1 , \qquad i = 1, 2 ,$$

we define continuous mappings from $[0, 1]$ into $\mathbb{H}^+(l)$ satisfying for $i = 1, 2$

$$D^i(0) = C^i , \qquad D^i(1) = Q^{l,r} \quad \text{and} \quad \operatorname{rank} D^i(\alpha) = r , \qquad 0 \leqq \alpha \leqq 1 .$$

The mapping $D^1(D^2)^-$ constructed according to (5.3.19) and (5.3.20) is then an arc in $\mathbb{H}^+(l, r)$ with initial point C^1 and endpoint C^2. □

Theorem 5.3.3. *Consider a parametric optimization problem (5.3.1) and let an index set $I \subset \{1, \dots, m\}$ and a non-negative integer $d \leqq n$ be given. The local stability set $\mathfrak{S}(I, d)$ is then arcwise connected.*

Proof. Assume $(C, p, b) \in \mathfrak{S}(I, d)$ and $y \in \operatorname{ri} \psi(C, p, b) \subset \Sigma^I(b)$ are arbitrarily chosen. Further, let A^I be the matrix formed from the vectors $a^{i\mathsf{T}}$, $i \in I$, $s = \operatorname{rank} A^I$ and $L^I = \{h \in \mathbb{E}_n \mid a^{i\mathsf{T}} h = 0, i \in I\}$; the zero vector of $\mathbb{E}_k$ (k is a positive integer) is here denoted by 0^k.

We first assume $s < n$. There then exists an orthonormal basis $\{g^1, \dots, g^{n-s}\}$ in L^I and we write

$$G = (g^1 \dots g^{n-s}); \tag{5.3.24}$$

the linear mapping corresponding to the matrix G is denoted by $\mathfrak{G}$.

We define a mapping

$$\beta^1\colon [0, 1] \to \mathbb{H}^{n,n} \times \mathbb{E}_n \times \mathbb{E}_m \tag{5.3.25}$$

by

$$C^1 = GDG^\mathsf{T} \quad \text{where} \quad D = G^\mathsf{T} C G , \tag{5.3.25a}$$

$$C^1(\alpha) = (1 - \alpha)\, C + \alpha C^1 , \qquad 0 \leqq \alpha \leqq 1 , \tag{5.3.25b}$$

$$p^1(\alpha) = (1 - \alpha)^2\, p + \alpha(-\bar{a} - 2(1 - \alpha)\, Cy - 2C^1 y) , \tag{5.3.25c}$$

where $0 \leqq \alpha \leqq 1$, $\bar{a} = \sum_{i \in I} a^i$, and

$$\beta^1(\alpha) = (C^1(\alpha), p^1(\alpha), b) \qquad 0 \leqq \alpha \leqq 1 . \tag{5.3.25d}$$

β^1 is obviously continuous on $[0, 1]$. Since C and C^1 are symmetric and positive semidefinite we also have $C^1(\alpha) \in \mathbb{H}^+(n)$ for all $0 \leqq \alpha \leqq 1$. The definition of $\mathfrak{S}(I, d)$ gives $I = \operatorname{ch} \psi(C, p, b)$ and since $y \in \psi(C, p, b)$ and $y_j > 0$, $j \notin I$, there exist $\bar{\mu}_i \geqq 0$, $i \in I$, such that $2Cy + p = - \sum_{i \in I} \bar{\mu}_i a^i$. Furthermore,

$$2C^1(\alpha)\, y + p^1(\alpha) = 2(1-\alpha)\, Cy + 2\alpha C^1 y + p^1(\alpha)$$
$$= (1-\alpha)^2 (2Cy + p) - \alpha\bar{a} = - \sum_{i \in I} \left((1-\alpha)^2 \bar{\mu}_i + \alpha\right) a^i$$

for all α satisfying $0 \leqq \alpha \leqq 1$, all conditions of Lemma 5.3.2 are thus fulfilled for y and all $\mu_i(\alpha) = (1-\alpha)^2 \bar{\mu}_i + \alpha \;\forall \alpha : 0 < \alpha \leqq 1$. Bearing $\beta^1(0) = (C, p, b)$ in mind

$$\operatorname{ch} \psi(\beta^1(\alpha)) = I \quad \text{and} \quad y \in \operatorname{ri} \psi(\beta^1(\alpha)) \tag{5.3.26}$$

then follow for all α satisfying $0 \leqq \alpha \leqq 1$.

In order to demonstrate $\beta^1(\alpha) \in \mathfrak{S}(I, d)$, $0 \leqq \alpha \leqq 1$, we must first confirm that $\dim \psi(\beta^1(\alpha)) = d$, $0 \leqq \alpha \leqq 1$, holds. Assume $\alpha \in [0, 1]$. Each $h \in L^I$ satisfying the system of equations $G^{\mathsf{T}} C^1(\alpha)\, h = 0^{n-s}$ also satisfies $h^{\mathsf{T}} C^1(\alpha)\, h = 0$ since $\{g^1, \dots, g^{n-s}\}$ is a basis of L^I and G is given by (5.3.24). Because of $C^1(\alpha) \in H^+(n)$ it follows from $h^{\mathsf{T}} C^1(\alpha)\, h = 0$ that $C^1(\alpha)\, h = 0^n$ holds, we then have

$$\tilde{L}^I(\alpha) =_{\mathrm{Df}} \{h \in L^I \mid C^1(\alpha)\, h = 0^n\}$$
$$= \{h \in L^I \mid G^{\mathsf{T}} C^1(\alpha)\, h = 0^{n-s}\}$$
$$= \mathfrak{G}(\{z \in E_{n-s} \mid G^{\mathsf{T}} C^1(\alpha)\, Gz = 0^{n-s}\}) . \tag{5.3.27}$$

From $G^{\mathsf{T}} C^1(\alpha)\, G = \alpha G^{\mathsf{T}} C G + (1-\alpha)\, G^{\mathsf{T}} G D G^{\mathsf{T}} G = D$, where we use the fact that $G^{\mathsf{T}} G = E^{(n)}$ (since G is an orthogonal matrix), (5.3.36), Lemma 5.3.1, and (5.3.27) we find

$$\psi(\beta^1(\alpha)) = \{x \in E_n \mid a^{i\mathsf{T}} x = a^{i\mathsf{T}} y,\, i \in I,\, a^{j\mathsf{T}} x \leqq b_j,\, j \in \bar{I},\, C^1(\alpha)\, x = C^1(\alpha)\, y\}$$
$$= \left(y + \tilde{L}^I(\alpha)\right) \cap \{x \in E_n \mid a^{j\mathsf{T}} x \leqq b_j,\, j \in \bar{I}\}\,^{1)}$$
$$= \left(y + \mathfrak{G}(\{z \in E_{n-s} \mid Dz = 0^{n-s}\})\right) \cap \{x \in E_n \mid a^{j\mathsf{T}} x \leqq b_j,\, j \in \bar{I}\}$$
$$= \psi(C, p, b) .$$

The number $\alpha \in [0, 1]$ was arbitrary, β^1 is thus an arc in $\mathfrak{S}(I, d)$.

We have $\dim \psi(C, p, b) = d$ and thus $\dim \tilde{L}^I(0) = n - d$. Since G has maximal rank (5.3.27) yields for $\alpha = 0$ $\operatorname{rank} D = n - d - s$.

By Lemma 5.3.3 there exists an arc

$$\tilde{C} : [0, 1] \to H^+(n - s, n - d - s)$$

satisfying $\tilde{C}(0) = D$ and $\tilde{C}(1) = Q^{n-s, n-s-d}$, in the degenerate case $n - d = s$ (which is equivalent to $D = 0$) we set $Q^{n-s, 0} =_{\mathrm{Df}} D = 0$. We now show that an arc is given in $\mathfrak{S}(I, d)$ by

$$\beta^2(\alpha) = \left(G\tilde{C}(\alpha)\, G^{\mathsf{T}}, -2G\tilde{C}(\alpha)\, G^{\mathsf{T}} y - \bar{a}, b\right) , \qquad 0 \leqq \alpha \leqq 1 . \tag{5.3.28}$$

By definition we immediately obtain $G\tilde{C}(\alpha)\, G^{\mathsf{T}} \in H^+(n)$, $\alpha \in [0, 1]$. Because of $y \in \Sigma^I(b)$ and $-\nabla_x f(y, G\tilde{C}(\alpha)\, G^{\mathsf{T}}, -2G\tilde{C}(\alpha)\, G^{\mathsf{T}} y - \bar{a}) = \bar{a}$ the conditions of Lemma 5.3.2 are fulfilled and thus $I = \operatorname{ch} \psi(\beta^2(\alpha))$, $0 \leqq \alpha \leqq 1$, follows. By Lemma 5.3.1 the dimension of $\psi(\beta^2(\alpha))$, $0 \leqq \alpha \leqq 1$, is determined by the dimension of

$$\hat{L}(\alpha) =_{\mathrm{Df}} \{h \in L^I \mid G\tilde{C}(\alpha)\, G^{\mathsf{T}} h = 0^n\} ,$$

therefore $\operatorname{rank} G\tilde{C}(\alpha) = \operatorname{rank} \tilde{C}(\alpha) = n - s - d$, $0 \leqq \alpha \leqq 1$, and

$$\hat{L}(\alpha) = \mathfrak{G}(\{z \in E_n \mid G\tilde{C}(\alpha)\, z = 0^n\}) , \qquad 0 \leqq \alpha \leqq 1 ,$$

---

1) $y + \tilde{L}^I(\alpha)$ is the Minkowski sum of $\{y\}$ and $\tilde{L}^I(\alpha)$.

give

$$\dim \psi(\beta^2(\alpha)) = \dim \hat{L}(\alpha) = \dim \{z \in E_n \mid G\tilde{C}(\alpha)\, z = 0^n\}$$
$$= (n-s) - (n-d-s) = d\,, \qquad 0 \leqq \alpha \leqq 1\,.$$

By coupling the arcs β^1 and β^2 according to (5.3.19) we then obtain an arc $\beta = \beta^1\beta^2$: $[0, 1] \to \mathfrak{S}(I, d)$ for the case $s < n$, this arc satisfies

$$\beta(0) = (C, p, b) \quad \text{and} \quad \beta(1) = (Q, -2Qy - \bar{a}, b) \tag{5.3.29}$$

where $Q = GQ^{n-s,n-d-s}G^{\mathsf{T}}$.

The considerations relating to β^1 may be extended to the case $s = n$ in which G is the zero vector of E_n: if $s = n$ we define β^1 by (5.3.25a) to (5.3.25d) where now $G = 0^n$; we then have $\beta^1(0) = (C, p, b)$ and $\beta^1(1) = (0, -\bar{a}, b)$ so that the coupled arc $\beta^1\beta^2$ fulfils the conditions in (5.3.29) if β^2 is defined by $\beta^2(\alpha) \equiv (0, -\bar{a}, b)$, $0 \leqq \alpha \leqq 1$.

Now let $(\hat{C}, \hat{p}, \hat{b})$ be a parameter triplet which is distinct from (C, p, b), we further choose a $\hat{y} \in \operatorname{ri} \psi(\hat{C}, \hat{p}, \hat{b})$. As in the first part of the proof there then exists an arc

$$\hat{\beta}: [0, 1] \to \mathfrak{S}(I, d)$$

satisfying $\hat{\beta}(0) = (C, \hat{p}, b)$ and $\hat{\beta}(1) = (Q, -2Q\hat{y} - \bar{a}, \hat{b})$ (in analogy to (5.3.29)). Defining

$$\begin{aligned} \gamma(\alpha) &= \big(Q, -2Qy(\alpha) - \bar{a}, b(\alpha)\big)\,, \\ y(\alpha) &= (1-\alpha y) + \alpha\hat{y}\,, \qquad b(\alpha) = (1-\alpha)\, b + \alpha\hat{b}\,, \end{aligned} \qquad 0 \leqq \alpha \leqq 1\,, \tag{5.3.30}$$

one easily confirms that the conditions of Lemma 5.3.2 are fulfilled for $y(\alpha)$, $0 \leqq \alpha \leqq 1$, and $\mu_i = 1$, $i \in I$, it follows that $I = \operatorname{ch} \psi(\gamma(\alpha))$, $0 \leqq \alpha \leqq 1$, holds. Then, by Lemma 5.3.1,

$$\operatorname{ri} \psi(\gamma(\alpha)) = \{x \in \Sigma^I(b(\alpha)) \mid Qx = Qy(\alpha)\}$$

for all $\alpha \in [0, 1]$ and thus $\dim \psi(\gamma(\alpha)) = \dim \psi(\gamma(0)) = d$, $0 \leqq \alpha \leqq 1$, also holds. $\beta\gamma\hat{\beta}^-$ coupled according to (5.3.19) and (5.3.20) is the desired arc: we have

$$\beta\gamma\hat{\beta}^-(0) = (C, p, b)\,, \qquad \beta\gamma\hat{\beta}^-(1) = (\hat{C}, \hat{p}, \hat{b})$$

and

$$\beta\gamma\hat{\beta}^-(\alpha) \in \mathfrak{S}(I, d) \qquad 0 \leqq \alpha \leqq 1\,,$$

thus completing the proof. □

Note 5.3.2. An essential feature of the arcwise connectedness of the stability sets $\mathfrak{S}(I, d)$ is the fact that this follows from the fully parametric character of the objective function: in other words all coefficients of the objective function may be varied independently of each other on $H^+(n) \times E_n$, the reader may easily find examples to confirm that Theorem 5.3.3 is in general not valid if for example an affine-linear dependence of the above coefficients is assumed. The fully parametric character was used in the proof of Lemma 3.3.3 and in conditions (5.3.25a) to (5.3.25d) and (5.3.28). Repeating the proof of Theorem 5.3.3 step for step with respect to the local stability sets

$$\tilde{\mathfrak{S}}(I, d) = \{(C, p, \tilde{b}) \mid \operatorname{ch} \psi(C, p, \tilde{b}) = I, \dim \psi(C, p, \tilde{b}) = d\}$$

corresponding to the problem $\min \{x^{\mathsf{T}}Cx + p^{\mathsf{T}}x \mid Ax \leqq b^0 + B\tilde{b}\}$ (introduced in (5.3.7) and (5.3.7a)) one finds that these sets are also arcwise connected. In constructing the arc $\beta\gamma\beta^-$ in the proof of Theorem 5.3.3 we only needed the property that the feasible parameter set is convex, which is also the case in problem (5.3.7) (cf. Note 5.3.1).

With Theorem 5.3.3 we have shown that the system of sets considered in (5.3.11) constitutes a decomposition of the solubility set $\mathfrak{A}$ of problem (5.3.1) into a finite number of arcwise connected sets. The remainding two theorems of this section are concerned with properties exhibited by the mappings φ and ψ of problem (5.3.1) on the components of this decomposition.

Theorem 5.3.4. *Consider problem* (5.3.1) *and assume* $I \subset \{1, \ldots, m\}$, $0 \leqq d \leqq n$ *and* $\mathfrak{S}(I, d) \neq \emptyset$. *Then*

(1) *the extreme value function* $\varphi|_{\mathfrak{S}(I,d)}$ *is continuous on* $\mathfrak{S}(I, d)$;

(2) $\psi|_{\mathfrak{S}(I,d)}$ *is l.s.c.-B on* $\mathfrak{S}(I, d)$;

(3) *a vector function* $\hat{x}|_{\mathfrak{S}(I,d)}$ *which is continuous on* $\mathfrak{S}(I, d)$ *exists such that*

$$\hat{x}|_{\mathfrak{S}(I,d)}(C, p, b) \in \psi_{\mathfrak{S}(I,d)}|(C, p, b)$$

holds for all $(C, p, b) \in \mathfrak{S}(I, d)$;

(4) *the optimal set mapping restricted to* $\mathfrak{S} = \mathfrak{S}(I, d) \cap \mathfrak{A}_0$[1]) *is H-continuous on* $\mathfrak{S}$.

Proof. Since $M|_{\mathfrak{B}}$ is by Theorem 5.3.2 (or Theorem 3.4.1) l.s.c.-B and quadratic convex functions are in particular also weakly analytic all propositions follow from Corollary 4.3.6.1. □

Note 5.3.3. Proposition (3) of the preceding theorem may be sharpened in the case $d = 0$ ($I \subset \{1, \ldots, m\}$ is arbitrary): The components $\hat{x}_i$, $i = 1, \ldots, n$, of the vector function $\hat{x}|_{\mathfrak{S}(I,0)}$ may be represented on the whole of $\mathfrak{S}(I, 0)$ by

$$\hat{x}_i(C, p, b) = \frac{\tau_i(C, p, b)}{\tau_0(C, p, b)}, \qquad i = 1, \ldots, n, \tag{5.3.31}$$

where $\tau_0, \tau_1, \ldots, \tau_n$ are polynomials in $c_{11}, \ldots, c_{1n}, \ldots, c_{n1}, \ldots, c_{nn}, p_1, \ldots, p_n, b_1, \ldots, b_m$ and $\tau_0(C, p, b) \neq 0 \ \forall (C, p, b) \in \mathfrak{S}(I, 0)$.

To prove this last we proceed as follows. By definition of $\mathfrak{S}(I, 0)$ and (5.3.13) we have

$$\operatorname{ri} \psi(C, p, b) \subset \{x \in E_n \mid a^{i\mathsf{T}}x = b_i, i, j \in I, a^{j\mathsf{T}}x < b_j, j \in \{1, \ldots, m\} \setminus I\}$$

where $(C, p, b) \in \mathfrak{S}(I, 0)$. Let the matrix A^I, the linear subspace L^I and the matrix G be given as in the proof of Theorem 5.3.3 and set $s = \operatorname{rank} A^I$. Further, let $y^I(b)$ be a vector function with the property

$$L^I(b) =_{\mathrm{Df}} \{x \in E_n \mid a^{i\mathsf{T}}x = b_i, i \in I\} = y^I(b) + L^I.$$

Obviously one can always find a function y^I which is linear in b. Furthermore, for each $(C, p, b) \in \mathfrak{S}(I, 0)$ the optimal solution $\hat{x}|_{\mathfrak{S}(I,0)}(C, p, b)$ is also a minimum with respect to the affine subspace $L^I(b)$. Setting $x = y^I(b) - G\xi$, $\xi \in E_{n-s}$ for $x \in L^I(b)$ we then find

$$\hat{x}_{\mathfrak{S}(I,0)}(C, p, b) = y^I(b) - G\xi(C, p, b)$$

where $\hat{\xi}(C, p, b)$ is (the only) optimal point of the problem

$$\min \{\xi^{\mathsf{T}}G^{\mathsf{T}}CG\xi - 2\xi^{\mathsf{T}}G^{\mathsf{T}}Cy^I(b) - \xi^{\mathsf{T}}G^{\mathsf{T}}p \mid \xi \in E_{n-s}\},$$

i.e. $\hat{\xi}(C, p, b) = (G^{\mathsf{T}}CG)^{-1}\left(G^{\mathsf{T}}Cy^I(b) + \frac{1}{2}G^{\mathsf{T}}p\right)$ and thus

$$\hat{x}|_{\mathfrak{S}(I,0)}(C, p, b) = y^I(b) - G(G^{\mathsf{T}}CG)^{-1}\left(G^{\mathsf{T}}Cy^I(b) + \tfrac{1}{2}G^{\mathsf{T}}p\right);$$

we have therefore found a representation of the form (5.3.31).

---

[1]) $\mathfrak{A}_0$ was introduced in Theorem 5.3.2(3).

Note 5.3.4. Consider the set

$$\mathfrak{S}(I) = \{(C, p, b) \in \mathfrak{A} \mid \operatorname{ch} \psi(C, p, b) = I\} = \bigcup_{d=0}^{n} \mathfrak{S}(I, d) \tag{5.3.32}$$

defined for arbitrary $I \subset \{1, \dots, m\}$. One easily confirms that

$$\mathfrak{S}(I) \subset \operatorname{cl} \mathfrak{S}(I, 0) \tag{5.3.33}$$

holds since $(C, p, b) \in \mathfrak{S}(I, d)$, $d \geqq 1$, and $\bar{x} \in \psi(C, p, b)$ imply that multipliers $\bar{u}_i \geqq 0$, $i \in I$, exist such that $2C\bar{x} + p = -\sum_{i \in I} a^i \bar{u}_i$. Hence

$$2(C + \varepsilon E^{(n)})\, \bar{x} + (p - 2\varepsilon\bar{x}) = -\sum_{i \in I} a^i \bar{u}_i$$

where $E^{(n)}$ is the unit (n, n)-matrix and $\varepsilon \in \mathbb{R}$; consequently we have

$$\bar{x} \in \psi(C + \varepsilon E^{(n)}, p - 2\varepsilon\bar{x}, b)$$

and

$$(C + \varepsilon E^{(n)}, p - 2\varepsilon\bar{x}, b) \in \mathfrak{S}(I, 0)$$

for all $\varepsilon > 0$.

It would thus appear reasonable to study the decomposition of $\mathfrak{A}$ into the sets $\mathfrak{S}(I)$, this is however not the case considering our objective of finding subsets of $\mathfrak{A}$ exhibiting "good" stability properties with respect to φ and ψ. As shown by Example 5.2.1 in general neither is $\varphi|_{\mathfrak{S}(I)}$ continuous on $\mathfrak{S}(I)$ nor can the vector function $\hat{x}|_{\mathfrak{S}(I,0)}$ considered in Note 5.3.3 be continuously extended to $\mathfrak{S}(I)$. The reader may easily confirm that in general $\mathfrak{S}(I)$ must neither be convex nor closed.

The qualitative results derived in this section do not enable one to establish methods for calculating local stability sets $\mathfrak{S}(I, d)$, an optimal vector function or the extreme value function; on the contrary, it becomes apparent that this is a very difficult problem. If algorithms for solving parametric quadratic problems are the objective then a further restriction of the problem class considered becomes necessary (see Section 5.5) and if one wishes to determine a continuous optimal vector function then this goal must be replaced by a weaker objective such as the calculation of so-called approximate selection functions (Chapter 6).

A special situation is of course that in which the constraint polyhedron of problem (5.3.1) is fixed ($b \equiv b^0$) and $\Sigma^I(b^0) = \{x^0\}$ is a vertex of $M(b^0)$. The local stability set $\mathfrak{S}(I, 0)$ is then the same as the local stability set $\mathfrak{K}^+(x^0)$ considered in Section 5.2. We thus specialize problem (5.3.1) by taking $b = b^0$ and also imposing the condition that the constraint polyhedron possesses vertices:

Consider the parametric problem

$$\min \{x^{\mathsf{T}} C x + p^{\mathsf{T}} x \mid x \in M_0\}\,, \qquad (C, p) \in \hat{\Lambda}\,, \tag{5.3.34}$$

subject to

$$M_0 = \{x \in E_n \mid Ax \leqq b^0\} \tag{5.3.34a}$$

where the (m, n)-matrix A and $b^0 \in E_m$ are fixed and $M_0 \neq \emptyset$, rank $A = n \leqq m$,

$$\hat{\Lambda} = \mathfrak{C}_f \quad (\mathfrak{C}_f \text{ defined according to (5.3.1c)})\,. \tag{5.3.34b}$$

$\hat{\mathfrak{A}}$ is the solubility set of problem (5.3.34), $\hat{\varphi}$ and $\hat{\psi}$ are the extreme value function and the optimal set mapping restricted to $\hat{\mathfrak{A}}$ respectively, and $\hat{\mathfrak{S}}(I)$ denotes the set

$$\hat{\mathfrak{S}}(I) = \{(C, p) \in \hat{\mathfrak{A}} \mid \operatorname{ri} \bar{\psi}(C, p) \subset \Sigma^I\}$$

if I is a subset of $\{1, \dots, m\}$ and $\Sigma^I = \Sigma^I(b^0)$ is a face of M_0. The rank condition imposed on A ensures that the polyhedron M_0 has vertices.

It follows from Theorem 5.3.1 that $\hat{\mathfrak{A}}$ is a convex cone of dimension $s(n) + n$. Local stability sets are defined by

$$\hat{\mathfrak{S}}(I, d) = \{(C, p) \in \hat{\mathfrak{A}} \mid \dim \psi(C, p) = d, \operatorname{ch} \psi(C, p) = I\}$$

which sets constitute a decomposition of $\hat{\mathfrak{A}}$ into a finite number of pairwise disjoint, arcwise connected subsets[1]). Besides the stability propositions of Theorems 5.3.2 and 5.3.4, these specialized to problem (5.3.34), we also have the following.

Theorem 5.3.5. *Let a parametric problem subject to the conditions* (5.3.34) *be given. If x^0 is a vertex of M_0 and $I = \{i \in \{1, \dots, m\} \mid x_i^0 = 0\}$, then*

(1) *the set* $\operatorname{cl} \hat{\mathfrak{S}}(I)$ *is a non-empty, convex and closed cone of dimension* $s(n) + n$, *and*

$$\operatorname{cl} \hat{\mathfrak{S}}(I) = \mathfrak{K}^+(x^0) = \{(C, p) \in \hat{\mathfrak{A}} \mid x^0 \in \hat{\psi}(C, p)\}$$

holds;

(2) *the extreme value function* $\hat{\varphi}$ *may be represented on* $\operatorname{cl} \hat{\mathfrak{S}}(I)$ *by* $\hat{\varphi}(C, p) = x^{0\mathsf{T}}(x^0 + p^{\mathsf{T}}x^0)$ *and it is thus linear on* $\operatorname{cl} \hat{\mathfrak{S}}(I)$.

Proof. By Theorem 5.2.7 (4) $\mathfrak{K}^+(x^0)$ is anon-empty, convex and closed cone with a dimension satisfying $s(n) + \dim K^p(x_0, M_0) \leqq \dim \mathfrak{K}^+(x^0)$. Since x^0 is a vertex of M_0 Theorem A.1.3 then gives $\dim K^p(x^0, M_0) = n$ and thus $\dim \mathfrak{K}^+(x^0) = s(n) + n$. By Theorem 5.2.7 (2) and because of $\operatorname{ri} \mathfrak{K}^+(x^0) = \operatorname{ri} \mathfrak{K}(x^0) \cap \operatorname{ri} \mathfrak{C}_f$ we have $\operatorname{ri} \mathfrak{K}^+(x^0) \subset \mathfrak{S}(I)$. On the other hand $\mathfrak{S}(I) \subset \mathfrak{K}^+(x_0)$ by definition and since $\mathfrak{K}^+(x^0)$ is closed $\operatorname{cl} \mathfrak{S}(I) = \mathfrak{K}^+(x^0)$ follows. This proves (1) and (2) is then immediate. □

If Σ^I is an arbitrary face of the convex polyhedron M_0 in (5.3.34a) then in general one cannot weaken the hypotheses of Theorem 5.3.4 and still obtain the corresponding continuity properties, this applies in particular to the necessary restriction of the extreme value function and the optimal set mapping to local stability sets $\hat{\mathfrak{S}}(I, d)$.

Example 5.2.1 already indicates that $\hat{\varphi}|_{\hat{\mathfrak{S}}(I)}$ (if Σ^I is a fixed face and the dimension may possibly be variable) is neither necessarily continuous nor need there exist a continuous optimal vector function on $\hat{\mathfrak{S}}(I)$. On the other hand the invariance alone of the dimension of the optimal set (variations of $\operatorname{ch} \hat{\psi}(C, p)$ are allowed) is not sufficient to ensure that a continuous optimal vector function exists, as is shown in the next example.

Example 5.3.2. Consider the parametric problem

$$\min \{(y - \alpha^2 x + \alpha)^2 \mid y \geqq 0\}, \qquad \alpha \in R.$$

Obviously

$$\bar{\psi}(\alpha) = \{(x, y) \mid y = \alpha^2 x - \alpha, y \geqq 0\} \qquad \forall \alpha \in R$$

holds for the optimal sets $\bar{\psi}(\alpha)$. Now, $\dim \bar{\psi}(\alpha) = 1$ $\forall \alpha \in R$, however, a continuous function $\bar{Z}_v\colon R \to E_2$ with $\bar{Z}_v(\alpha) \in \bar{\psi}(\alpha)$ $\forall \alpha \in R$ does not exist for any $v \in \bar{\psi}(0)$; since for each $v \in \bar{\psi}(0)$ there exists an $\alpha_0 > 0$ such that $\|v - w\| \geqq 1$ holds for all $w \in \bar{\psi}(\alpha)$ where $\alpha \in (0, \alpha_0)$.

To conclude this section we consider the case of parametric quadratic problems which are not necessarily convex.

[1]) cf. Theorem 5.3.3 und Note 5.3.2.

Note 5.3.5.

We replace condition (5.3.1c) of problem (5.31) by

$$\Lambda = \mathbb{H}_{s(n)} \times \mathbb{E}_n \times \mathbb{E}_m \tag{5.3.35}$$

and likewise substitute

$$\hat{\Lambda} = \mathbb{H}_{s(n)} \times \mathbb{E}_n \tag{5.3.36}$$

for the parameter set $\mathfrak{C}_f$ in problem (5.3.34).

In the problem given by $\{(5.3.1.), (5.3.1a), (5.3.1b), (5.3.35)\}$ $\mathfrak{A}^N$ is the solubility set, ψ^N the optimal set mapping and φ^N the extreme value function, the corresponding notation for problem $\{(5.3.34), (5.3.34a), (5.3.36)\}$ is given by $\hat{\mathfrak{A}}^N$, $\hat{\psi}^N$ and $\hat{\varphi}^N$. Further, let $\mathfrak{D}$ denote the set $\mathfrak{D} = \mathbb{H}_{s(n)} \times \mathbb{E}_n \times \mathfrak{B}$. Since convexity of the objective function at x is not needed in the propositions (1) and (2) of Theorem 5.3.2 we have

(i) $\varphi^N|_{\mathfrak{D}}$ and $\psi^N|_{\mathfrak{D}}$ are upper semicontinuous and closed respectively on $\mathfrak{A}^N$.

If the set $M(b)$ is bounded for each $b \in \mathfrak{B}$ then Theorem 4.2.2 may be applied since $M|_{\mathfrak{B}}$ is H-continuous, thus

(ii) if the recession cone $\{h \in \mathbb{E}_n \mid Ah \leqq 0\}$ of the constraint polyhedron given by (5.3.1a) or (5.3.34a) only consists of the zero vector then $\varphi^N|_{\mathfrak{D}}$ is continuous on $\mathfrak{A}^N$ and $\psi^N|_{\mathfrak{D}}$ is u.s.c.-B on $\mathfrak{A}^N$.

The solubility set $\mathfrak{A}^N$ is in general not convex as is shown by the following example. By convex combination (with the factor $\frac{1}{2}$) of the coefficients of

$$\min \{x^2 - y^2 - 2x \mid x - y = 1, x \geqq 0, y \geqq 0\}$$

and

$$\min \{-x^2 + y^2 \mid x - y = -1, x \geqq 0, y \geqq 0\}$$

which are obviously soluble, we obtain the problem

$$\min \{-x \mid x - y = 0, x \geqq 0, y \geqq 0\}$$

which does not have an optimal solution. If Theorems 5.2.2 and 5.3.1 are taken into account we however have

(iii) $\hat{\mathfrak{A}}^N$ is a convex cone of full dimension with a vertex at $0 \in \mathbb{H}_{s(n)} \times \mathbb{E}_n$; the extreme value function $\hat{\varphi}^N$ is concave and continuous on int $\hat{\mathfrak{A}}^N$.

5.4. Linear Complementarity Problems with Parameters in the Right-Hand Sides of the Constraints

In this section we study a class of linear complementarity problems in which the right-hand sides of the constraints are parameter-dependent. The investigations centre about the "fully" parametric problem

Determine for each $\lambda \in \mathbb{E}_n$ the set

$$\psi(\lambda) = \{(x, y) \in M(\lambda) \mid x^{\mathsf{T}} y = 0\}\,^{1)}$$
$$\text{where}\quad M(\lambda) = \{(x, y) \in \mathbb{E}_{2n} \mid -Kx + y = \lambda, x \geqq 0, y \geqq 0\}\,. \tag{5.4.1}$$

From the theoretical point of view we can deduce comparatively comprehensive results on this problem and from which by simple considerations the most important pro-

1) Here and in the following we write (x, y) and actually mean the vector $\binom{x}{y}$. Other combined column vectors are dealt with in like fashion in this and the following section.

perties of the following problem follow:

Determine for each $t \in E_d$ the set

$$\psi_d(t) = \{(x, y) \in M_d(t) \mid x^{\mathsf{T}} y = 0\}$$

where

$$M_d(t) = \{(x, y) \in E_{2n} \mid -Kx + y = q(t), x^{\mathsf{T}} \geqq 0, y \geqq 0\}; \tag{5.4.2}$$

$q(t)$ is affine linear in the parameter vector $t \in E_d$:

$$q(t) = q^0 + \sum_{i=1}^{d} q^i t_i , \qquad q^1, \dots, q^d \text{ linearly independent.} \tag{5.4.3}$$

We treat the single-parameter case ($d = 1$) separately since relatively speaking, this case is the most amenable as far as effective numerical calculations are concerned.

In all problem classes considered we assume

$$K \in \tilde{H}^+(n) \cup H^P(n) \tag{5.4.4}$$

where $\tilde{H}^+(n)$ is the class of all positive semidefinite (and not necessarily symmetric) (n, n)-matrices and $H^P(n)$ the class of (n, n)-matrices whose main minors are all positive. As indicated in the text some results are also valid for more general classes of matrices. The theoretical results presented in this section are of direct practical significance since they provide a foundation for solution procedures based on the simplex technique (see Section 6.2).

The present investigations are concerned with the feasible parameter set

$$\mathfrak{B} = \{\lambda \mid M(\lambda) \neq \emptyset\} , \tag{5.4.5}$$

the solubility set

$$\mathfrak{A} = \{\lambda \mid \psi(\lambda) \neq \emptyset\} , \tag{5.4.6}$$

and the solution set mapping

$$\psi : \lambda \to \psi(\lambda) . \tag{5.4.7}$$

We also introduce a concept of a local stability set which is suited to our purpose here, deduce properties and representations thereof and demonstrate that a decomposition or partitioning of the solubility set $\mathfrak{A}$ into local stability sets is possible.

We first present some consequences relating to $\mathfrak{B}$ and ψ of material given in the Appendix.

By Theorem A3.3 (a) we know that under condition (5.4.4) the set $\psi(\lambda)$ is a closed face of the convex polyhedron $M(\lambda)$ for each $\lambda \in \mathfrak{A}$ or may even be identical to $M(\lambda)$ for certain λ. We denote the characteristic index set pair of the set $\psi(\lambda)$ by $(\hat{I}(\lambda), \hat{J}(\lambda))$ for $\lambda \in \mathfrak{A}$ (cf. (3.2.8)); we may then write

$$\psi(\lambda) = \{(x, y) \in M(\lambda) \mid x_i = 0, i \in \hat{I}(\lambda), y_j = 0, j \in \hat{J}(\lambda)\} \tag{5.4.8}$$

and

$$\operatorname{ri} \psi(\lambda) = \{(x, y) \in \psi(\lambda) \mid x_i > 0, i \notin \hat{I}(\lambda), y_j > 0, j \notin \hat{J}(\lambda)\} . \tag{5.4.9}$$

Since $\hat{I}(\lambda), \hat{J}(\lambda) \subset \{1, \dots, n\}$ the pair $(\hat{I}(\lambda), \hat{J}(\lambda))$ only takes on a finite number of values on the solubility set and which we denote by

$$(I^1, J^1), \dots, (I^s, J^s) . \tag{5.4.10}$$

We note that the relation $I^r \cup J^r = \{1, \dots, n\}$ holds for $r = 1, \dots, S$ and that in the case $K \in H^A(n)$ we even have in addition $I^r \cap J^r = \emptyset$. Properties of the set $\mathfrak{B}$

may be derived from Theorem 5.3.1. If furthermore we exploit the particular structure of the constraints we arrive at the following theorem if Theorems A.1.2 and A.1.3 are invoked.

Theorem 5.4.1. *The set $\mathfrak{B}$ is a convex polyhedral cone of dimension n. It may be represented in the form*

$$\mathfrak{B} = -Z^P \tag{5.4.11}$$

where

$$Z = \{u \in E_n \mid K^\mathsf{T} u \leqq 0, u \geqq 0\} . \tag{5.4.12}$$

Proof. See the preceding remarks. □

Note 5.4.1. If $u^1, \dots, u^G$ are the edge vectors of the cone (5.4.12) we obtain by Theorem A.1.2 a description of $\mathfrak{B}$ in the form

$$\mathfrak{B} = \{\lambda \mid \lambda^\mathsf{T} u^l \geqq 0, l = 1, \dots, G\} \tag{5.4.13}$$

(see also Note 5.2.1).

Theorem 5.4.2. *Assume* (5.4.4) *is fulfilled. Then the following is true:*

(1) $\mathfrak{A} = \mathfrak{B}$[1]);

(2) *the mapping $\psi|_\mathfrak{A}$ is u.s.c.-H on $\mathfrak{A}$*[2]);

(3) *the solution set $\psi(\lambda)$ is bounded for $\lambda \in \operatorname{int} \mathfrak{A}$ and unbounded if $\lambda \in \operatorname{bd} \mathfrak{A}$*[3]).

Proof. The first proposition follows from Theorem A.3.1, the second from Theorem 3.4.1 and the fact that $\mathfrak{A}$ is closed and ψ is a finite union of upper semicontinuous (H) maps, and the third is a consequence of Theorem A.3.4 if one takes the representation (5.4.11) of $\mathfrak{B}$ and proposition (1) of the present theorem into account and then derives a representation of $\operatorname{int} \mathfrak{A} = \operatorname{ri} \mathfrak{A}$ using Theorem A.1.1. □

Note 5.4.2. For $K \in H^P(n)$ the relation $\mathfrak{A} = E_n$ and the H-continuity of ψ follow from Theorem A.3.3 (1), $\psi|_\mathfrak{A}$ is however in general not H-continuous on the entire set $\mathfrak{A}$.

We now take an index set pair (I^r, J^r) of (5.4.10) and introduce in a natural fashion the following concept of a local stability set of problem (5.4.1):

$$\mathfrak{A}^r = \{\lambda \in \mathfrak{A} \mid \hat{I}(\lambda) = I^r, \hat{J}(\lambda) = J^r\} \tag{5.4.14}$$

where $\mathfrak{A}^r$ is the local stability set corresponding to the index set pair (I^r, J^r). It follows immediately from the definition that the local stability sets $\mathfrak{A}^1, \dots, \mathfrak{A}^s$ constitute a decomposition of the solubility set $\mathfrak{A}$.

The following theorem gives characteristic properties of the local stability sets and of the mapping ψ restricted to the same, we however first define

$$B^r = (-k^i_{i \notin I^r}, e^j_{j \notin J^r}) \tag{5.4.15}$$

---

1) The set $\mathfrak{A}$ thus has the property derived in Theorem 5.4.1 for $\mathfrak{B}$. By Theorem A.3.1 this even holds if $K \in H^L(n) \cup H^Z(n)$.

2) According to P. Kleinmann [1] this holds for much more general linear complementarity problems having parameter-dependent right-hand sides.

3) By Theorem A.3.4 condition (5.4.4) may be weakened.

as the submatrix of $(-K, E)$ consisting of the columns $-k^i$, $i \notin I^r$, and e^i, $j \notin J^r$, and further

$$Z^r = \{u \in \mathbb{E}_n \mid (B^r)^\mathsf{T} u \geqq 0\}\,, \tag{5.4.16}$$

$$T^r(\lambda) = \{(x, y) \in M(\lambda) \mid x_i = 0, i \in I^r, y_j = 0, j \in J^r\}\,, \tag{5.4.17}$$

$$L^r = \{(x, y) \in \mathbb{E}_{2n} \mid -Kx + y = 0, x_i = 0, i \in I^r, y_j = 0, j \in J^r\} \tag{5.4.18}$$

and

$$d(r) = \dim L^r\,. \tag{5.4.19}$$

Theorem 5.4.3. *Assume that* (5.4.4) *holds. Then*

(1) *the set* $\mathfrak{A}^r$ *in* (5.4.14) *is the relative interior of a convex polyhedral cone and may be represented in the form*

$$\mathfrak{A}^r = -\operatorname{ri}(Z^r)^p;$$

(2) $\psi(\lambda) = T^r(\lambda)$ *and* $\dim \psi(\lambda) = d(r)$ *for all* $\lambda \in \mathfrak{A}^r$;

(3) $\operatorname{cl} \mathfrak{A}^r = \{\lambda \mid T^r(\lambda) \neq \emptyset\}$ *and* $T^r(\lambda) \subset \psi(\lambda)$ *for all* $\lambda \in \operatorname{cl} \mathfrak{A}^r$;

(4) $\dim \mathfrak{A}^r + d(r) + |I^r| + |J^r| = 2n$;

(5) *the mapping* $\psi|_{\mathfrak{A}^r}$ *is H-continuous on* $\mathfrak{A}^r$.

Proof. (1) Employing Theorems A.1.1 and A.1.2 we obtain the representations

$$-(Z^r)^p = \{\lambda \mid \lambda = \sum_{i \notin I^r} - k^i x_i + \sum_{j \notin J^r} e^j y_j,\ x_i \geqq 0, i \notin I^r, y_j \geqq 0, j \notin J^r\} \tag{5.4.20}$$

and

$$-\operatorname{ri}(Z^r)^p = \{\lambda \mid \lambda = \sum_{i \notin I^r} - k^i x_i + \sum_{j \notin J^r} e^j y_j,\ x_i > 0, i \notin I^r, y_j > 0, j \notin J^r\}\,. \tag{5.4.21}$$

If $\lambda \in \mathfrak{A}^r$ then for all $(x, y) \in \operatorname{ri} \psi(\lambda)$ we have $x_i > 0$ for $i \notin I^r$ and $y_i > 0$ for $j \notin J^r$ (cf. (5.4.9)). From (5.4.21) we, however, then find $-\operatorname{ri}(Z^r)^p \subset \mathfrak{A}^r$. Assume then $\lambda \in -\operatorname{ri}(Z^r)^p$, i.e.

$$\lambda = \sum_{i \notin I^r} - k^i \bar{x}_i + \sum_{j \notin J^r} e^j \bar{y}_j \quad \text{where } \bar{x}_i > 0\,, \quad i \notin I^r\,, \quad \bar{y}_j > 0\,, \quad j \notin J^r\,,$$

$(\bar{x}, \bar{y})$ denotes the point in $\mathbb{E}_{2n}$ obtained by inserting the zero components $\bar{x}_i = 0$, $i \in I^r$, $\bar{y}_j = 0$, $j \in J^r$. Because of $\emptyset \neq T^r(\lambda) \subset \psi(\lambda)$ ($T^r(\lambda)$ defined by (5.4.17)) and (5.4.21) we have $(\hat{I}(\lambda), \hat{J}(\lambda)) \subset (I^r, J^r)$.

Now assume $\lambda \notin \mathfrak{A}^r$, then $(\hat{I}(\lambda), \hat{J}(\lambda)) \neq (I^r, J^r)$ and $\tilde{x}_i > 0$ for an $i \in I^r$ or $\tilde{y}_i > 0$ for a $j \in J^r$ hold for an arbitrary point $(\tilde{x}, \tilde{y}) \in \operatorname{ri} \psi(\lambda)$. The manner of construction of $\mathfrak{A}^r$, I^r, J^r given in (5.4.10) and (5.4.14) implies $\mathfrak{A}^r \neq \emptyset$. Given $\mu \in \mathfrak{A}^r$ and $(x^0, y^0) \in \operatorname{ri} \psi(\mu)$ we define the point

$$(x(t), y(t)) = (x^0, y^0) + t((\tilde{x}, \tilde{y}) - (\bar{x}, \bar{y}))\,.$$

For all t we then have $-Kx(t) + y(t) = \mu$ and $x(t)^\mathsf{T} y(t) = 0$. One may also find a $t^0 > 0$ with the property $x(t) \geqq 0$ and $y(t) \geqq 0$ for all $t \in [0, t_0]$. We then obtain $(x(t), y(t)) \in \psi(\mu)$ for $t \in [0, t_0]$. By the construction of $(\tilde{x}, \tilde{y})$ there exists an index $i \in I^r$ such that $x_i(t) > 0$ or an index $j \in J^r$ such that $y_i(t) > 0$ holds for $t > 0$. This, however, contradicts $\mu \in \mathfrak{A}^r$.

(2) $\psi(\lambda) = T^r(\lambda)$ for $\lambda \in \mathfrak{A}^r$ follows from the definition (5.4.14) of the set $\mathfrak{A}^r$ and the relations (5.4.8), (5.4.9), and (5.4.21). The same also imply that the pair (I^r, J^r) is for $\lambda \in \mathfrak{A}^r$ the characteristic index set pair of the closed face $T^r(\lambda)$ of $M(\lambda)$, from which $\dim \psi(\lambda) = \dim T^r(\lambda) = d(r)$ for $\lambda \in \mathfrak{A}^r$ follows.

(3) $\operatorname{cl} \mathfrak{A}^r = \{\lambda \mid T^r(\lambda) \neq \emptyset\}$ follows directly from (5.4.20) and (1) above, and $T^r(\lambda) \subset \psi(\lambda)$ is a consequence of $I^r \cup J^r = \{1, \ldots, n\}$.

(4) We further have

$$d(r) = 2n - (|I^r| + |J^r| + \text{rank } B^r)$$

and

$$\dim \{u \in E_n \mid (B^r)^\mathsf{T} u = 0\} = n - \text{rank } B^r$$

and the assertion follows from (5.4.20) by Theorem A.1.3.

(5) This follows from $\psi(\lambda) = T^r(\lambda)$ for $\lambda \in \mathfrak{A}^r$ if Corollary 3.4.1.1 is invoked. □

Note 5.4.3. If $u^1, \dots, u^{F(r)}$ is a fundamental system of the cone Z^r in (5.4.16) and $v^1, \dots, v^{G(r)}$ is a basis of the latter's vertex set then by Theorem A.1.2 we obtain the representation

$$\mathfrak{A}^r = \{\lambda \mid \lambda^\mathsf{T} u^l > 0, l = 1, \dots, F(r), \lambda^\mathsf{T} v^k = 0, k = 1, \dots, G(r)\} \tag{5.4.22}$$

(see also Note 5.2.1).

Note 5.4.4. If $K \in H^P(n)$ the relation

$$\dim \mathfrak{A}^r + |I^r| + |J^r| = 2n$$

holds because of $\dim \psi(\lambda) = 0$ for all $\lambda \in E_n$, and if $K \in H^A(n)$ then

$$\dim \mathfrak{A}^r + d(r) = n$$

for all r follows from $|I^r| + |J^r| = n$.

Because of proposition (5) there exists for each $r = 1, \dots, S$ a selection function which is defined and continuous on $\mathfrak{A}^r$:

$$\left(x^r(\lambda), y^r(\lambda)\right) \in \psi(\lambda) \quad \text{(cf. Theorem 2.3.1)}\,.$$

Since the mapping ψ is in general not l.s.c.-H on the entire set $\mathfrak{A}$ a selection function for ψ which is continuous on the whole of $\mathfrak{A}$ does not necessarily exist (except in the case $K \in H^P(n)$). However the following somewhat weaker proposition does hold and it plays an important role in finding suitable solution procedures.

Note 5.4.5. The graph G of the mapping ψ

$$G = \{(x, y, \lambda) \in E_{3n} \mid \lambda \in \mathfrak{A}, (x, y) \in \psi(\lambda)\} \tag{5.4.23}$$

is the union of a finite number of convex polyhedra

$$N^r = \{(x, y, \lambda) \in E_{3n} \mid (x, y) \in M(\lambda), x_i = 0, i \in I^r, y_j = 0, j \in J^r\}\,,$$
$$r = 1, \dots, S\,,$$

and is arcwise connected.

Moreover, for any two elements $(x^0, y^0, \lambda^0), (x^1, y^1, \lambda^1) \in G$ one can find a piece-wise linear vector function

$$\left(x(\cdot), y(\cdot), \lambda(\cdot)\right): [0, 1] \to G$$

satisfying $\left(x(0), y(0), \lambda(0)\right) = (x^0, y^0, \lambda^0)$ and $\left(x(1), y(1), \lambda(1)\right) = (x^1, y^1, \lambda^1)$ as well as two non-decreasing functions

$$\tau_1(\cdot), \tau_2(\cdot): [0, 1] \to [0, 1]$$

satisfying $\tau_1(t) \leq \tau_2(t)$ and $\lambda(\tau) = t\lambda^1 + (1 - t)\lambda^0$ for all $t \in [0, 1]$ and all $\tau \in [\tau_1(t), \tau_2(t)]$.

Proof. The sets N^r are obviously convex polyhedra and the relation

$$G = \bigcup_{r=1}^{S} N^r$$

follows from Theorem 5.4.3. The arcwise connectedness of G results from the upper semicontinuity of ψ and the above representation of G. Since the local stability sets $\mathfrak{A}^r$ constitute a decomposition of the set $\mathfrak{A}$ one can find a finite number of values t_k, $k = 1, \dots, L$, with the property $t_0 = 0 < t_1 < t_2 < \dots < t_L = 1$ and a finite number of local stability sets $\mathfrak{A}^k$, $k = 1, \dots, L$, such that

$$[\lambda^0, \lambda^1] \cap \mathrm{cl}\, \mathfrak{A}^k = \{\lambda \mid \lambda = t\lambda^1 + (1-t)\,\lambda^0,\ t_{k-1} \leqq t \leqq t_k\}$$

holds.

Assume that for $k = 1, \dots, L$ the points $(\bar{x}^k, \bar{y}^k) \in T^k(t_{k-1}\lambda^1 + (1 - t_{k-1})\,\lambda^0)$ and $(\bar{\bar{x}}^k, \bar{\bar{y}}^k) \in T^k(t_k\lambda^1 + (1 - t_k)\,\lambda^0)$ are arbitrary. The polygon formed by connecting the points

$$\begin{array}{ll}
(\bar{x}^1, \bar{y}^1, t_0\lambda^1 + (1-t_0)\,\lambda^0)\,, & (\bar{\bar{x}}^1, \bar{\bar{y}}^1, t_1\lambda^1 + (1-t_1)\,\lambda^0)\,,\\
(\bar{x}^2, \bar{y}^2, t_1\lambda^1 + (1-t_1)\,\lambda^0)\,, & (\bar{\bar{x}}^2, \bar{\bar{y}}^2, t_2\lambda^1 + (1-t_2)\,\lambda^0)\,,\\
\dots & \dots\\
(\bar{x}^L, \bar{y}^L, t_{L-1}\lambda^1 + (1-t_{L-1})\,\lambda^0)\,, & (\bar{\bar{x}}^L, \bar{\bar{y}}^L, t_L\lambda^1 + (1-t_L)\,\lambda^0)
\end{array}$$

is by Theorem 5.4.3 (2) and (3) and because of the convexity of $\psi(\lambda)$ contained entirely in G. This polygon may be described by a piece-wise linear vector function having the above properties. □

In the following $\mathfrak{A}^1, \dots, \mathfrak{A}^T$, $T \leqq S$, are the local stability sets of highest possible dimension n.

Theorem 5.4.4. *Assume that* (5.4.4) *holds, then*

(1') *if $\mathfrak{A}^r$ is a local stability set of dimension n* (i.e. $r \in \{1, \dots, T\}$) *it follows that the corresponding matrix B^r* (5.4.15) *is a basis matrix of* $(-K, E)$ *and that*

$$\psi(\lambda) = \left\{\begin{pmatrix} x^r(\lambda) \\ y^r(\lambda) \end{pmatrix}\right\} \quad \textit{for} \quad \lambda \in \mathfrak{A}^r\,, \tag{5.4.24}$$

$$\psi(\lambda) \supset \left\{\begin{pmatrix} x^r(\lambda) \\ y^r(\lambda) \end{pmatrix}\right\} \quad \textit{for} \quad \lambda \in \mathrm{bd}\, \mathfrak{A}^r\,, \tag{5.4.25}$$

and

$$\mathfrak{A}^r = \left\{\lambda \,\middle|\, \begin{pmatrix} x^r(\lambda) \\ y^r(\lambda) \end{pmatrix}_{\mathrm{BV}} > 0\right\} \tag{5.4.26}$$

hold for the corresponding basis solution

$$\begin{pmatrix} x^r(\lambda) \\ y^r(\lambda) \end{pmatrix}_{\mathrm{BV}} = (B^r)^{-1}\,\lambda\,, \qquad \begin{pmatrix} x^r(\lambda) \\ y^r(\lambda) \end{pmatrix}_{\mathrm{NBV}} = 0; \tag{5.4.27}$$

(vector of the basis variables) (vector of the non-basis variables)

(1'') *if on the other hand $\tilde{B}$ is a basis matrix of* $(-K, E)$ *whose index set pair* $(\tilde{I}, \tilde{J})$ *of the non-basis variables satisfies* $\tilde{I} \cup \tilde{J} = \{1, \dots, n\}$,
then there exists one and only one index $r \in \{1, \dots, T\}$ with the property $\tilde{I} = I^r$, $\tilde{J} = J^r$, and thus $\tilde{B} = B^r$, such that (5.4.24), (5.4.25), *and* (5.4.26) *are fulfilled*;

(2) *the convex polyhedral cones* $\mathrm{cl}\, \mathfrak{A}^r$, $r = 1, \dots, T$, *constitute a partitioning of the set* $\mathfrak{A}$;

(3) *if the index set pairs* (I^{r_1}, J^{r_1}) *and* (I^{r_2}, J^{r_2}) *differ by one and only one index for* $r_1, r_2 \in \{1, \dots, T\}$ *then* $\mathrm{cl}\, \mathfrak{A}^{r_1} \cap \mathrm{cl}\, \mathfrak{A}^{r_2}$ *is a common* $(n-1)$*-dimensional closed face of* $\mathrm{cl}\, \mathfrak{A}^{r_1}$ *and* $\mathrm{cl}\, \mathfrak{A}^{r_2}$.

Proof. (1′) It follows from (5.4.21) and $I^r \cap J^r = \{1, \dots, n\}$ that $\dim \mathfrak{A}^r = n$ holds if and only if $d(r) = 0$ and $I^r \cup J^r = \emptyset$. The set $\psi(\lambda)$, where $\lambda \in \mathfrak{A}^r$ and $r \in \{1, \dots, T\}$ thus consists of just one non-degenerate basic solution of the convex polyhedron $M(\lambda)$, and the matrix B^r is a basis matrix of $(-K, E)$. The set $T^r(\lambda)$ consists of the point given by (5.4.27), Theorem 5.4.3 (2) and (3) then gives the relations (5.4.24), (5.4.25), and (5.4.26).

(1″) Let $\tilde{B}$ be an arbitrary basis matrix of $(-K, E)$ satisfying the above condition. We first note that the set $\{\lambda \mid \tilde{B}^{-1}\lambda > 0\}$ is non-empty and of dimension n since $\operatorname{rank} \tilde{B} = \operatorname{rank} \tilde{B}^{-1} = n$. Because of $\tilde{I} \cup \tilde{J} = \{1, \dots, n\}$ we obviously have

$$\begin{pmatrix} \tilde{x}(\lambda) \\ \tilde{y}(\lambda) \end{pmatrix} \in \psi(\lambda) \quad \text{for all} \quad \lambda \in \{\lambda \mid \tilde{B}^{-1}\lambda \geqq 0\}$$

for the corresponding basis solution given by

$$\begin{pmatrix} \tilde{x}(\lambda) \\ \tilde{y}(\lambda) \end{pmatrix}_{\mathrm{BV}} = \tilde{B}^{-1}\lambda\,, \qquad \begin{pmatrix} \tilde{x}(\lambda) \\ \tilde{y}(\lambda) \end{pmatrix}_{\mathrm{NBV}} = 0\,,$$

and therefore $(\hat{I}(\lambda), \hat{J}(\lambda)) \subset (\tilde{I}, \tilde{J})$ for $\lambda \in \{\lambda \mid \tilde{B}^{-1}\lambda > 0\}$. However, $|\tilde{I}| + |\tilde{J}| = n$ and $|\hat{I}(\lambda)| + |\hat{J}(\lambda)| \geqq n$ for all $\lambda \in \mathfrak{A}$ also hold and thus $\hat{I}(\lambda) = \tilde{I}, \hat{J}(\lambda) = \tilde{J}$ for $\lambda \in \{\lambda \mid \tilde{B}^{-1}\lambda > 0\}$ follows. Hence there exists one and only one index $r \in \{1, \dots, T\}$ such that $\tilde{I} = I^r$ and $\tilde{J} = J^r$ and thus $\tilde{B} = B^r$ hold, so that by (1′) the conditions (5.4.24), (5.4.25) and (5.4.26) are satisfied.

(2) follows from the fact that (i), the sets $\mathfrak{A}^1, \dots, \mathfrak{A}^S$ constitute a decomposition of $\mathfrak{A}$, (ii) their closures are convex polyhedral cones, and (iii) their number is finite.

(3) This follows from the representation (5.4.20) of the set $\mathrm{cl}\, \mathfrak{A}^r = -(Z^r)^p$ and the fact that the matrix (5.4.15) is a basis matrix of $(-K, E)$. □

Note 5.4.6. The union of the local stability sets $\mathfrak{A}^{T+1}, \dots, \mathfrak{A}^S$ of dimension smaller than n is the union of all faces of the n-dimensional convex polyhedral cones $\mathrm{cl}\, \mathfrak{A}^1, \dots, \mathrm{cl}\, \mathfrak{A}^T$.

We now turn our attention to the question of how the results obtained for the fully parametric problem (5.4.1) may be used to characterize the multiparametric problem (5.4.2). To this end we consider the invers $t = q^{-1}(\lambda)$ of the affine-linear mapping $\lambda = q(t)$ in (5.4.3), the former is defined on the d-dimensional affine subspace

$$L_d = \{\lambda \in \mathbb{E}_n \mid \lambda = q(t), t \in \mathbb{E}_d\}$$

and obviously establishes an isomorphism between L_d and $\mathbb{E}_d$. We thus for example find that the feasible parameter set of (5.4.2) is $q^{-1}(\mathfrak{B} \cap L_d)$, the solubility set is given by $q^{-1}(\mathfrak{A} \cap L_d)$ and $q^{-1}(\mathfrak{A}^r \cap L_d)$ is a local stability set providing $\mathfrak{A}^r \cap L_d \neq \emptyset$. Finally, for each $t \in \mathbb{E}_d$ the solution set of (5.4.2) is given by $\psi(q^{-1}(t))$. Using these relations important results established in the theorems on problem (5.4.1) may be extended to problem (5.4.2). Some of these results are however no longer valid as follows from the fact that instead of $\mathfrak{A}$ and $\mathfrak{A}^r$ we have to consider $\mathfrak{A} \cap L_d$ and $\mathfrak{A}^r \cap L_d$. In order not to merely repeat parts of the above theorems with different notation we now list the changes that arise in applying the previous results to problem (5.4.2). We refer to the fact that in F. NOŽIČKA et al. [2] the application of theorems on linear fully parametric problems to the multiparametric case according to the basic idea given above is presented in much detail.

1. The characteristic parameter sets of problem (5.4.2) are in general, assuming $q^0 \neq 0$, not cones as is the case for $\mathfrak{B}$, $\mathfrak{A}$, and cl $\mathfrak{A}^r$ in problem (5.4.1), but rather convex polyhedra.

2. The solubility set of (5.4.2) does not need to be a set of maximal dimension (which here is d), as is the case for $\mathfrak{A}$, and may even be empty.

3. The dimensional proposition of Theorem 5.4.3 (4) cannot be extended in a suitable manner to the present problem.

4. The number of local stability sets of problem (5.4.2) may be smaller than the number S of those of problem (5.4.1).

5. Theorem 5.4.4 may not be applied to problem (5.4.2). In general there is no unique relation existing between the local stability sets of highest possible dimension and the basis matrices of the matrix $(-K, E)$ considered in that theorem. All that may be inferred is the following: If the solution set consists of one and only one non-degenerate vertex point of $M_d(t)$ on a local stability set of (5.4.2) then this local stability set has the dimension d, and in this sense the relations (5.4.24), (5.4.25), and (5.4.26) may be extended to the problem considered.

Finally, we consider in somewhat more detail the important special case of a single-parametric problem of type (5.4.2), i.e. a problem of the form

Determine for each $t \in R$ the set

$$\psi_1(t) = \{(x, y) \in M_1(t) \mid x^\mathsf{T} y = 0\}$$

where

$$M_1(t) = \{(x, y) \in E_{2n} \mid -Kx + y = q^0 + q^1 t, x \geqq 0, y \geqq 0\}\,. \tag{5.4.28}$$

The particular features of this problem largely follow from the fact that the parameter sets of interest are convex subsets of the one-dimensional Euclidean space and may thus be relatively easily represented and computed. As already indicated at the beginning of the section the one-parameter case in the present as well as other classes of parametric optimization problems is also the case most efficiently dealt with numerically. In the following we provide foundations for a solution procedure given in Section 6.2.

The solubility set of problem (5.4.28) either is empty, consists of one element or is a bounded or unbounded interval. Since the first two possibilities are not of interest for what follows, we assume

$$\dim \mathfrak{A}_1 = 1\,. \tag{5.4.29}$$

One may then find numbers $\underline{t}, \bar{t} \in \bar{R}$[1]) such that

$$\mathfrak{A}_1 = \{t \in R \mid \underline{t} \leqq t \leqq \bar{t}\} \tag{5.4.30}$$

holds.

The local stability sets of problem (5.4.28) are either open intervals or sets consisting of one element: the corresponding boundary points. There thus exists a finite number of parameter values $t_r \in \bar{R}$, $r = 0, \ldots, P$, such that

$$\underline{t} = t_0 < t_1 < \ldots < t_P = \bar{t}$$

[1]) $\bar{R} =_{\mathrm{Df}} R \cup \{-\infty\} \cup \{+\infty\}$

and if the local stability sets are numbered accordingly

$$\mathfrak{A}_1^r = (t_{r-1}, t_r)\,, \qquad r = 1, \dots, P\,, \tag{5.4.31}$$

are the local stability sets of dimension one and

$$\mathfrak{A}_0^r = \{t_r\}\,, \qquad r = 0, \dots, P, \qquad t_r \text{ finite}\,, \tag{5.4.32}$$

are those of zero dimension.

We denote the characteristic index set pair corresponding to a local stability set $\mathfrak{A}_1^r$ (5.4.31) or $\mathfrak{A}_0^r$ (5.4.32) by (I_1^r, J_1^r) and (I_0^r, J_0^r) respectively. We also introduce the following notation:

$$N = \{(x, y, t) \in E_{2n+1} \mid -Kx + y - q^1 t = q^0, x \geqq 0, y \geqq 0\}\,, \tag{5.4.33}$$

$$G_1 = \{(x, y, t) \in N \mid x^{\mathsf{T}} y = 0\}\,, \tag{5.4.34}$$

$$N_1^r = \{(x, y, t) \in N \mid x_i = 0, i \in I_1^r, y_j = 0, j \in J_1^r\}\,, \qquad r = 1, \dots, P\,, \tag{5.4.35a}$$

$$N_0^r = \{(x, y, t) \in N \mid x_i = 0, i \in I_0^r, y_j = 0, j \in J_0^r\}\,,$$
$$r = 0, \dots, P\,,\ t_r \text{ finite}\,. \tag{5.4.35b}$$

The set N is a convex polyhedron and is the graph of the constraint set mapping M_1 of (5.4.28). Each of the sets N_1^r and N_0^r is a closed face of N and one of them may also be identical to the entire set N. The set G_1 is the graph of the solution set mapping of problem (5.4.28) and we have

$$G_1 = \bigcup_{r=1}^{P} N_1^r \cup \bigcup_{\substack{r=0\\ t_r \text{ finite}}}^{P} N_0^r\,. \tag{5.4.36}$$

The following also holds:

$$\underline{t} = \inf\{t \mid (x, y, t) \in N\} = \inf\{t \mid (x, y, t) \in G_1\}\,, \tag{5.4.37a}$$

$$\bar{t} = \sup\{t \mid (x, y, t) \in N\} = \sup\{t \mid (x, y, t) \in G_1\}\,, \tag{5.4.37b}$$

$$t_{r-1} = \inf\{t \mid (x, y, t) \in N_1^r\}\,, \qquad r = 1, \dots, P\,, \tag{5.4.37c}$$

$$t_r = \sup\{t \mid (x, y, t) \in N_1^r\}\,, \qquad r = 1, \dots, P\,, \tag{5.4.37d}$$

$$t_r = \inf\{t \mid (x, y, t) \in N_0^r\} = \sup\{t \mid (x, y, t) \in N_0^r\}$$
$$r = 0, \dots, P\,, \quad t_r \text{ finite}\,, \tag{5.4.37e}$$

and furthermore

$$N_1^r \cap T^r \subset N_0^r\,, \qquad r > 0\,, \tag{5.4.38a}$$

$$N_1^{r+1} \cap T^r \subset N_0^r\,, \qquad r < P\,, \tag{5.4.38b}$$

where $T^r = \{(x, y, t) \mid t = t_r\}$.

The mapping $\psi_1/_{\mathfrak{A}_1}$ is lower semicontinuous at all points $t \in \mathfrak{A}_1$ with $t \neq t_r$ and also at all points t_r for which equality holds in (5 4.38a) and (5.4.38b).

The following lemmas needed later in this volume are concerned with properties of the vertices of the convex polyhedron N (5.4.33).

Lemma 5.4.1. *A point $(x^0, y^0, t^0) \in N$ is a vertex of N if and only if*

(1) *of the $2n$ variables x_i, y_j at least $n + 1$ have the value zero, and*

(2) *the columns of the matrix $(-K, E)$ corresponding to positive variables x_i, y_j are, together with the vector q^1, linearly independent.*

Proof. This follows in analogous fashion to a corresponding criterion for convex polyhedra given in equation form (see F. NOŽIČKA et al. [1]) if one takes into account that the sign of the variable t is not restricted. □

As in the case of convex polyhedra in equation form we differentiate between degenerate and non-degenerate vertices of N and also note that to each vertex there corresponds at least one basis. A characteristic feature with respect to the latter is that the variable t must always be one of the n basis variables. The number of non-basic variables is $n + 1$.

The following lemma gives the conditions under which N actually does possess a vertex.

Lemma 5.4.2. *The set N possesses vertices if and only if $q^1 \neq 0$.*

Proof. By Lemma 5.4.1 N does not have a vertex if $q^1 = 0$. Assume therefore $q^1 \neq 0$. If N did not possess vertices then it would contain a straight line which can be represented in the form

$$g = \{(x, y, t) \in E_{2n} \mid (x, y, t) = (x^0, y^0, t^0) + \beta(x^1, y^1, t^1), \beta \in \mathbb{R}\},$$

where $(x^1, y^1, t^1) \neq 0$. If $g \subset N$ then $-Kx^1 + y^1 - q^1t^1 = 0$ and $x^1 = y^1 = 0$ are necessarily true, it follows, however, that $t^1 \neq 0$ and $q^1t^1 = 0$ hold, which contradict the assumption $q^1 \neq 0$. □

We thus find that the set N has vertices if and only if (5.4.28) is a genuinely parametric problem, i.e. if the parameter t actually does appear in the problem. This will be assumed in the following (cf. 5.4.3)).

If $q^1 \neq 0$ as assumed then not only N but also each of its closed faces and in particular the sets N_1^r and N_0^r (5.4.35) possess vertices; the vertices of the latter are of course also vertices of N.

The next lemma characterizes the vertices of N contained in G_1. These are of special interest in our considerations.

Lemma 5.4.3. *Let (x^0, y^0, t^0) be a vertex of N contained in the set G_1. Then to this vertex there corresponds a basis with the property*

$$\begin{aligned}&\textit{for each } i \in \{1, \dots, n\} \textit{ not more than one of the variables } x_i \textit{ or } y_i\\ &\textit{is contained in the basis and there exists one and only one index}\\ &i^* \in \{1, \dots, n\} \textit{ such that both } x_{i^*} \textit{ and } y_{i^*} \textit{ are non-basis variables.}\end{aligned} \tag{5.4.39}$$

Proof. If (x^0, y^0, t^0) is a vertex of N then it follows from Lemma 5.4.1 that (x^0, y^0) must be a degenerate vertex of the convex polyhedron $M_1(t^0)$. If furthermore (x^0, y^0, t^0) is a vertex of N contained in the set G_1 then we also have $(x^0, y^0) \in \psi_1(t^0)$. By Theorem A.3.5 we know that then a basis corresponding to the vertex (x^0, y^0) must exist and have the property that for each $i \in \{1, \dots, n\}$ one and only one of the variables x_i or y_i is a basis variable. Since the vertex is degenerate and by Lemma 5.4.1 the columns of $(-K, E)$ corresponding to the positive variables x_i, y_j and the column q^1 are linearly independent, the existence of an $i^* \in \{1, \dots, n\}$ with $x_{i^*}^0 = y_{i^*}^0 = 0$ follows such that the basis variables corresponding to the vertex (x^0, y^0) of $M_1(t^0)$ with the exception of x_{i^*} and y_{i^*}, respectively, together with the variable t constitute a basis corresponding to the vertex (x^0, y^0, t^0) of N. This basis obviously has the property (5.4.39). □

We note that Lemma 5.4.3 is trivial, if the vertex (x^0, y^0, t^0) is non-degenerate. The next lemma is concerned with regular vertices of N.

Lemma 5.4.4. *Let (x^0, y^0, t^0) be a regular vertex of N and let it be contained in G_1. Then there exist two and only two edges of N incident to (x^0, y^0, t^0) which are contained in G_1.*

Proof. An edge of N which is incident to (x^0, y^0, t^0) is characterized by the fact that in its relative interior exactly one of the non-basis variables corresponding to (x^0, y^0, t^0) attains a positive value and the basis variables retain their positive values since (x^0, y^0, t^0) is by hypothesis non-degenerate. Now, if this edge is contained in G_1 then by Lemma 5.4.3 this can only be the case if the corresponding non-basis variable is either x_{i^*} or y_{i^*}. There are thus two such edges. □

As is the case for the mapping ψ in (5.4.7) the mapping ψ_1 is also in general not H-continuous on the entire solubility set so that a selection function $(\hat{x}(t), \hat{y}(t)) \in \psi_1(t)$ which is continuous on $\mathfrak{A}_1$ need not necessarily exist. The following corollary which is closely related to Note 5.4.5 holds however.

Corollary 5.4.4.1. *Let $(\hat{x}, \hat{y}, \hat{t})$ be an arbitrary vertex of the convex polyhedron N and let it be contained in G_1.*

There then exists an edge polygon running through $(\hat{x}, \hat{y}, \hat{t})$ and contained in G_1, i.e. there exists a sequence of oriented closed edges $f_0, \ldots, f_L$ of N which are contained in G_1 and which may be represented in the form

$$f_0 = \{(x, y, t) \mid (x, y, t) = (x^1, y^1, t^1) + \beta(x^0, y^0, t^0), \beta \geqq 0\} \tag{5.4.40a}$$

where $t_0 = 0$ if $\underline{t} > -\infty$ and $t^0 < 0$ if $\underline{t} = -\infty$,

$$f_L = \{(x, y, t) \mid (x, y, t) = (x^L, y^L, t^L) + \beta(x^{L+1}, y^{L+1}, t^{L+1}), \beta \geqq 0\} \tag{5.4.40b}$$

where $t^{L+1} = 0$ if $\bar{t} < \infty$ and $t^{L+1} > 0$ if $\bar{t} = \infty$

$$f_k = \operatorname{conv} \{(x^k, y^k, t^k), (x^{k+1}, y^{k+1}, t^{k+1})\}, \qquad k = 1, \ldots, L-1, \tag{5.4.40c}$$

where the points (x^k, y^k, t^k), $k = 1, \ldots, L$, are vertex points of N, one of which is the point $(\hat{x}, \hat{y}, \hat{t})$, and the vectors (x^0, y^0, t^0) and $(x^{L+1}, y^{L+1}, t^{L+1})$ are direction vectors of unbounded edges of N with the property

$$\underline{t} = \inf_{f_0} \{t\} \leqq t^1 \leqq t^2 \leqq \ldots \leqq t^L \leqq \sup_{f_L} \{t\} = \bar{t}\,. \tag{5.4.41}$$

Proof. The basic idea used to construct the desired edge polygon may already be found in Note 5.4.5. Here however we can exploit the relations (5.4.36), (5.4.37), and (5.4.38) directly. Because of (5.4.36) the given vertex $(\hat{x}, \hat{y}, \hat{t})$ must also be a vertex of one of the sets N_1^r or N_0^r (5.4.35). Starting with one of the sets N_1^r or N_0^r we construct the edge polygon step by step. If $(\hat{x}, \hat{y}, \hat{t})$ is a vertex of a set N_1^r then we can find an edge subpolygon which is contained in N_1^r, runs through $(\hat{x}, \hat{y}, \hat{t})$ (in the above sense), has the monotonicity condition assumed in (5.5.41) and which connects optimal vertex points of both linear optimization problems

$$\min \{t \mid (x, y, t) \in N_1^r\} \quad \text{and} \quad \max \{t \mid (x, y, t) \in N_1^r\}$$

assuming that both problems are soluble, or if this is not the case, which then has the property that the value of t is not bounded on the first or last edge. If on the other hand $(\hat{x}, \hat{y}, \hat{t})$ is a vertex of a set N_0^r then one can find an edge subpolygon which is contained in N_0^r, runs through $(\hat{x}, \hat{y}, \hat{t})$, fulfills the monotonicity condition of (5.4.41) and connects optimal vertex points of the linear optimization problems

$$\max \{t \mid (x, y, t) \in N_1^r\} \quad \text{and} \quad \min \{t \mid (x, y, t) \in N_1^{r+1}\}$$

if $0 < r < P$ or whose first ($r = 0$) or last ($r = P$) edge is unbounded (see Theorem 5.4.2 (3)).

The entire edge polygon may be completed in like fashion by taking the endpoints of the previous subpolygon as the initial points of the next. □

The previous corollary may be sharpened if all vertex points of the convex polyhedron are regular.

Corollary 5.4.4.2. *Let all vertex points of the convex polyhedron N be regular. Then*

(1) $\dim N_1^r = 1$ *for* $r = 1, \dots, P$,
$\dim N_0^r = 1$ *for all* $r \in \{0, \dots, P\}$ *for which* t_r *is finite and* $\psi_1|_{\mathfrak{A}_1}$ *is not H-continuous at* t_r,
$\dim N_0^r = 0$ *for all* $r \in \{0, \dots, P\}$ *for which* t_r *is finite and* $\psi_1|_{\mathfrak{A}_1}$ *is H-continuous at* t_r;

(2) *the set* G_1 *consists of one and only one edge polygon as described in Corollary* 5.4.4.1 *and the sets* N_1^r, $r = 1, \dots, P$, *and* N_0^r, $r = 0, \dots, P$, t_r *finite*, $\dim N_0^r = 1$, *represent the edges* f_k;

(3) $\dim \psi_1(t) = 0$ *holds at all continuity points of the mapping* $\psi_1|_{\mathfrak{A}_1}$ *and likewise* $\dim \psi_1(t) = 1$ *at all points at which the mapping is not continuous; the set* $\psi_1(t)$ *is unbounded at the (finite) boundary points and only so at those points.*

Proof. Follows from Lemma 5.4.4 and Corollary 5.4.4.1. □

5.5. Convex Quadratic Optimization Problems Having Parameters in the Linear Part of the Objective Function and in the Right-Hand Sides of the Constraints

We consider the problem

$$\min \{\tfrac{1}{2} x^{\mathsf T} C x + p^{\mathsf T} x \mid Ax \leqq b, x \geqq 0\}, \qquad (p, b) \in E_n \times E_m, \tag{5.5.1}$$

which depends on the parameters p and b. We assume

$$C \in H^+(n) = \tilde{H}^+(n) \cap H^s(n) \tag{5.5.2}$$

where $H^+(n)$ is the class of all symmetric and positive semidefinite (n, n)-matrices (cf. (5.3.4)).

It should be noted that if $C = 0$ we have the special case of linear parametric optimization. Problem (5.5.1) is itself a special case of problem (5.3.1) if one disregards the fact that in the former non-negativity conditions are for technical reasons imposed on all variables. If the matrix C is fixed we find that much sharper and more practicable results may be obtained as compared to Section 5.3.

For each (p, b) the Lagrange dual problem is

$$\max \{h(p, b, u) \mid u \in N(p)\} \tag{5.5.3}$$

where

$$h(p, b, u) = \inf \{\tfrac{1}{2} x^{\mathsf T}(Cx + p^{\mathsf T} x + u^{\mathsf T} Ax \mid x \geqq 0\} - b^{\mathsf T} u$$

and

$$N(p) = \{u \in E_m \mid u \geqq 0, \inf_{x \geqq 0} \{\tfrac{1}{2} x^{\mathsf T} C x + p^{\mathsf T} x + u^{\mathsf T} Ax\} > -\infty\}\,.$$

$\psi_Q(p, b)$ denotes the optimal set of (5.5.1) and $\psi_D(p, b)$ the optimal set of (5.5.3).

Applying Theorems A.4.1 and A.4.2 we find that the problems (5.5.1) and (5.5.3) are closely related to the following parameter-dependent linear complementarity problem:

Determine for each $(p, b) \in E_{n+m}$ the set

$$\psi(p, b) = \{(x, u, y, z) \in M(p, b) \mid x^{\mathsf{T}}y + u^{\mathsf{T}}z = 0\}$$

where $M(p, b) = \{(x, u, y, z) \in E_{2n+2m} \mid -Cx - A^{\mathsf{T}}u + y = p,$

$$Ax + z = b, x \geqq 0, u \geqq 0, y \geqq 0, z \geqq 0\} . \tag{5.5.4}$$

This problem is of the type (5.4.1) studied in some detail in the previous section.

The matrix K in (5.4.1) here has the special structure

$$K = \begin{pmatrix} C & A^{\mathsf{T}} \\ -A & 0 \end{pmatrix} \tag{5.5.5}$$

and because of (5.5.2) we have $K \in \tilde{H}^+(n + m)$.

Denoting

$$\tilde{\psi}(p, b) = \{(x, u) \in E_{n+m} \mid \exists (y, z) \quad \text{with} \quad (x, u, y, z) \in \psi(p, b)\} \tag{5.5.6}$$

we conclude by Theorem A.4.2 that

$$\tilde{\psi}(p, b) = \psi_Q(p, b) \times \psi_D(p, b) \tag{5.5.7}$$

holds. It immediately follows that

$$\dim \psi(p, b) = \dim \tilde{\psi}(p, b) = \dim \psi_Q(p, b) + \dim \psi_D(p, b) \tag{5.5.8}$$

is also true.

All results of the previous section may be applied to problem (5.5.4), because of the special form of the matrix K (5.5.5) some of them may be obtained in a sharper form. In the following we employ all of these results in investigating (5.5.1), which is the real problem of interest.

Since the variables of problem (5.5.4) are collected in four subvectors and the set $\psi(p, b)$ is a closed face of the polyhedron $M(p, b)$ it is appropriate to describe $\psi(p, b)$ using a four-tuple of index sets $(\hat{I}_1(p, b), \hat{I}_2(p, b), \hat{J}_1(p, b), \hat{J}_2(p, b))$ where $\hat{I}_1(p, b)$, $\hat{J}_1(p, b) \subset \{1, \dots, n\}$, $\hat{I}_2(p, b), \hat{J}_2(p, b) \subset \{1, \dots, m\}$ and $\hat{I}_1(p, b) \cup \hat{J}_1(p, b) = \{1, \dots, n\}$, $\hat{I}_2(p, b) \cup \hat{J}_2(p, b) = \{1, \dots, m\}$ are defined such that

$$\psi(p, b) = \{(x, u, y, z) \in M(p, b) \mid x_i = 0 \ i \in \hat{I}_1(p, b), \quad u_j = 0 \ j \in \hat{I}_2(p, b),$$
$$y_i = 0 \ i \in \hat{J}_1(p, b), \quad z_j = 0 \ j \in \hat{J}_2(p, b)\} \tag{5.5.9}$$

and

$$\operatorname{ri} \psi(p, b) = \{(x, u, y, z) \in \psi(p, b) \mid x_i > 0 \ i \notin \hat{I}_1(p, b), \quad u_j > 0 \ j \notin \hat{I}_2(p, b),$$
$$y_i > 0 \ i \notin \hat{J}_1(p, b), \quad z_j > 0 \ j \notin \hat{J}_2(p, b)\} \tag{5.5.10}$$

hold. As in (5.4.10)

$$\{(I_1^r, I_2^r, J_1^r, J_2^r), r = 1, \dots, S\} \tag{5.5.11}$$

is the finite system of all values attained by the above four-tuple on the solubility set of problem (5.5.4).

The theorem that now follows establishes characteristic features of the solubility set

$$\mathfrak{A} = \{(p, b) \mid \psi_Q(p, b) \neq \emptyset\} \tag{5.5.12}$$

the optimal set mapping

$$\psi_Q \colon (p, b) \to \psi_Q(p, b) \tag{5.5.13}$$

and the extreme value function of problem (5.5.1)

$$\varphi(p, b) = \min \{\tfrac{1}{2} x^{\mathsf{T}} C x + p^{\mathsf{T}} x \mid Ax \leqq b, x \geqq 0\} , \qquad (p, b) \in \mathfrak{A} . \tag{5.5.14}$$

Theorem 5.5.1. *If (5.5.2) is assumed then*

(1) *the set $\mathfrak{A}$ (5.5.12) is an $(n + m)$-dimensional convex polyhedral cone and it may be represented as the product*

$$\mathfrak{A} = \mathfrak{B}_D \times \mathfrak{B}_Q \tag{5.5.15}$$

of the n-dimensional convex polyhedral cone

$$\mathfrak{B}_D = -(Z_1)^p \tag{5.5.16}$$

with the m-dimensional convex polyhedral cone

$$\mathfrak{B}_Q = -(Z_2)^p \tag{5.5.17}$$

where

$$Z_1 = \{v \in E_n \mid Av \leqq 0, Cv = 0, v \geqq 0\} \tag{5.5.18}$$

and

$$Z_2 = \{w \in E_m \mid A^{\mathsf{T}} w \geqq 0, w \geqq 0\} ; \tag{5.5.19}$$

(2) *the mapping $\psi_Q|_{\mathfrak{A}}$ is u.s.c-H on $\mathfrak{A}$;*

(3) *the solution set $\psi_Q(p, b)$ is bounded if $(p, b) \in \operatorname{int} \mathfrak{B}_D \times \mathfrak{B}_Q$ and unbounded if $(p, b) \in \operatorname{bd} \mathfrak{B}_D \times \mathfrak{B}_Q$;*

(4) *the extreme value function φ is continuous on $\mathfrak{A}$, convex on $\mathfrak{B}_Q$ for each fixed $p^0 \in \mathfrak{B}_D$ and concave on $\mathfrak{B}_D$ for each fixed $b^0 \in \mathfrak{B}_Q$.*

Proof. (1) It follows from Theorem A.4.1 that the solubility set $\mathfrak{A}$ of (5.5.1) is identical to that of (5.5.3), Theorems 5.4.1 and 5.4.2 may thus be applied. The cone Z in (5.4.12) may here be represented in the form

$$Z = \left\{ (v, w) \in E_{n+m} \,\middle|\, \begin{array}{ll} Cv - A^{\mathsf{T}} w \leqq 0, & v \geqq 0 , \\ Av \qquad\qquad \leqq 0, & w \geqq 0 \end{array} \right\} .$$

It suffices to show that $Z = Z_1 \times Z_2$ since it then follows by Theorems 5.4.1 and 5.4.2 that

$$\mathfrak{A} = (-Z_1)^p \times (-Z_2)^p$$

and thus (5.5.15) hold.

The relation $Z_1 \times Z_2 \subset Z$ is trivial and is independent of the conditions imposed on C.

In order to demonstrate $Z \subset Z_1 \times Z_2$ we exploit condition (5.5.2). First observe that the inequalites $Av \leqq 0$, $v \geqq 0$, $w \geqq 0$ occur in the representations of both Z and $Z_1 \times Z_2$. One must then show that the conditions $Cv = 0$ and $A^{\mathsf{T}} w \geqq 0$ are satisfied for $(v, w) \in Z$. Since $Cv - A^{\mathsf{T}} w \leqq 0$ and $v \geqq 0$ we have $v^{\mathsf{T}} C v - v^{\mathsf{T}} A^{\mathsf{T}} w \leqq 0$. Taking $Av \leqq 0$ and $v \geqq 0$ into account $-v^{\mathsf{T}} A^{\mathsf{T}} w \geqq 0$ follows and by (5.5.2) $v^{\mathsf{T}} C v \geqq 0$. Hence $v^{\mathsf{T}} C v = 0$ and again because of $C \in H^+(n)$ we have $Cv = 0$, which then gives the inequality $A^{\mathsf{T}} w \geqq 0$.

(2) By Theorem 5.4.2 the solution set mapping $\psi|_{\mathfrak{A}}$ of (5.5.4) and consequently also $\tilde{\psi}: (p, b) \to \tilde{\psi}(p, b)$ are, both restricted to $\mathfrak{A}$, u.s.c.-H. Relation (5.5.7) then implies that $\psi_Q|_{\mathfrak{A}}$ is u.s.c.-H on $\mathfrak{A}$ (this however also follows from Theorem 4.5.5).

(3) Assume that $(p^0, b^0) \in \mathfrak{A}$ is arbitrary and $x^0 \in \operatorname{aff} \psi_Q(p^0, b^0)$. By Theorem A.4.3 (1) we have the representation

$$\psi_Q(p^0, b^0) = \{x \mid Ax \leqq b^0, x \geqq 0, Cx = Cx^0, p^{0\mathsf{T}} x = p^{0\mathsf{T}} x^0\} .$$

The recession cone $O^+\psi_Q(p^0, b^0)$ of the convex polyhedron $\psi_Q(p^0, b^0)$ may thus be written in the form

$$O^+\psi_Q(p^0, b^0) = \{v \in E_n \mid Av \leqq 0, v \geqq 0, Cv = 0, p^{0T}v = 0\} .$$

The set $\psi_Q(p^0, b^0)$ is bounded if and only if $O^+\psi_Q(p^0, b^0) = \{0\}$ is true. Since $(p^0, b^0) \in \mathfrak{A}$ it follows from (1) that $p^{0T}v \geqq 0$ for all $v \in Z_1$ and consequently $\psi_Q(p^0, b^0)$ is bounded if and only if $p^{0T}v > 0$ holds for all $v \in Z_1$ satisfying $v \neq 0$. This is, however, equivalent to the condition

$$(p^0, b^0) \in \operatorname{int} \mathfrak{B}_D \times \mathfrak{B}_Q .$$

(4) Follows from Theorems 4.3.8 and 4.5.2. □

Note 5.5.1. If the matrix C is in addition positive definite then we have $Z_1 = E_n$ and $\psi_Q|_{\mathfrak{A}}$ is H-continuous on $\mathfrak{A}$.

Note 5.5.2

(1) The solubility set of the dual problem (5.5.3) is also identical to the set $\mathfrak{A}$ given by (5.5.12).

(2) The mapping ψ_D: $(p, b) \to \psi_D(p, b)$ restricted to $\mathfrak{A}$ is likewise u.s.c.-H on $\mathfrak{A}$.

(3) The function φ (5.5.14) is simultaneously also the extreme value function of problem (5.5.3).

(4) The set $\psi_D(p, b)$ is bounded for $(p, b) \in \mathfrak{B}_D \times \operatorname{int} \mathfrak{B}_Q$ and unbounded for $(p, b) \in \mathfrak{B}_D \times \operatorname{bd} \mathfrak{B}_Q$.

(5) The set $\mathfrak{B}_Q$ in (5.5.17) is the feasible parameter set of problem (5.5.1) and the set $\mathfrak{B}_D$ (5.5.16) the feasible parameter set of problem (5.5.3).

Note 5.5.3. If the matrix C is not positive semidefinite one may show that the set $\mathfrak{A}$ is a closed cone (cf. B. C. Eaves [3]), this cone is, however, not necessarily convex as the following example demonstrates:

$$C = \begin{pmatrix} -1 & 0 & 0 & 0 \\ 0 & 1 & 0 & 0 \\ 0 & 0 & 0 & 0 \\ 0 & 0 & 0 & 0 \end{pmatrix}, \quad A = \begin{pmatrix} 1 & -1 & 1 & 0 \\ 1 & -1 & 0 & 1 \end{pmatrix}, \quad p^1 = p^2 = \begin{pmatrix} 4 \\ 0 \\ 0 \\ 0 \end{pmatrix},$$

$$b^1 = \begin{pmatrix} 2 \\ 4 \end{pmatrix}, \quad b^2 = \begin{pmatrix} 4 \\ 2 \end{pmatrix}.$$

Consider the following optimization problems

$$\min \{Q(x, p^1) \mid x \in M(b^1)\} = \min \{Q(x, p^2) \mid x \in M(b^2)\}$$
$$= \min \{-x_1^2 + x_2^2 + 4x_1 \mid x_1 - x_2 \leqq 2, x_1 \geqq 0, x_2 \geqq 0\}$$

and

$$\min \{Q(x, \tfrac{1}{2} p^1 + \tfrac{1}{2} p^2) \mid x \in M(\tfrac{1}{2} b^1 + \tfrac{1}{2} b^2)\}$$
$$= \min \{-x_1^2 + x_2^2 + 4x_1 \mid x_1 - x_2 \leqq 3, x_1 \geqq 0, x_2 \geqq 0\} .$$

One observes that the first two problems (in this case they are equivalent) possess a solution whereas the third problem formed by convex combination of the parameter values has no optimal solution.

Proposition (1) of Theorem 5.5.1 can, however, be generalized if a parametric problem in equation form

$$\min \{\tfrac{1}{2} x^T Cx + p^T x \mid Ax = b, x \geqq 0\}$$

is under consideration and the matrix C is an *A-flat matrix*, i.e. if the conditions

$$\text{(i)}\quad Ay = 0\,, \qquad y \geqq 0 \Rightarrow y^{\mathrm{T}}Cy \geqq 0$$

and

$$\text{(ii)}\quad Ay = 0\,, \qquad y^{\mathrm{T}} \geqq 0\,, \qquad y^{\mathrm{T}}Cy = 0 \Rightarrow (C + C^{\mathrm{T}})\, y = 0$$

are fulfilled. Subject to these conditions the set $\mathfrak{A}$ is also an $(n + m)$-dimensional convex polyhedral cone which may be represented as the product of the feasible sets of the primal and the dual problem (B. C. Eaves [3], J. Guddat [3], J. Guddat and P. Hohmuth [1]). In particular this also applies to the somewhat better known copositive plus matrices, i.e. to the matrices satisfying

$$\text{(i)}\quad y \geqq 0 \Rightarrow y^{\mathrm{T}}Cy \geqq 0\,,$$

$$\text{(ii)}\quad y \geqq 0\,, \qquad y^{\mathrm{T}}Cy = 0 \Rightarrow (C + C^{\mathrm{T}})\, y = 0\,.$$

One notes that the set of all positive semidefinite matrices is a subset of the set of all copositive plus matrices.

Here we define a local stability set $\mathfrak{A}^r$ for the parametric linear complementarity problem (5.5.4) by

$$\mathfrak{A}^r = \{(p, b) \in \mathfrak{A} \mid \hat{I}_1(p, b) = I_1^r,\, \hat{I}_2(p, b) = I_2^r,\, \hat{J}_1(p, b) = J_1^r,\, \hat{J}_2(p, b) = J_2^r\} \tag{5.5.20}$$

where we employ the four-tuple given by (5.5.11). Because of the particular structure of the matrix K (5.5.5) we also replace (5.4.17) and (5.4.16) by

$$\begin{aligned} T^r(p, b) = \{(x, y, z) \in M(p, b) \mid\, & x_i = 0,\, i \in I_1^r,\, u_j = 0,\, j \in I_2^r, \\ & y_i = 0,\, i \in J_1^r,\, z_j = 0,\, j \in J_2^r\} \end{aligned} \tag{5.5.21}$$

and

$$\begin{aligned} Z^r = \{(v, w) \in E_{n+m} \mid\, & c^{i\mathrm{T}}v - a^{i\mathrm{T}}w \leqq 0,\, i \in I_1^r,\, a_jv \leqq 0,\, j \notin I_2^r, \\ & v_i \geqq 0\;\, i \notin J_1^r,\, w_j \geqq 0\;\, j \notin J_2^r\} \end{aligned} \tag{5.5.22}$$

where c^i is the i-th column of C, a^i the i-th column of A, and a_j the j-th row of A.

By Theorem 5.4.3 we have $\mathfrak{A}^r = -\mathrm{ri}\,(Z^r)^p$ and hence the set (5.5.20) may be represented in the form

$$\begin{aligned} \mathfrak{A}^r = \{(p, b) \mid\, & p^{\mathrm{T}}v^k + b^{\mathrm{T}}w^k > 0,\, k = 1, \dots, F(r),\, p^{\mathrm{T}}y^l + b^{\mathrm{T}}z^l = 0, \\ & l = 1, \dots, G(t)\} \end{aligned} \tag{5.5.23}$$

where (v^k, w^k), $k = 1, \dots, F(r)$, is a fundamental system of the cone Z^r (5.5.22) and (y^l, z^l), $l = 1, \dots, G(r)$, is a basis of its vertex set.

We note that the local stability sets $\mathfrak{A}^r$ may not, in contrast to the solubility set $\mathfrak{A}$ be represented as a Cartesian product (see Example 5.5.4). Besides the representation (5.5.23) the set $\mathfrak{A}^r$ has all the other properties following from Theorem 5.4.3, together with the relations (5.5.7) and (5.5.8) this can be used to prove the next theorem which characterizes the mapping ψ_Q and the function φ on $\mathfrak{A}^r$. We, however, first introduce the notation

$$L_Q(r) = \{x \mid x_i = 0,\, i \in I_1^r,\, a_jx = 0,\, j \in J_2^r,\, Cx = 0\} \tag{5.5.24}$$

and

$$d_Q(r) = \dim L_Q(r) \tag{5.5.25}$$

as abbreviations.

Theorem 5.5.2

(1) $\dim \psi_Q(p, b) = d_Q(r)$ *is true for all* $(p, b) \in \mathfrak{A}^r$.

(2) *The mapping* $\psi_Q|_{\mathfrak{A}^r}$ *is H-continuous on* $\mathfrak{A}^r$.

(3) *The function* φ *is a quadratic function of the vector* (p, b) *on* $\mathfrak{A}^r$.

Proof. (1) For $(p, b) \in \mathfrak{A}^r$ and using Theorems A.4.3 (2) we obtain the representation

$$\begin{aligned}\psi_Q(p, b) = \{x \mid x_i = 0,\ i \in I_1^r,\ a_j x = b_j,\ j \in J_2^r,\ x_i \geqq 0,\ i \notin I_1^r,\\ a_j x \leqq b_j,\ j \notin J_2^r,\ Cx = Cx^0(p, b)\}\end{aligned} \tag{5.5.26}$$

and because of the representation given by (5.5.10)

$$\begin{aligned}\mathrm{ri}\,\psi_Q(p, b) = \{x \mid x_i = 0,\ i \in I_1^r,\ a_j x = b_j,\ j \in J_2^r,\ x_i > 0,\ i \notin I_1^r,\\ a_j x < b_j,\ j \notin J_2^r,\ Cx = Cx^0(p, b)\}\end{aligned} \tag{5.5.27}$$

assuming $x^0(p, b) \in \mathrm{aff}\,\psi_Q(p, b)$. Now the first proposition of the theorem follows immediately from (5.5.27).

(2) This follows from Corollary 3.4.1.1 if we use (5.5.26) and calculate $x^0(p, b)$ above according to the formula

$$\begin{pmatrix} x^0(p, b) \\ u^0(p, b) \\ y^0(p, b) \\ z^0(p, b) \end{pmatrix} = (D^r)^+ \begin{pmatrix} p \\ b \end{pmatrix} \tag{5.5.28}$$

where $(D^r)^+$ is the pseudo-inverse of the coefficient matrix of the system of linear equations

$$\begin{gathered} -Cx - A^{\mathsf{T}}u + y = p\,, \qquad Ax + z = b\,, \\ x_i = 0\,,\quad i \in I_1^r\,,\qquad u_j = 0\,,\quad j \in I_2^r\,,\qquad y_i = 0\,,\quad i \in J_1^r\,,\qquad z_j = 0\,,\quad j \in J_2^r \end{gathered} \tag{5.5.29}$$

which for $(p, b) \in \mathfrak{A}^r$ describes the non-empty set $\mathrm{aff}\,\psi_Q(p, b)$.

(3) Because of $\varphi(p, b) = \frac{1}{2}\, x^{\mathsf{T}}Cx + p^{\mathsf{T}}x$ for all $x \in \mathrm{aff}\,\psi_Q(p, b)$ this proposition follows if $x^0(p, b)$ in (5.5.28) is inserted into the objective function. □

Note 5.5.4. One also confirms that $\dim \psi_D(p, b)$ is equal to $d_D(r) = d(r) - d_Q(r)$ for all $(p, b) \in \mathfrak{A}^r$ and that $\psi_D|_{\mathfrak{A}^r}$ is H-continuous on $\mathfrak{A}^r$.

Note 5.5.5. According to Theorem 5.4.4 the closures of the local stability sets $\mathfrak{A}^1, \ldots, \mathfrak{A}^T$ of maximal dimension (which here is equal to $n + m$) constitute a partitioning of the solubility set and have the properties given in that theorem. The relation $\dim \psi(p, b) = 0$ for $(p, b) \in \mathfrak{A}^r$, $r = 1, \ldots, T$, here gives

$$\dim \psi_Q(p, b) = \dim \psi_D(p, b) = 0$$

because of (5.5.8), i.e. problems (5.5.1) and (5.5.3) possess for $(p, b) \in \mathfrak{A}^r$, $r = 1, \ldots, T$, uniquely determined solutions $x^r(p, b)$ and $u^r(p, b)$, respectively, these solutions depend linearly on the vector (p, b).

Because of Theorem 5.5.2 it appears reasonable to consider the local stability sets $\mathfrak{A}^r$ of problem (5.5.4) also as local stability sets of the parametric quadratic problem (5.5.1).

One recalls that in Sections 4.3 and 5.3 local stability sets for parametric optimization problems were introduced and examined with the help of characteristic index sets (cf. (4.3.22), (4.3.23), (5.3.9), and (5.3.32)). Let $I(p, b) \subset \{1, \dots, n\}$ and $J(p, b) \subset \{1, \dots, m\}$ be the characteristic index set pair with respect to the optimal set $\psi_Q(p, b)$, i.e.

$$x_i^0 = 0\,, \quad i \in I(p, b)\,, \quad a_j x^0 = b_j\,, \quad j \in J(p, b) \quad \text{for all} \quad x^0 \in \psi_Q(p, b)\,.$$

The local stability set mentioned above may then, for each given index set pair $I \subset \{1, \dots, n\}$ and $J \subset \{1, \dots, m\}$ be defined here by

$$\mathfrak{S}(I, J) = \{(p, b) \in \mathfrak{A} \mid I(p, b) = I, J(p, b) = J\}\,. \tag{5.5.30a}$$

The reader may find the following alternative more obvious (cf. (5.3.13)): Defining

$$\Sigma^{(I,J)}(b) =_{\mathrm{Df}} \{x \in E_n \mid x_i = 0, i \in I, x_i > 0, i \notin I, a_j x = b_j, j \in J,$$
$$a_j x < b_j, j \notin J\}$$

for a given index set pair we find that $\Sigma^{(I,J)}(b)$ is either a face of the polyhedron $M(b) = \{x \in E_n \mid Ax \leqq b, x \geqq 0\}$ or is equal to ri $M(b)$, assuming that b is chosen such that $\Sigma^{(I,J)}(b)$ is non-empty; we then have

$$\mathfrak{S}(I, J) = \{(p, b) \in \mathfrak{A} \mid \operatorname{ri} \psi(p, b) \subset \Sigma^{(I,J)}(b)\}\,. \tag{5.5.30b}$$

As is the case for $\mathfrak{A}^r$ the sets $\mathfrak{S}(I, J)$ constitute a decomposition of the solubility set $\mathfrak{A}$. The following statement demonstrates that both systems of sets differ only in unessential features.

Note 5.5.6. To each non-empty set $\mathfrak{S}(I, J)$ (5.5.30) there corresponds one and only one index $r \in \{1, \dots, S\}$ (5.5.11) such that

$$\mathfrak{A}^r \subset \mathfrak{S}(I, J) \subset \operatorname{cl} \mathfrak{A}^r \tag{5.5.31}$$

holds; we also have $I = I_1^r$ and $J = J_2^r$.

It follows then from (5.5.31) that the local stability sets $\mathfrak{A}^r$ and $\mathfrak{S}(I, J)$ can only differ by boundary points. $\mathfrak{S}(I, J)$ can be obtained from the union of $\mathfrak{A}^r$ and certain faces of its closure. The sets $\mathfrak{S}(I, J)$ are in general neither relatively open nor closed. The number of non-empty sets $\mathfrak{S}(I, J)$ is smaller than or equal to S, the number of sets $\mathfrak{A}^r$ in (5.5.11). We also refer to the examples given towards the end of this section. Here we note that in the definition of a local stability set for problem (5.3.1)

$$\min \{x^{\mathrm{T}}Cx + p^{\mathrm{T}}x \mid Ax \leqq b\}\,, \qquad (C, p, b) \quad \text{variable}\,,$$

the assumption that the dimension of the optimal set $\psi(C, p, b)$ is constant was essential in obtaining strong stability statements relating to the extreme value function and the optimal set mapping (Theorems 5.3.3 and 5.3.4), this property is however automatically given in the problem class (5.5.1) (see Theorem 5.5.2).

Some special results on local stability sets follow if $C = 0$, i.e. in the case of linear parametric optimization problems; the most important of these are summarized in the following corollaries. We note that in this case the matrix K (5.5.5) is antisymmetric, or in other words $K \in H^A(n + m)$. In Section 5.4 we referred to various special properties of parametric complementarity problems involving an antisymmetric matrix K, these properties will now be employed.

Corollary 5.5.2.1. *Assume* $C = 0$ *in* (5.5.1) *and* $r \in \{1, \dots, S\}$. *Then*

(1) *the set* $\mathfrak{A}^r$ *given in* (5.5.20) *may be represented in the form*

$$\mathfrak{A}^r = \mathfrak{A}_D^r \times \mathfrak{A}_Q^r \tag{5.5.32}$$

where

$$\mathfrak{A}_D^r = -\mathrm{ri}\,(Z_1^r)^p\,, \tag{5.5.33}$$

$$\mathfrak{A}_Q^r = -\mathrm{ri}\,(Z_2^r)^p\,, \tag{5.5.34}$$

and Z_1^r *and* Z_2^r *are given by*

$$Z_1^r = \{v \in E_n \mid a_j v \leqq 0, j \in J_2^r, v_i \geqq 0, i \in I_1^r\}\,, \tag{5.5.35}$$

$$Z_2^r = \{w \in E_m \mid a^{i\mathsf{T}} w \geqq 0, i \in I_1^r, w_j \geqq 0, j \notin J_2^r\}\,; \tag{5.5.36}$$

(2) *for all* $(p, b) \in \mathfrak{A}^r$ *we have*

$$\psi_Q(p, b) \equiv \psi_Q(b) = \{x \mid x_i = 0, i \in I_1^r, a_j x = b_j, j \in J_2^r, x_i \geqq 0, i \notin I_1^r, \\ a_j x \leqq b_j, j \notin J_2^r\}\,,$$

$$\dim \psi_Q(p, b) = d_Q(r)$$

where $d_Q(r) = \dim \{x \mid x_i = 0, i \in I_1^r, a_j x = 0, j \in J_2^r\}$ *and further*

$$\psi_D(p, b) \equiv \psi_D(p) = \{u \mid u_j = 0, j \notin J_2^r, a^{i\mathsf{T}} u = p_i, i \notin I_1^r, u_j \geqq 0, \\ j \in J_2^r, a^{i\mathsf{T}} u \geqq p_i, i \in I_1^r\}\,,$$

$$\dim \psi_D(p, b) = d_D(r)$$

where $d_D(r) = \dim \{u \mid u_j = 0, j \notin J_2^r, a^{i\mathsf{T}} u = 0, i \notin I_1^r\}$;

(3) $\dim \mathfrak{A}_Q^r + d_D(r) = m$, $\dim \mathfrak{A}_D^r + d_Q(r) = n$;

(4) *the extreme value function* $\varphi(p, b)$ *is linear on* $\mathfrak{A}_Q^r$ *for fixed* $p \in \mathfrak{A}_D^r$ *and linear on* $\mathfrak{A}_D^r$ *for fixed* $b \in \mathfrak{A}_Q^r$, *respectively.*

Proof. All propositions follow by specializing Theorems 5.4.3 and 5.5.2 if one takes $Z^r = Z_1^r \times Z_2^r$ (Z^r is given by (5.5.22)) and the relations $J_1^r = \{1, \dots, n\} \setminus I_1^r$ and $I_2^r = \{1, \dots, m\} \setminus J_2^r$ into account. □

Corollary 5.5.2.2. *Assume* $C = 0$ *in* (5.5.1). *Then*

(1) *if the local stability set* $\mathfrak{A}^r$ *is of (highest possible) dimension* $n + m$ *it follows that the matrix*

$$A^r = (a_{ij})_{i \in J_2^r, j \in J_1^r} \tag{5.5.37}$$

is quadratic and regular; if the matrix (5.5.37) *is suitably complemented by zero columns to form the matrix*

$$B^r = (b_{ij}^r)_{i \in J_2^r, j = 1, \dots, n}$$

where

$$b_{ij}^r = \begin{cases} a_{ij} & \text{if} \quad i \in J_2^r\,, \qquad j \in J_1^r\,, \\ 0 & \text{if} \quad i \in J_2^r\,, \qquad j \in \{1, \dots, n\} \setminus J_1^r \end{cases} \tag{5.5.38}$$

then we have for all $(p, b) \in \mathfrak{A}^r$

$$\psi_Q(p, b) \equiv \psi_Q(b) = \{(x^r(b))\}\,,$$

$$\psi_D(p, b) \equiv \psi_D(p) = \{(u^r(p))\}\,,$$

and

$$\varphi(p, b) = p^{\mathsf{T}} (B^r)^+ b$$

as well as

$$\mathfrak{B}_D^r = \{p \mid u_i^r(p) > 0, i \in J_2^r\}$$

and

$$\mathfrak{B}_Q^r = \{b \mid x_j^r(b) > 0, j \in J_1^r\}$$

for the points given by

$$x^r(b) = (B^r)^+ b \quad \text{and} \quad u^r(p) = [-p^{\mathsf{T}}(B^r)^+]^{\mathsf{T}} \tag{5.5.39}$$

where $(B^r)^+$ *is the pseudo-inverse of the matrix* B^r, *and these points depend linearly on* b *and* p *respectively;*
(2) *if on the other hand the matrix*

$$\tilde{A} = (a_{ij})_{i \in \tilde{J}_2, j \in \tilde{J}_1}$$

is quadratic and regular for a suitable index set pair $\tilde{J}_1 \subset \{1, \dots, n\}$, $\tilde{J}_2 \subset \{1, \dots, m\}$ *it follows that there exists one and only one index* $r \in \{1, \dots, T\}$ *such that* $\tilde{J}_1 = J_1^r$, $\tilde{J}_2 = J_2^r$, $\tilde{A} = A^r$, *and* (1) *hold.*

Proof. This may be obtained by specializing Theorem 5.4.4. □

One notes that in the case $C = 0$ the sets $\mathfrak{S}(I, J)$ of (5.5.30) are closed, thus in (5.5.31)

$$\mathfrak{S}(I, J) = \operatorname{cl} \mathfrak{A}^r \quad \text{holds}.$$

We would like to mention the fact that further investigations on linear parametric optimization problems involving methods which are in part different from the above, lead to a number of additional results, this is, however, outside the scope of this volume: we refer the reader to F. Nožička et al. [2].

Finally, we remark that some important results obtained for problem (5.5.1) may also be applied to multiparametric problems of the type

$$\min \{\tfrac{1}{2} x^{\mathsf{T}} C x + p(t)^{\mathsf{T}} x \mid Ax \leqq b(t), x \geqq 0\}, \tag{5.5.40}$$

where $p(t)$ and $b(t)$ exhibit an affine-linear dependence on the parameter vector $t \in \mathbb{E}_d$. As this question was considered in Section 5.4 in some detail in the context of multiparametric linear complementarity problems we will not discuss the implications for problem (5.5.40) here.

Special cases of multiparametric problems arise if in (5.5.1) only p or b but not both are parameter-dependent, i.e. if we consider the problems

$$\begin{aligned} &\min \{\tfrac{1}{2} x^{\mathsf{T}} C x + p^{\mathsf{T}} x \mid Ax \leqq b, x \geqq 0\}, \\ &p \in \mathbb{E}_n \text{ variable}, \qquad b \in \mathbb{E}_m \text{ fixed}, \end{aligned} \tag{5.5.41}$$

or

$$\begin{aligned} &\min \{\tfrac{1}{2} x^{\mathsf{T}} C x + p^{\mathsf{T}} x \mid Ax \leqq b, x \geqq 0\}, \\ &b \in \mathbb{E}_m \text{ variable}, \qquad p \in \mathbb{E}_n \text{ fixed}. \end{aligned} \tag{5.5.42}$$

These problems are in a certain sense again fully parametric and may without difficulty be examined in the same way as was the case for problem (5.5.1); similar strong results may be obtained, with the exception that the closures of the local stability sets in the case $C \neq 0$ are no longer cones, but rather only convex polyhedra. In order to ensure that the solubility set is non-empty in problem (5.5.41) one must, however, assume that the primal constraint set $\{x \mid Ax \leqq b, x \geqq 0\}$ which here is parameter-independent, is non-empty or alternatively that this holds for the then parameter-independent dual constraint set $\{u \mid u \geqq 0, \inf_{x \geqq 0} \{\tfrac{1}{2} x^{\mathsf{T}} C x + p^{\mathsf{T}} x + u^{\mathsf{T}} A x\} > -\infty\}$ in the case (5.5.42). The only additional result which does not follow immediately from the above theorems on problem (5.5.1) is the fact that the optimal set mapping of (5.5.42) is not only u.s.c.-H but also l.s.c.-H.

We recall that a local stability set $\mathfrak{R}(x^0)$ with respect to a fixed point $x^0 \in M$ was introduced in Section 5.2 (5.2.7) for parametric optimization problems with fixed constraint sets M. Such sets $\mathfrak{R}(x^0)$ may also be defined for problem (5.5.41) and the results of Section 5.2 also hold, this applies in particular to the properties given by Theorem 5.2.5. The set $\mathfrak{R}(x^0)$ now has the form

$$\mathfrak{R}(x^0) = \{-Cx^0\} - K^p(x^0, M) ,$$

i.e. we add the negative cone corresponding to $K^p(x^0, M)$ to the image point of x^0 under the linear mapping $(-C)$ (see Examples 5.5.1 and 5.5.2).

The number of sets $\mathfrak{R}(x^0)$, where $x^0 \in M$, is also in general not finite in the case of the parametric quadratic problem (5.5.41). A finite number is only guaranteed if in (5.5.41) $C = 0$, in this case there is a one-to-one correspondence between sets $\mathfrak{R}(x^0)$ and closed faces of M as well as M itself. If $C \neq 0$ and x^0 is a vertex of M then by Theorem 5.3.5 $\mathfrak{R}(x^0)$ is equal to the closure of the local stability sets $\mathfrak{S}(I^0, J^0)$ corresponding to this vertex, and assuming that (I^0, J^0) is the corresponding index set pair, it is by Note 5.5.6 also equal to the closure of a certain local stability set $\mathfrak{A}_b^{r_0}$ of problem (5.5.41) defined in analogy to (5.5.20) (see Examples 5.5.1 and 5.5.2).

If $\Sigma^{(I^0, J^0)}$ is not a vertex of M one easily finds the relation existing between $\mathfrak{R}(x^0)$ and $\mathfrak{S}(I^0, J^0)$ or $\mathfrak{A}_b^{r_0}$ assuming $x^0 \in \Sigma^{(I^0, J^0)}$ and that in addition the matrix C is positive definite. This special case is examined in the following corollary, in which ψ_b denotes the optimal set mapping of problem (5.5.41).

Corollary 5.5.2.3. *Let $M \neq \emptyset$ and C be positive definite. Then*
(1) *ψ_b is continuous on E_n;*
(2) *$\psi_b^{-1}(x^0) = \mathfrak{R}(x^0)$ for all $x \in M$;*
(3) *$\psi_b^{-1}(M) = E_n$.*

Assume furthermore that $\Sigma^{(I,J)}$ is non-empty and $\mathfrak{S}(I, J)$ is the corresponding local stability set in (5.5.3). *Then*
(4) $\psi_b^{-1}(\Sigma^{(I,J)}) = \mathfrak{S}(I, J)$;
(5) $\mathfrak{S}(I, J) = \bigcup_{x^0 \in \Sigma^{(I,J)}} \mathfrak{R}(x^0)$.

Proof. See the preceding remarks and Example 5.3.1. □

The following examples demonstrate the decomposition and partitioning theorems and the relations existing between the various types of local stability sets. In particular the correspondence between the faces of the convex polyhedron M and the local stability sets of problem (5.5.41) is illustrated. We therefore choose the concept of a local stability set given in (5.5.30a) and (5.5.30b); the relation of this to the local stability sets introduced in (5.5.20) is examined in Note 5.5.6. The interested reader may easily find the particular form of these parameter sets in the concrete examples given below.

Example 5.5.1. $\min \{Q(x, p) \mid x \in M\}$, $p \in E_2$, where

$$Q(x, p) = x_1^2 + x_2^2 + p_1x_1 + p_2x_2$$

and

$$M = \{x \in E_2 \mid x_1 + x_2 \leqq 1, x_1 \geqq 0, x_2 \geqq 0\} .$$

We decompose the polyhedron into ri M and its boundary faces:

$$\begin{aligned} &\Sigma_1 = \text{ri } M, \quad \Sigma_2 = \{x \in E_2 \mid 0 < x_1 < 1, x_2 = 0\}, \\ &\Sigma_3 = \{x \in E_2 \mid 0 < x_2 < 1, x_1 = 0\}, \\ &\Sigma_4 = \{x \in E_2 \mid x_1 + x_2 = 1, x_2 < 0, x_2 < 0\}, \\ &\Sigma_5 = \{(0, 0)\}, \quad \Sigma_6 = \{(1, 0)\}, \quad \Sigma_7 = \{(0, 1)\}. \end{aligned} \tag{5.5.43}$$

In order to construct the corresponding local stability sets $\mathfrak{S}(\Sigma_j)$, $j = 1, \dots, 7$, we must by Corollary 5.5.2.3 find the images of the sets Σ_j, $j = 1, \dots, 7$, under the linear mapping $\pi = -2C$ with $C = \begin{pmatrix} 1 & 0 \\ 0 & 1 \end{pmatrix}$ and then add to these the negative polar cone corresponding to the direction cone $K(x^0, M)$ if $x^0 \in \Sigma_j$, $j \in \{1, \dots, 7\}$, is chosen arbitrarily. The matrix C is regular, i.e. there is a one-to-one correspondence between the boundary faces of M and the boundary faces of the image polyhedron $\pi(M) = \{p \in E_2 \mid p = -2Cx, x \in M\}$ (cf. fig. 5.1). This also applies to the relative interior of both polyhedra.

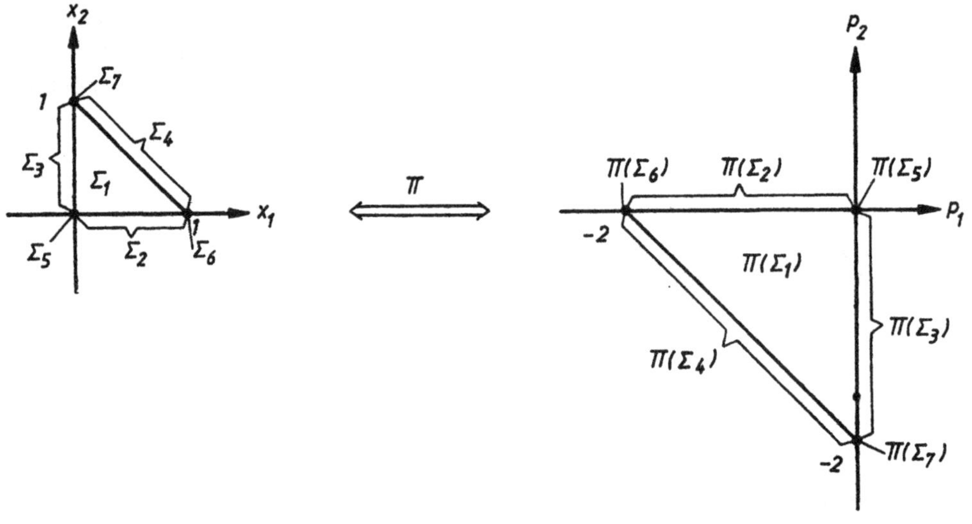

Fig. 5.1

If the corresponding negative polar cones assigned to the boundary faces of $\pi(M)$ are added to these boundary faces then we obtain the local stability sets $\mathfrak{S}(\Sigma_i)$, $j = 2, \dots, 7$. Obviously $\mathfrak{S}(\Sigma_1) = \pi(\Sigma_1)$. We thus have

$$\begin{aligned} &\mathfrak{S}(\Sigma_1) = \{p \in E_2 \mid p_1 + p_2 > -2, p_1 < 0, p_2 < 0\}, \\ &\mathfrak{S}(\Sigma_2) = \{p \in E_2 \mid -2 < p_1 < 0, p_2 \geqq 0\}, \\ &\mathfrak{S}(\Sigma_3) = \{p \in E_2 \mid p_1 \geqq 0, -2 < p_2 < 0\}, \\ &\mathfrak{S}(\Sigma_4) = \{p \in E_2 \mid p_1 - p_2 < 2, p_1 - p_2 < -2, p_1 + p_2 \leqq -2\}, \\ &\mathfrak{S}(\Sigma_5) = \{p \in E_2 \mid p_1 \geqq 0, p_2 \geqq 0\}, \\ &\mathfrak{S}(\Sigma_6) = \{p \in E_2 \mid p_1 \leqq -2, p_1 - p_2 \geqq -2\}, \\ &\mathfrak{S}(\Sigma_7) = \text{j}\{p \in E_2 \mid -p_1 + p_2 \geqq -2, p_2 \leqq -2\}. \end{aligned}$$

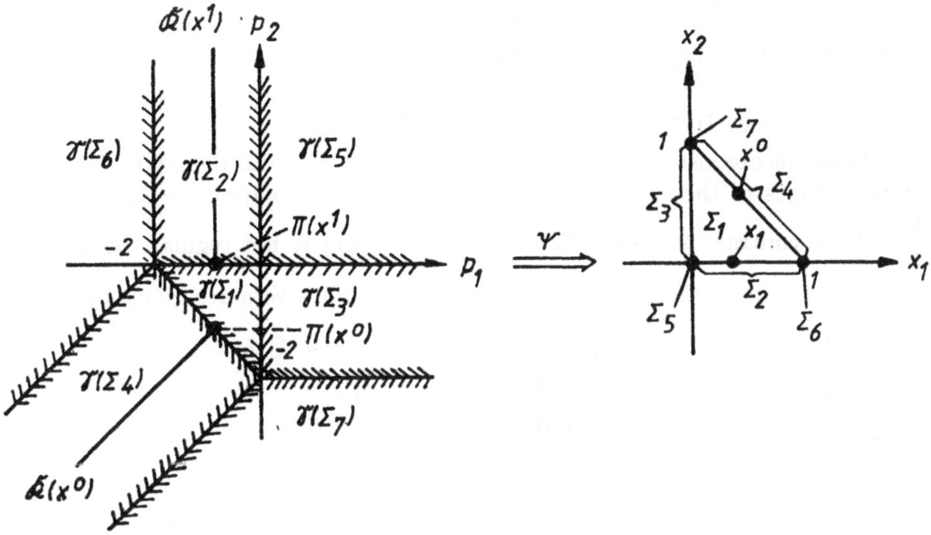

Fig. 5.2

The optimal set mapping is given by

$$\psi_b(p) = \begin{cases} \left\{\left(-\frac{p_1}{2}, -\frac{p_2}{2}\right)\right\} & \text{if } p \in \operatorname{cl} \mathfrak{S}(\Sigma_1)\,, \\ \left\{\left(-\frac{p_1}{2}, 0\right)\right\} & \text{if } p \in \operatorname{cl} \mathfrak{S}(\Sigma_2)\,, \\ \left\{\left(0, -\frac{p_2}{2}\right)\right\} & \text{if } p \in \operatorname{cl} \mathfrak{S}(\Sigma_3)\,, \\ \left\{\left(-\frac{p_1}{4} + \frac{p_2}{4} + \frac{1}{2}, \frac{p_1}{4} - \frac{p_2}{4} + \frac{1}{2}\right)\right\} & \text{if } p \in \operatorname{cl} \mathfrak{S}(\Sigma_4)\,, \\ \{(0, 0)\} & \text{if } p \in \operatorname{cl} \mathfrak{S}(\Sigma_5)\,, \\ \{(1, 0)\} & \text{if } p \in \operatorname{cl} \mathfrak{S}(\Sigma_6)\,, \\ \{(0, 1)\} & \text{if } p \in \operatorname{cl} \mathfrak{S}(\Sigma_7) \end{cases}$$

(cf. fig. 5.2). The extreme value function then follows:

$$\varphi_b(p) = \begin{cases} -\frac{p_1^2}{4} - \frac{p_2^2}{4} & \text{if } p \in \operatorname{cl} \mathfrak{S}(\Sigma_1)\,, \\ -\frac{p_1^2}{4} & \text{if } p \in \operatorname{cl} \mathfrak{S}(\Sigma_2)\,, \\ -\frac{p_2^2}{4} & \text{if } p \in \operatorname{cl} \mathfrak{S}(\Sigma_3)\,, \\ -\frac{p_1^2}{3} - \frac{p_2^2}{8} + \frac{p_1 p_2}{4} + \frac{p_1}{2} + \frac{p_2}{2} + \frac{1}{2} & \text{if } p \in \operatorname{cl} \mathfrak{S}(\Sigma_4)\,, \\ 0 & \text{if } p \in \operatorname{cl} \mathfrak{S}(\Sigma_5)\,, \\ p_1 + 1 & \text{if } p \in \operatorname{cl} \mathfrak{S}(\Sigma_6)\,, \\ p_2 + 1 & \text{if } p \in \operatorname{cl} \mathfrak{S}(\Sigma_7)\,. \end{cases}$$

Example 5.5.2. $\min \{Q(x, p) \mid x \in M\}$, $p \in E_2$, where

$$Q(x, p) = x_1^2 + 2x_1x_2 + x_2^2 + p_1x_1 + p_2x_2$$

and

$$M = \{x \in E_2 \mid x_1 + x_2 \leqq 1, x_1 \geqq 0, x_2 \geqq 0\} .$$

As in (5.5.4) we decompose the polyhedron M into ri$M$ and its boundary faces and consider the images of the sets $\Sigma_1, \dots, \Sigma_7$ under the linear mapping $\pi = -2C$ with $C = \begin{pmatrix} 1 & 1 \\ 1 & 1 \end{pmatrix}$. In contrast to Example 5.5.1 this mapping is not regular. Images and inverse images are illustrated in fig. 5.3. The local stability sets are given by

$$\mathfrak{S}(\Sigma_1) = \{p \in E_2 \mid p_2 - p_1 = 0, -2 < p_1 < 0, -2 < p_2 < 0\} ,$$
$$\mathfrak{S}(\Sigma_2) = \{p \in E_2 \mid p_1 - p_2 \leqq 0, -2 < p_1 < 0\} ,$$
$$\mathfrak{S}(\Sigma_3) = \{p \in E_2 \mid p_1 - p_2 \geqq 0, -2 < p_2 < 0\} ,$$

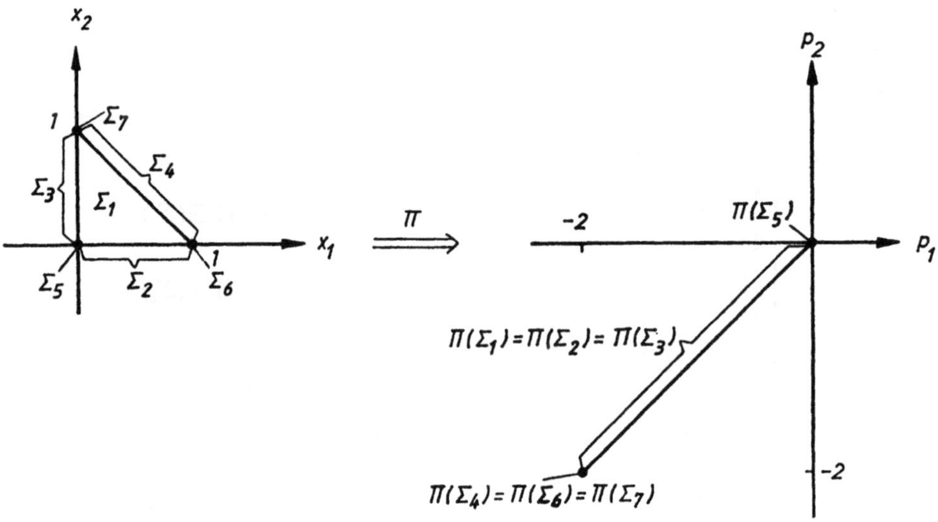

Fig. 5.3

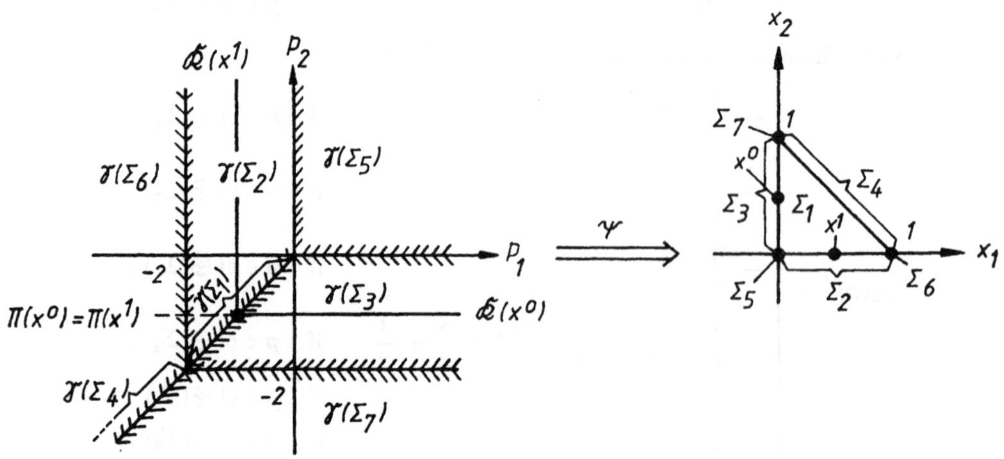

Fig. 5.4

$$\mathfrak{S}(\Sigma_4) = \{p \in E_2 \mid p_1 \leqq -2, p_2 \leqq -2, p_1 - p_2 = 0\} ,$$
$$\mathfrak{S}(\Sigma_5) = \{p \in E_2 \mid p_1 \geqq 0, p_2 \geqq 0\} ,$$
$$\mathfrak{S}(\Sigma_6) = \{p \in E_2 \mid p_1 - p_2 \leqq 0, p_2 \leqq -2\} \setminus \{(-2, -2)\} ,$$
$$\mathfrak{S}(\Sigma_7) = \{p \in E_2 \mid p_1 - p_2 \geqq 0, p_2 \leqq -2\} \setminus \{(-2, -2)\}$$

(cf. fig. 5.4). The optimal set mapping ψ_b is given by

$$\psi_b(p) = \begin{cases} \pi(p) \cap \operatorname{cl} \Sigma_1 & \text{if} \quad p \in \operatorname{cl} \mathfrak{S}(\Sigma_1) , \\ \left\{\left(-\frac{p_1}{2}, 0\right)\right\} & \text{if} \quad p \in \operatorname{cl} \mathfrak{S}(\Sigma_2) , \\ \left\{\left(0, -\frac{p_2}{2}\right)\right\} & \text{if} \quad p \in \operatorname{cl} \mathfrak{S}(\Sigma_3) , \\ \operatorname{cl} \Sigma_4 & \text{if} \quad p \in \operatorname{cl} \mathfrak{S}(\Sigma_4) , \\ \{(0, 0)\} & \text{if} \quad p \in \operatorname{cl} \mathfrak{S}(\Sigma_5) , \\ \{(1, 0)\} & \text{if} \quad p \in \operatorname{cl} \mathfrak{S}(\Sigma_6) , \\ \{(0, 1)\} & \text{if} \quad p \in \operatorname{cl} \mathfrak{S}(\Sigma_7) . \end{cases}$$

From this we obtain the extreme value function

$$\varphi_b(p) = \begin{cases} -\frac{1}{4} p_1^2 & \text{if} \quad p \in \mathfrak{S}(\Sigma_1) \cup \mathfrak{S}(\Sigma_2), \\ -\frac{1}{4} p_2^2 & \text{if} \quad p \in \mathfrak{S}(\Sigma_3) , \\ p_1 & \text{if} \quad p \in \mathfrak{S}(\Sigma_4) , \\ 0 & \text{if} \quad p \in \mathfrak{S}(\Sigma_5) , \\ p_1 + 1 & \text{if} \quad p \in \mathfrak{S}(\Sigma_6) , \\ p_2 + 1 & \text{if} \quad p \in \mathfrak{S}(\Sigma_7) . \end{cases}$$

We note that the mapping ψ_b is not lower semicontinuous at $\hat{p} = (-2, -2)$. In both examples there exists a partitioning of the solubility set $\mathfrak{A}_b = E_2$ into convex polyhedra, these polyhedra are closures of certain local stability sets, namely those of highest dimension. We also observe that the local stability sets are not necessarily either relatively open or closed. The last two examples have in the first case a fixed objective function and in the second parameter-dependent objective function and constraint vector.

Example 5.5.3

$$\min \{Q(x) \mid x \in M(b)\} , \qquad b \in R , \tag{5.5.44}$$

where

$$Q(x) = x_1^2 + x_2^2$$

and

$$M(b) = \{x \in E_2 \mid x_1 + x_2 = b, x_1 \geqq 0, x_2 \geqq 0\} .$$

$\mathfrak{A}_p = [0, +\infty)$ is obviously the solubility set. We consider the four sets $I_1 = \{1\}$, $I_2 = \{2\}$, $I_3 = \{1, 2\}$, and $I_4 = \emptyset$ which are all subsets of the index set I_3. We then have

$$\Sigma^{I_1}(b) = \{x \in E_2 \mid x_1 + x_2 = b, x_1 = 0, x_2 > 0\} ,$$
$$\Sigma^{I_2}(b) = \{x \in E_2 \mid x_1 + x_2 = b, x_1 > 0, x_2 = 0\} ,$$
$$\Sigma^{I_3}(b) = \{(0, 0)\} , \qquad \Sigma^{I_4}(b) = \operatorname{ri} M(b) .$$

Since $Q(x)$ is strictly convex the problem (5.5.44) possesses for each $b \in \mathfrak{A}_p$ a uniquely determined optimal point. We obtain $\mathfrak{S}(I_1) = \mathfrak{S}(I_2) = \emptyset$, $\mathfrak{S}(I_3) = \{0\}$, $\mathfrak{S}(I_4) = (0, +\infty)$.

Example 5.5.4. $\min \{Q(x, p) \mid x \in M(b)\}$, $p \in E_2$, $b \in R$, where

$$Q(x, p) = x_1^2 + x_2^2 + p_1x_1 + p_2x_2$$

and

$$M(b) = \{x \in E_2 \mid x_1 + x_2 = b, x_1 \geqq 0, x_2 \geqq 0\} .$$

Here the solubility set is given by

$$\mathfrak{A} = \left\{\begin{pmatrix} p \\ b \end{pmatrix} \in E_3 \,\middle|\, p \in E_2, b \in [0, +\infty)\right\}.$$

We consider the index sets $I_1 = \{1\}$, $I_2 = \{2\}$, $I_3 = \{1, 2\}$, and $I_4 = \emptyset$ and obtain the sets $\Sigma^{I_j}(b)$, $j = 1, \dots, 4$, given in the previous example. The closures of the local stability sets are then

$$\operatorname{cl} \mathfrak{S}(I_1) = \left\{\begin{pmatrix} p \\ b \end{pmatrix} \in E_3 \,\middle|\, p_1 - p_2 + 2b \geqq 0, 0 \leqq b < \infty\right\},$$

$$\operatorname{cl} \mathfrak{S}(I_2) = \left\{\begin{pmatrix} p \\ b \end{pmatrix} \in E_3 \,\middle|\, p_1 - p_2 + 2b \leqq 0, 0 \leqq b < \infty\right\},$$

$$\operatorname{cl} \mathfrak{S}(I_3) = \left\{\begin{pmatrix} p \\ b \end{pmatrix} \in E_3 \mid p \in E_2, b = 0\right\},$$

$$\operatorname{cl} \mathfrak{S}(I_4) = \left\{\begin{pmatrix} p \\ b \end{pmatrix} \in E_3 \,\middle|\, p_1 - p_2 + 2b \geqq 0, p_1 - p_2 - 2b \leqq 0, 0 \leqq b < \infty\right\}.$$

Further,

$$\psi(p, b) = \begin{cases} \{(0, b)\} & \text{if } \begin{pmatrix} p \\ b \end{pmatrix} \in \operatorname{cl} \mathfrak{S}(I_1)\,, \\ \{(b, 0)\} & \text{if } \begin{pmatrix} p \\ b \end{pmatrix} \in \operatorname{cl} \mathfrak{S}(I_2)\,, \\ \{(0, 0)\} & \text{if } \begin{pmatrix} p \\ b \end{pmatrix} \in \operatorname{cl} \mathfrak{S}(I_3)\,, \\ \left\{\left(-\frac{p_1}{4} + \frac{p_2}{4} + \frac{b}{2}, \frac{p_1}{4} - \frac{p_2}{4} - \frac{b}{2}\right)\right\} & \text{if } \begin{pmatrix} p \\ b \end{pmatrix} \in \operatorname{cl} \mathfrak{S}(I_4) \end{cases}$$

and

$$\varphi(p, b) = \begin{cases} p_2 + b^2 & \text{if } \begin{pmatrix} p \\ b \end{pmatrix} \in \operatorname{cl} \mathfrak{S}(I_1)\,, \\ p_1 + b^2 & \text{if } \begin{pmatrix} p \\ b \end{pmatrix} \in \operatorname{cl} \mathfrak{S}(I_2)\,, \\ 0 & \text{if } \begin{pmatrix} p \\ b \end{pmatrix} \in \operatorname{cl} \mathfrak{S}(I_3)\,, \\ -\frac{p_1^2}{8} - \frac{p_2^2}{8} + \frac{p_1p_2}{4} + \frac{p_1b}{2} + \frac{p_2b}{2} + \frac{b^2}{2} & \text{if } \begin{pmatrix} p \\ b \end{pmatrix} \in \operatorname{cl} \mathfrak{S}(I_4)\,. \end{cases}$$

5.6. Linear Parametric Optimization Problems with Integer Conditions and Parameters in the Objective Function and in the Right-Hand Sides of the Linear Constraints

Consider the parametric problem

$$(\mathrm{P}_s(p, \lambda)) \quad \inf \{p^\mathsf{T} x \mid x \in G(\lambda)\} , \qquad (p, \lambda) \in E_{n+m} , \tag{5.6.1}$$

where

$$G(\lambda) = \{x \in E_n \mid Ax \leqq \lambda, x_1, \dots, x_s \in \mathbb{Z}\}$$

according to (3.4.3) and the number s of integer variables satisfies $0 < s \leqq n$. Problem (5.6.1) follows from the parametric problem $(\mathrm{P}_s(p, \lambda))$ considered in Section 4.5 if D in the latter is the zero (n, n)-matrix. The aim of the present considerations is to identify and characterize subsets of the solubility set $\mathfrak{A}_s$ of $(\mathrm{P}_s(p, \lambda))$ on which the restricted extreme value function $\varphi_s(p, \lambda)$ is continuous and the restricted optimal set mapping ψ_s is u.s.c.-H; we further wish to determine whether the solubility set may be decomposed into such subsets, which we call *local stability sets*. These investigations are motivated by the fact that a mixed-integer or pure-integer linear optimization problem can only exhibit stable behaviour under perturbations of the coefficients on such a set.

The following is based on Sections 3.4 and 4.5 and the same notation is employed. The theorems given here are also valid if no integer conditions are imposed. One could therefore also consider $s = 0$, however since the results then obtaining are known and frequently hold in much sharper form we confine ourselves here to the case $0 < s \leqq n$, i.e. there is at least one integer variable, and refer the reader to Section 5.2 for the case $s = 0$.

We first characterize the feasible parameter set

$$\mathfrak{B}_s =_{\mathrm{Df}} \{\lambda \in E_m \mid G(\lambda) \neq \emptyset\} \tag{5.6.2}$$

and show that it can be decomposed into subsets on which the restricted constraint set mapping $G\colon E_m \to 2^{E_n}$ with images $G(\lambda)$ is continuous.

For convenience we decompose the matrices needed in the definition of $G(\lambda)$ according to its integer and arbitrarily real sub-matrices: If s, $0 < s \leqq n$, is the number of integer variables then the subvector z consists of the integer variables $x_1, \dots, x_s$ and the subvector y of the remaining variables $x_{s+1}, \dots, x_n$; the (m, s)-matrix B and the $(m, n - s)$-matrix C likewise partition the (m, n)-matrix A. The constraint set then has the form

$$G(\lambda) = \{x \in E_n \mid x = (z, y), Bz + Cy \leqq \lambda, z \in \mathbb{Z}^s\} . \tag{5.6.3}$$

As before $M(\lambda)$ denotes the convex polyhedron assigned to $G(\lambda)$,

$$M(\lambda) = \{x \in E_s \mid Ax \leqq \lambda\} \tag{5.6.4}$$

and for which the feasible parameter set $\mathfrak{B}_0$ is by Theorem 5.3.1 an m-dimensional cone with vertex set $\{Az \mid z \in E_n\}$. We obviously have

$$\mathfrak{B}_s \subset \mathfrak{B}_0 \tag{5.6.5}$$

for all s satisfying $0 \leqq s \leqq n$.

Theorem 5.6.1. *The feasible parameter set $\mathfrak{B}_s$ is for all s, $0 < s \leqq n$, a connected set equal to the union of a countable family of convex polyhedral cones each of which is ob-*

tained by a translation of the m-dimensional cone

$$K =_{\mathrm{Df}} \{\vartheta \in E_m \mid \vartheta \geqq Cy\} . \tag{5.6.6}$$

Proof. Let $G(g, \lambda)$ be the polyhedron defined in (3.4.16), i.e.

$$G(z, \lambda) = \{x \in E_n \mid x = (z, y), Cy \leqq \lambda - Bz\} \tag{5.6.7}$$

where $z \in Z^s$ and $\lambda \in E_m$ are fixed, we then have

$$G(z, \lambda) \neq \emptyset \Leftrightarrow \lambda - Bz \in K \tag{5.6.8}$$

from which

$$\mathfrak{B}_s = \bigcup_{z \in Z^s} (Bz + K) \tag{5.6.9}$$

follows since $G(\lambda) \neq \emptyset$ is the union of polyhedra $G(z, \lambda) \neq \emptyset$.

We must now demonstrate that $\mathfrak{B}_s$ is connected. Assume that the cone K possesses a face of smallest dimension (the proof would otherwise be complete). One easily confirms that

$$(Bz + K) \cap (B\bar{z} + K) \neq \emptyset \tag{5.6.10}$$

holds for any two points z and $\bar{z}$ of Z^s. If λ' and λ'' are arbitrary points contained in $\mathfrak{B}_s$ then the convexity of K and (5.6.10) imply the existence of a polygon connecting λ' and λ'' which is contained entirely in $\mathfrak{B}_s$. □

One may easily find examples to show that the feasible parameter set is in general not convex.

Theorem 5.6.2. *If the matrix $A = (B, C)$ has only rational elements then the feasible parameter set $\mathfrak{B}_s$ is closed for $0 < s \leqq n$.*

Proof. If $\{\lambda^t\} \subset \mathfrak{B}_s$ is an arbitrary sequence with the property $\lambda^t \to \lambda^0$ then one must show that $G(\lambda^0)$ is non-empty. By Theorem 3.4.2 there exists for each $t = 1, 2, \ldots$ a representation

$$G(\lambda^t) = K_G(\lambda^t) + U_G$$

where $K_G(\lambda^t)$ are compact sets (3.4.14) and U_G is given by (3.4.4). It obviously suffices to demonstrate $K_G(\lambda^0) \neq \emptyset$ which may be established as in the proof of Theorem 3.4.2. □

The rationality of the matrix A is essential for the validity of the last theorem as is illustrated by the example given in Section 3.4.

In order to ensure that the point-to-set mapping defined by $G(\lambda)$ (5.6.3) is continuous (or in particular l.s.c.-B) we must by Theorem 3.4.3 impose the condition (3.4.15)

$$P_z G(\lambda) = P_z G(\lambda^0) \qquad \forall \lambda \in \Lambda$$

where as in Section 3.4 P_z denotes the projection of the set $G(\lambda)$ onto the subspace spanned by the unit vectors corresponding to the integer subvector $z \in Z^s$.

We now wish to determine the subsets of $\mathfrak{B}_s$ on which this condition is satisfied. For convenience we define

$$T(\lambda) =_{\mathrm{Df}} P_z G(\lambda) \subset Z^s \tag{5.6.11}$$

for a fixed $\lambda \in \mathfrak{B}_s$. The projection may be finite or countable according to whether $G(\lambda)$ is bounded with respect to the integer variables or not.

Definition. Let $\lambda^0 \in \mathfrak{B}_s$ be a fixed parameter point, the set

$$Q(\lambda^0) =_{\mathrm{Df}} \{\lambda \in \mathfrak{B}_s \mid T(\lambda) = T(\lambda^0)\} \tag{5.6.12}$$

is then called a *local stability set* with respect to the parameter point $\lambda^0 \in \mathfrak{B}_s$.

The local stability set $Q(\lambda^0)$ contains λ^0 and is thus always non-empty; it may be represented as

$$Q(\lambda^0) = \bigcap_{z \in T(\lambda^0)} [Bz + K] \setminus \bigcup_{z \in \mathbb{Z}^s \setminus T(\lambda_0)} [Bz + K] \tag{5.6.12a}$$

since $T(\lambda) = T(\lambda^0)$ is equivalent to

$$G(z, \lambda) \neq \emptyset \qquad \forall z \in T(\lambda^0),$$

$$G(z, \lambda) = \emptyset \qquad \forall z \in \mathbb{Z}^s \setminus T(\lambda^0)$$

from which (5.6.12a) follows if (5.6.8) is taken into account.

In the following two lemmas we show that if the matrix A is rational then regardless of the cardinality of the projection $T(\lambda^0)$ there exist two finite sets $\Omega' \subset T(\lambda^0)$ and $\Omega'' \subset \mathbb{Z}^s \setminus T(\lambda^0)$ (if $T(\lambda^0) = \mathbb{Z}^s$ then $\Omega'' = \emptyset$) such that $Q(\lambda^0)$ may be represented in the form (5.6.12a) if Ω' and Ω'' are substituted for $T(\lambda^0)$ and $\mathbb{Z}^s \setminus T(\lambda^0)$ respectively.

Lemma 5.6.1. *Let the matrix $A = (B, C)$ be rational, $\lambda^0 \in \mathfrak{B}_s$ be an arbitrary fixed parameter point and s satisfy $0 < s \leqq n$ and be otherwise arbitrary. The set*

$$D(\lambda^0) =_{\mathrm{Df}} \bigcap_{z \in T(\lambda^0)} [Bz + K] \tag{5.6.13}$$

is then an m-dimensional convex polyhedral cone. Furthermore, there then exists a $\vartheta^0 \in \mathfrak{B}_s$ such that

$$D(\lambda^0) = \vartheta^0 + K \subset \mathfrak{B}_s \tag{5.6.14}$$

holds; K is the polyhedral cone given by (5.6.6).

Proof. We can assume that K has a face (of smallest dimension), since otherwise the lemma is trivial. If the projection $T(\lambda^0)$ is finite then the set $D(\lambda^0)$ defined by (5.6.13) is an m-dimensional convex polyhedral cone because of (5.6.10). Assume then that $T(\lambda^0)$ is not finite. The following is true for each subset $\Omega \subset T(\lambda^0)$

$$D(\lambda^0) \subset \bigcap_{z \in \Omega} [Bz + K]\,. \tag{5.6.15}$$

We show that a finite subset $\Omega' \subset T(\lambda^0)$ exists such that the inclusion (5.6.15) may be reversed.

Because of the rationality of A the set $G(\lambda^0)$ may be represented as

$$G(\lambda^0) = K_G(\lambda^0) + U_G \tag{5.6.16}$$

according to (3.4.12); $K_G(\lambda^0)$ and U_G are given by (3.4.14) and (3.4.4) respectively. The projection $\Omega' =_{\mathrm{Df}} P_z K_G(\lambda^0)$ of the compact set $K_G(\lambda^0)$ onto the subspace spanned by the integer variables then has the following properties

$$\Omega' \neq \emptyset\,, \qquad \Omega' \subset T(\lambda^0)\,, \qquad \Omega' \quad \text{finite}\,.$$

We now show that the reverse of (5.6.15) is true for $\Omega = \Omega'$. Assume that

$$\tilde{\lambda} \in \bigcap_{z \in \Omega'} [Bz + K]$$

is chosen arbitrarily, which by (5.6.8) implies

$$G(z, \tilde{\lambda}) \neq \emptyset \qquad \forall z \in \Omega'\,. \tag{5.6.17}$$

We must demonstrate that $\tilde{\lambda}$ is contained in $D(\lambda^0)$, which because of (5.6.13) and (5.6.8) is equivalent to

$$G(z, \tilde{\lambda}) \neq \emptyset \qquad \forall z \in T(\lambda^0) . \tag{5.6.18}$$

Now assume that $\tilde{z} \in T(\lambda^0)$ is chosen arbitrarily. If $\tilde{z} \in \Omega'$ then nothing remains to be said. Assume therefore $\tilde{z} \in T(\lambda^0) \setminus \Omega'$. By definition of $T(\lambda^0)$ (5.6.11) we have $G(\tilde{z}, \lambda^0) \neq \emptyset$. We then choose an $\tilde{x} = (\tilde{z}, \tilde{y})$ in $G(\tilde{z}, \lambda^0)$. From $G(\tilde{z}, \lambda^0) \subset G(\lambda^0)$ and (5.6.16) it follows that points $\bar{x} = (\bar{z}, \bar{y}) \in K_G(\lambda^0)$ and $\bar{u} = (\bar{v}, \bar{w}) \in U_G$ exist (u is partitioned in a like manner to x into v and w) such that

$$\tilde{x} = \bar{x} + \bar{u} , \qquad \tilde{z} = \bar{z} + \bar{v} , \qquad \bar{z} \in \Omega' , \tag{5.6.19}$$

hold. By (5.6.17) $G(\bar{z}, \tilde{\lambda})$ is non-empty and we may choose a point $\hat{x} = (\bar{z}, \hat{y}) \in G(\bar{z}, \tilde{\lambda})$. We then consider the point x defined as

$$x = (z, y) , \qquad x = \hat{x} + \bar{u} , \qquad z = \bar{z} + \bar{v} , \qquad y = \hat{y} + \bar{w} . \tag{5.6.20}$$

By the definitions of the polyhedron $G(z, \lambda)$ (5.6.7) and U_G (3.4.4) we have

$$C\hat{y} \leqq \tilde{\lambda} - B\bar{z} , \qquad C\bar{w} \leqq -B\bar{v}$$

from which

$$C(\hat{y} + \bar{w}) \leqq \tilde{\lambda} - B(\bar{z} + \bar{v})$$

follows. Together with (5.6.19) this however implies $z = \tilde{z}$, thus $x \in G(\tilde{z}, \tilde{\lambda})$, and consequently $G(\tilde{z}, \tilde{\lambda})$ is non-empty. Since $\tilde{z} \in T(\lambda^0)$ was arbitrary we have demonstrated the validity of (5.6.18). Hence

$$D(\lambda^0) = \bigcap_{z \in \Omega'} [Bz + K] .$$

As Ω' is finite it follows that $D(\lambda^0)$ is an m-dimensional convex polyhedral cone.

One now need only show that the equality (5.6.14) holds. K has a face of smallest dimension and thus each cone $Bz + K$ has a vertex set, which also applies to $D(\lambda^0)$. (5.6.14) holds for each point ϑ^0 contained in the latter. □

Lemma 5.6.2. *Let the matrix $A = (B, C)$ be arbitrary, s satisfy $0 < s \leqq n$ and otherwise be arbitrary and $Q(\lambda^0)$ be the local stability set corresponding to an arbitrary fixed parameter point $\lambda^0 \in \mathfrak{B}_s$. Furthermore, assume $T(\lambda^0) \neq Z^s$. There then exists a finite set $\Omega'' \subset Z^s \setminus T(\lambda^0)$ such that*

$$Q(\lambda^0) = D(\lambda^0) \setminus \bigcap_{z \in \Omega''} D^z \tag{5.6.21}$$

holds, where $D(\lambda^0)$ is the m-dimensional convex polyhedral cone in (5.6.13) *and*

$$D^z =_{\mathrm{Df}} Bz + K , \qquad z \in Z^s . \tag{5.6.22}$$

Proof. Because of (5.6.22) and (5.6.13) we may also write (5.6.12a) in the form

$$Q(\lambda^0) = \bigcap_{z \in Z^s \setminus T(\lambda^0)} (D(\lambda^0) \setminus D^z) . \tag{5.6.23}$$

Since $Q(\lambda^0)$ is non-empty

$$D(\lambda^0) \setminus D^z \neq \emptyset \qquad \forall z \in Z^s \setminus T(\lambda^0) \tag{5.6.24}$$

also holds, which implies $D(\lambda^0) \subset D^z$. Assume that $\vartheta^0 \in \mathfrak{B}_s$ is a point satisfying (5.6.14) and define

$$\tilde{D}(\lambda^0) =_{\mathrm{Df}} \vartheta^0 - K . \tag{5.6.25}$$

We then prove the relation

$$Bz \in \tilde{D}(\lambda^0) \Leftrightarrow D(\lambda^0) \subset D^z . \tag{5.6.26}$$

If $D(\lambda^0) \subset D^z$ then there exists a $\vartheta \in K$ such that

$$\vartheta^0 = Bz + \vartheta \Rightarrow Bz = \vartheta^0 - \vartheta \in \tilde{D}(\lambda^0) .$$

On the other hand $Bz \in \tilde{D}(\lambda^0)$ implies the existence of a $\vartheta \in K$ with the property

$$Bz = \vartheta^0 - \vartheta \Rightarrow \vartheta^0 \in D^z \Rightarrow D(\lambda^0) \subset D^z$$

thus proving (5.6.26).

Because of (5.6.24) and (5.6.26) we find that

$$\mathbb{Z}^s \setminus T(\lambda^0) \subset \mathbb{Z}^s \setminus \{z \in \mathbb{Z}^s \mid Bz \in \tilde{D}(\lambda^0)\} \tag{5.6.27}$$

must hold. The inclusion

$$\bigcap_{z \in \Omega''} (D(\lambda^0) \setminus D^z) \subset \bigcap_{z \in \mathbb{Z}^s \setminus T(\lambda^0)} (D(\lambda^0) \setminus D^z) \tag{5.6.28}$$

is valid for arbitrary $\Omega'' \subset \mathbb{Z}^s \setminus T(\lambda^0)$. Hence we need only demonstrate the existence of a finite set Ω'' contained in $\mathbb{Z}^s \setminus T(\lambda^0)$ and for which the reverse of (5.6.28) holds. Because of the rationality of the matrix C there exists a rational matrix F such that the polyhedral cone K in (5.6.6) has the form

$$K = \{\lambda \in E_m \mid F\lambda \geqq 0\} \tag{5.6.29}$$

which gives the representation

$$D(\lambda^0) = \{\lambda \in E_m \mid F\lambda \geqq f^0\} \tag{5.6.30}$$

for the cone $D(\lambda^0)$. Assume that F is a (p, m)-matrix such that (5.6.29) and (5.6.30) do not contain redundant inequalities. If we denote the rows of F by f^i, $i = 1, \dots, p$, then the hyperplanes

$$H_i = \{\lambda \in E_m \mid f^{i\mathsf{T}}\lambda = f_i^0\} , \qquad i = 1, \dots, p , \tag{5.6.31}$$

are at the same time boundary hyperplanes of $D(\lambda^0)$ and $\tilde{D}(\lambda^0)$. We define

$$\mathfrak{C} =_{\mathrm{Df}} \{\lambda \in E_m \mid \lambda = Bz, z \in \mathbb{Z}^s\} \setminus \tilde{D}(\lambda^0) \tag{5.6.32}$$

and consider for each $i = 1, \dots, p$ the sets

$$H_i^+ \cap \mathfrak{C} =_{\mathrm{Df}} \{\lambda \in E_m \mid f^{i\mathsf{T}}\lambda > f_i^0\} \cap \mathfrak{C} \tag{5.6.33}$$

of which only the non-empty sets are here of interest, therefore define

$$I =_{\mathrm{Df}} \{i \in \{1, \dots, p\} \mid H_i^+ \cap \mathfrak{C} \neq \emptyset\} \tag{5.6.34}$$

and let $i \in I$ be an arbitrary index. There then exists a $\lambda^i \in H_i^+ \cap \mathfrak{C}$ with the property

$$d(\lambda^i, H_i) = \min \{d(\lambda, H_i) \mid \lambda \in H_i^+ \cap \mathfrak{C}\} > 0 \tag{5.6.35}$$

as we now prove indirectly. Assume then that no $\lambda^i \in H_i^+ \cap \mathfrak{C}$ exists with the property (5.6.35). One can then however find a sequence $\{\lambda^t\} \subset H_i^+ \cap \mathfrak{C}$ satisfying

$$d(\lambda^t, H_i) \to \delta \geqq 0$$

which is equivalent to

$$|f^{i\mathsf{T}}\lambda^t - \tilde{f}_i| \to 0 \tag{5.6.36}$$

($\tilde{f}_i$ may be implied from δ and f_i^0). Because of (5.6.32) and (5.6.33) each λ^t has the form $\lambda^t = Bz^t$ where $z^t \in \mathbb{Z}^s$. We denote the elements of the matrix B by b_{jk} and the components of f^i by f^i_j. We further define

$$b^i_k =_{\mathrm{Df}} \sum_{j=1}^{m} f^i_j b_{jk} , \qquad i = 1, \dots , p , \qquad k = 1, \dots , s .$$

Relation (5.6.36) is then equivalent to

$$\forall \varepsilon > 0 \quad \exists t' : \left| \sum_{k=1}^{s} b^i_k z^t_k - \tilde{f}_i \right| < \varepsilon \qquad \forall t \geqq t'$$

where $z^t_k \in \mathbb{Z}$, $k = 1, \dots , s$, $t = 1, 2, \dots$ This however contradicts KRONECKER's approximation theorem (see for instance O. PERRON [1]) since B and F are rational which thus also applies to b^i_k. Consequently, there exists a $\lambda^t \in H^+_i \cap \mathfrak{C}$ satisfying (5.6.35).

We define

$$\mathfrak{C}_i =_{\mathrm{Df}} \{\lambda \in H^+_i \cap \mathfrak{C} \mid d(\lambda, H_i) = d(\lambda^i, H_i)\} .$$

Now, if there exists a $\tilde{\lambda} \in \mathfrak{C}_i$ such that $\tilde{\lambda} \notin D(\lambda^0)$ then we choose a $z \in \mathbb{Z}^s$ satisfying $\tilde{\lambda} = Bz$ and denote it by z^i. If on the other hand $\mathfrak{C}_i \subset D(\lambda^0)$ then we determine a $\hat{\lambda} \in \mathfrak{C}_i$ with the property

$$d(\hat{\lambda}, \vartheta^0) = \min \{d(\lambda, \vartheta^0) \mid \lambda \in \mathfrak{G}_i\} .$$

We now choose a $z \in \mathbb{Z}^s$ satisfying $\hat{\lambda} = Bz$ which we call z^i. The choice of z^i then gives in both coses

$$D(\lambda^0) \setminus D^{z^i} \subset D(\lambda^0) \setminus D^z \ \forall \ z \in \mathbb{Z}^s : \lambda = Bz \in H^+_i, \cap \mathfrak{C} . \tag{5.6.38}$$

If we proceed in this fashion for all $i \in I$ (5.6.34) and define

$$\Omega'' =_{\mathrm{Df}} \{z^i \mid i \in I\}$$

then Ω'' is obviously finite and (5.6.21) holds because of (5.6.38). □

Corollary 5.6.2.1. *If in Lemma* 5.6.2 *the condition that A is rational is deleted then in general a finite set Ω'' with the property* (5.6.21) *does not exist.*

Proof. This follows directly from the proof of Lemma 5.6.2 since KRONECKER's approximation theorem implies that parameter points λ^i contained in $H^+_i \cap \mathfrak{C}$, $i \in I$ arbitrary, may lie arbitrarily close to the hyperplane H_i. □

Before stating the following theorem we note the definition of a *star-shaped* subset of E_m: A subset $S \subset E_m$ is said to be star-shaped if there exists a point $z \in S$ such that

$$s(x, z) = \{y \in E_m \mid y = \alpha x + (1 - \alpha) z, \alpha \in (0, 1)\} \subset S \qquad \forall x \in S$$

holds.

Theorem 5.6.3. *If the matrix $A = (B, C)$ is rational and s is arbitrary apart from $0 < s \leqq n$ then the following statements are true:*

(1) *if $\lambda^0 \in \mathfrak{B}_s$ then the local stability set $Q(\lambda^0)$* (5.6.12) *has the properties*

(a) $\lambda \in Q(\lambda^0) \Rightarrow Q(\lambda) = Q(\lambda^0)$;

(b) $\lambda \in \mathfrak{B}_s \setminus Q(\lambda^0) \Rightarrow Q(\lambda) \cap Q(\lambda^0) = \emptyset$;

(c) *$Q(\lambda^0)$ is star-shaped and in general neither open nor closed;*

(2) *the feasible parameter set $\mathfrak{B}_s$ may be decomposed into a countable system of disjoint local stability sets $Q(\lambda)$.*

Proof. (1) Property (a) follows from the definition of $Q(\lambda^0)$. If (b) were false then there would exist a $\hat{\lambda} \in Q(\lambda) \cap Q(\lambda^0)$ which according to (5.6.12) implies $T(\hat{\lambda}) = T(\lambda) = T(\lambda^0)$. Since then $\lambda \in Q(\lambda^0)$ follows we have a contradiction. In order to prove (c) we take Lemmas 5.6.1 and 5.6.2 into account and write $Q(\lambda^0)$ in the form

$$Q(\lambda^0) = (\vartheta^0 + K) \setminus \bigcup_{z \in \Omega''} D^z .$$

We then show that the line segment $s(\lambda, \vartheta^0)$ is contained in $Q(\lambda^0)$ for arbitrary $\lambda \in Q(\lambda^0)$. We can again assume that K is a cone with a face of smallest dimension, since otherwise the proposition is trivial. First we demonstrate that

$$\vartheta^0 \in \lambda - K \tag{5.6.39}$$

and

$$(\lambda - K) \cap \bigcup_{z \in \Omega''} D^z = \emptyset \tag{5.6.40}$$

hold for arbitrary $\lambda \in Q(\lambda^0)$. This means that each cone $\lambda - K$ separates the point ϑ^0 of the vertex set of $D(\lambda^0)$ (5.6.14) from the set $\bigcup_{z \in \Omega''} D^z$. Each $\lambda \in Q(\lambda^0)$ is contained in $D(\lambda^0)$, hence there exists a $\vartheta \in K$ with the property $\lambda = \vartheta^0 + \vartheta$, and (5.6.39) follows. If (5.6.40) were false then one could find a $\tilde{z} \in \Omega''$ and a $\tilde{\vartheta} \in K$ such that $B\tilde{z} + \tilde{\vartheta} \in \lambda - K$. This would then imply the existence of a $\vartheta \in K$ satisfying $B\tilde{z} + \tilde{\vartheta} = \lambda - \vartheta$ which gives $\lambda = B\tilde{z} + \tilde{\vartheta} + \vartheta$ and thus $\lambda \in \bigcup_{z \in \Omega''} D^z$. This last contradicts $\lambda \in Q(\lambda^0)$, (5.6.40) is therefore correct. We next show that

$$Q(\lambda^0) \cap (\lambda - K) = D(\lambda^0) \cap (\lambda - K) \tag{5.6.41}$$

holds for arbitrary $\lambda \in Q(\lambda^0)$. From (5.6.21) one then has

$$Q(\lambda^0) \cap (\lambda - K) = [D(\lambda^0) \cap (\lambda - K)] \setminus \left[\left(\bigcup_{z \in \Omega''} D^z \cap (\lambda - K)\right) \cap D(\lambda^0)\right].$$

Since λ is contained in $Q(\lambda^0)$ (5.6.40) and thus (5.6.41) follow. Because of (5.6.39) we have

$$\vartheta^0 \in (\lambda - K) \cap D(\lambda^0) .$$

Since K contains the zero point of $\mathbb{E}_n$ we also find $\lambda \in (\lambda - K) \cap D(\lambda^0)$. The convexity of $(\lambda - K) \cap D(\lambda^0)$ implies that the line segment $s(\lambda, \vartheta^0)$ is contained in the same. Because of (5.6.41) $s(\lambda, \vartheta^0)$ lies entirely in $Q(\lambda^0)$.

The second part of (c) follows immediately from the representation (5.6.21) of $Q(\lambda^0)$ and the fact that λ^0 is always contained in $Q(\lambda^0)$.

(2) This follows from (b) of (1), Lemma 5.6.2 and the fact that the system of finite subsets of $\mathbb{Z}^s$ is countable. □

The following corollary follows immediately from Theorem 3.4.3 and the properties of the local stability set $Q(\lambda^0)$ corresponding to the parameter point λ^0.

Corollary 5.6.3.1. *If the matrix $A = (B, C)$ is rational and s, $0 < s \leqq n$, is arbitrary then the restriction $G|_{Q(\lambda^0)}$ of the point-to-set mapping defined by*

$$G(\lambda) = \{x \in \mathbb{E}_n \mid Ax \leqq \lambda, x_1, \dots, x_s \in \mathbb{Z}\}$$

is continuous.

Proof. See the preceding remarks. □

Theorem 5.6.4. *Let the matrix $A = (B, C)$ be rational and s, $0 < s \leqq n$, be arbitrary. The solubility set $\mathfrak{A}_s$ of $(\mathrm{P}_s(p, \lambda))$ is then non-empty, closed and connected, and*

$$\mathfrak{A}_s = -U^p \times \mathfrak{B}_s \tag{5.6.42}$$

holds, where U^p is the polar cone of the recession cone U in (3.4.3) and $\mathfrak{B}_s$ the feasible parameter set in (5.6.2).

Proof. Since $(P_s(0, 0))$ is soluble $\mathfrak{A}_s \neq \emptyset$ follows. We first show that, if A is rational,

$$U_G^p = U^p \tag{5.6.43}$$

holds for U_G of (3.4.3). Since $U_G \subset U$ holds for all s satisfying $0 < s \leqq n$ we need only demonstrate $U_G^p \subset U^p$. If $u \in U$ is chosen arbitrarily we may write

$$u = \sum_{i=0}^{n} \alpha_i u^i\,, \quad \alpha_i \geqq 0\,, \quad u^i \in U_G\,, \quad i = 0, \dots, n\,, \quad \sum_{i=0}^{n} \alpha_i = 1$$

because of the rationality of A (which implies conv $U_G = U$, cf. H. Noltemeier [1]). For an arbitrary $v \in U_G^p$ we then have

$$v^\top u = v^\top \left(\sum_{i=0}^{n} \alpha_i u^i \right) = \sum \alpha_i (v^\top u^i) \leqq 0$$

which verifies (5.6.43).

Now consider (5.6.42). If $(p, \lambda) \in \mathfrak{A}_s$ then $G(\lambda)$ is non-empty because this is the case for $\psi_s(p, \lambda)$, which implies $\lambda \in \mathfrak{B}_s$. For arbitrary $\bar{x} \in \psi_s(p, \lambda)$

$$p^\top \bar{x} \leqq p^\top x \qquad \forall x \in G(\lambda)$$

holds. Since the point $x + \alpha u$ is always contained in $G(\lambda)$ if $x \in G(\lambda)$, $\alpha \in Z_r$ and $u \in U_G$ are arbitrary, it follows that $p^\top u \geqq 0$ must hold for all $u \in U_G$ and this implies $p \in -U_G^p$. We have $\mathfrak{A}_s \subset -U^p \times \mathfrak{B}_s$ from (5.6.43). Assume now $(p, \lambda) \in -U^p \times \mathfrak{B}_s$. An arbitrary $\tilde{x} \in G(\lambda)$ may be represented as

$$\tilde{x} = \bar{x} + u\,, \qquad \bar{x} \in K_G(\lambda)\,, \qquad u \in U_G\,,$$

and we then have

$$p^\top \tilde{u} = p^\top \bar{x} + p^\top u \geqq p^\top \bar{x} \geqq \inf_{x \in K_G(\lambda)} p^\top x\,.$$

Since $K_G(\lambda)$ is non-empty and compact and is a subset of $G(\lambda)$ it follows that $\psi_s(p, \lambda) \neq \emptyset$ and thus $\mathfrak{A}_s \supset -U^p \times \mathfrak{B}_s$. Since U^p is a convex polyhedral cone the remainding assertions follow from Theorems 5.6.1 and 5.6.2. □

Theorem 5.6.5. *Let the matrix $A = (B, C)$ be rational and s, $0 < s \leqq n$, be arbitrary. The following statements are then true:*

(1) *The restriction of the extreme value function $\varphi_s|_{\mathfrak{A}_s}$ is continuous on the solubility set $\mathfrak{A}_s$;*

(2) *if $Q(\lambda^0)$ is the local stability set corresponding to an arbitrary fixed parameter point $\lambda^0 \in \mathfrak{B}_s$ then the restriction of the extreme value function $\varphi_s|_{-U^p \times Q(\lambda^0)}$ is continuous on the set $-U^p \times Q(\lambda^0)$, and likewise the restriction of the optimal set mapping $\psi_s|_{-U^p \times Q(\lambda^0)}$ is u.s.c.-H on $-U^p \times Q(\lambda^0)$.*

Proof. (1) follows immediately from Theorem 4.5.2 and (2) holds since the conditions of Corollary 4.5.5.1 are fulfilled. □

Note 5.6.1. The sets $-U^p \times Q(\lambda^0)$ are the largest subsets of the solubility set $\mathfrak{A}_s$ on which the problem $(P_s(p, \lambda))$ is stable (continuity of the extreme value function and upper semicontinuity (H) of the optimal set mapping). Because of Theorem 5.6.3 the stability sets $-U^p \times Q(\lambda^0)$ are star-shaped and in general neither open nor closed.

The solubility set $\mathfrak{A}_s$ may be decomposed into a countable system of disjoint stability sets $-U^p \times Q(\lambda^0)$.

Note 5.6.2. Additional difficulties arise in the case of quadratic parametric optimization problems with integer conditions as considered in Section 4.5, since little is known on the properties of the set on which the infimum is finite, in contrast to the linear integer case.

6. On Procedures for Analysing Parametric Optimization Problems

6.1. Introduction

It is well-known that there are many areas of application of parametric optimization, here we merely note the fields of vector optimization (see for example W. DINKELBACH [1], J. FOCKE [1], J. GUDDAT [5], and M. ZELENY [1]) and stochastic optimization (B. BEREANU [1], [2], [5], M. M. FABER [1], R.-R. REDETZKI [1], [2], and K. TAMMER [7], [8], [11]). The practical exploitation of this potential however involves the need for efficient procedures for the analysis of parameter-dependent optimization problems. It is first necessary to agree on what one expects from such solution procedures. Recalling the investigations presented in the preceding chapter one could, at least for certain classes of parametric optimization problems, consider the objective of calculating a finite decomposition or partitioning of the solubility set. We will see that although achieving such a goal is conceivable it is not always necessary or indeed possible. The conditions to be met by a solution procedure may be seen as depending on

(i) the theoretical results obtained on the structure of characteristic parameter sets of the problem and on the existence of a finite decomposition or partitioning,

(ii) available computing technology, and

(iii) the practical needs involved in the concrete application underlying the problem.

As already noted in the General Introduction and in the introductory section of Chapter 5, the original aim of parametric optimization is to find one or even all solutions, if such exist, corresponding to each parameter point. Since a parametric optimization problem constitutes a class of infinitely many optimization problems this aim may in general only be realized if one can divide this class into a finite number of subclasses such that for each subclass one may find the solution set in a closed form. It is for this reason that decomposition and partitioning theorems play such an important role. In Chapter 5 we observed that the current limits are given by the parametric linear complementarity problems dealt with there if one wishes to calculate a finite partitioning of the solubility set and the optimal sets $\psi(\lambda)$ corresponding to the partitioning sets, or likewise a linear selection function $\hat{x}(\lambda) \in \psi(\lambda)$ and the extreme value function $\varphi(\lambda)$. We first note that procedures for analysing fully parametric linear optimization problems with parameters in the objective function and the right-hand sides of the constraints may be found in F. NOŽIČKA et al. [2]. These procedures are not based on the simplex technique so that they are not affected by difficulties arising from a possible degeneracy, however other problems appear. Above all one should note that computing expense and storage capacity demands become extremely large in even not very high dimensioned problems, as was found by K. Wendler [2]. If instead of a fully parametric problem one wishes to solve a single-parametric or multi-parametric problem, which arise more frequently than the former in practical applications, then it is for this reason recommendable to avoid the indirect course involving the associated fully parametric problem. Most authors propose the use of the simplex technique for solving single-parametric or multi-parametric problems in both the linear and quadratic case as well as in linear complementarity problems (see for example W. Dinkelbach [1], T. Gál [4], [6], and C. van de Panne [3]). In Section 6.2 we present two procedures for solving single-parametric and multi-parametric problems which are based on this method. In principle these procedures are not new. They were however previously founded on the inner rules of the simplex method, whereas here we derive them to a large extent from the theoretical results obtained in Section 5.4 on the parametric problems to be solved. The first procedure is appropriate for solving multi-parametric problems. In the special case of linear parametric optimization with parameters in the objective function and the right-hand sides of the constraints this procedure has been known for some time (see for instance W. Dinkelbach [1], T. Gal [6]). The generalization to parameter-dependent convex quadratic optimization problems as well as parametric linear complementarity problems (this last however only in abbreviated form) may be found in C. van de Panne [3].

Using this procedure it is in principle possible to calculate the complete partitioning of the solubility set, however multi-parametric problems of large dimension will at the current state of computing technology probably prove intractable. This is however not necessary in many applications, thus in vector optimization and decision problems of stochastic optimization referred to above one is not interested in a partitioning as much as in finding solutions for particular parameter points in an efficient manner (taking theoretical results into account as far as possible). We therefore propose reducing the analysis of multi-parametric optimization problems to that of a sequence of single-parametric problems. This idea has already been realized in U. Kaussmann [1] for linear multi-parametric optimization problems, and now we give a brief description of the principle in our case.

We consider in the parameter space a finite number of given lines of the form

$$g(\lambda^0, \lambda^1) = \{\lambda \mid \lambda = \lambda^0 + t(\lambda^1 - \lambda^0), t \in R\} \tag{6.1.1}$$

where λ^0 and λ^1, $\lambda^0 \neq \lambda^1$,are fixed. Substituting (6.1.1) in the initial problem (P_λ) (General Introduction) we obtain a single-parametric problem $\left(\mathrm{P}_t(\lambda^0, \lambda^1)\right)$. The actual choice of the lines $g(\lambda^0, \lambda^1)$ in (6.1.1) depends on the concrete application. In particular possibilities are then opened of determining the pair of points (λ^0, λ^1) using a dialogue procedure. Using this approach we find that single-parametric problems are of central importance. Efficient algorithms for calculating a partitioning for single-parametric linear complementarity problems with a positive semidefinite matrix may also be given in cases of practical interest. The second procedure given in Section 6.2 is exclusively intended for single-parametric problems and the basic idea is due originally to P. Wolfe [1] who, in the third phase of his method for solving convex quadratic optimization problems, found it necessary to solve a special single-parametric quadratic auxiliary problem. A more suitable version with respect to the computing effort involved may be found in, for example, L. Collatz and W. Wetterling [1]. The same basic idea underlies the well-known procedure of C. E. Lemke [1] for solving linear complementarity problems, in which procedure a special single-parametric linear complementarity problem is involved in a subsidiary role. Other authors have shown that this procedure of LEMKE and consequently as its centrepiece the procedure for solving the auxiliary parametric problem always gives a solution in a larger class of matrices than that considered by LEMKE (B. C. EAVES [1], R. CHANDRASEKARAN [1], R. SAIGAL [1]). Detailed expositions are given by H. J. LÜTHI [1] and C. VAN DE PANNE [3]. This last monograph moreover contains a description of a procedure for solving single-parametric linear complementarity problems in general form.

A procedure which is also based on the simplex technique and is designed to solve certain single-parameter piece-wise linear complementarity problems is that of M. KOJIMA, N. NISHINO, and T. SEKINE [1], which also applies to a procedure of M. KOJIMA [2] which determines a piece-wise linear approximation for a continuously parameter-dependent stationary point of a not necessarily convex non-linear single-parametric optimization problem.

Proposals for dealing with single-parametric non-linear complementarity problems have been given by N. MEGIDDO [2]. We further note a procedure of I. ZSIGMOND [1] for linear problems with one parameter in the constraint matrix and for which a computer program exists, as well as an algorithm of A. M. GEOFFRION [2], [3] for solving strictly convex single-parameter problems entailing analytic functions. In the case of quadratic problems in which the parameter also occurs in the objective function matrix there also exist proposals for solution procedures (see J. C. G. BOOT [1], K. RITTER [1], J. GUDDAT and K. TAMMER [1], K. TAMMER [2], [10]).

Studies by F. GUERRA [1] confirm that at the present state of computing technology most of the last-named procedures are hardly suited to calculating a partitioning of the solubility set for larger problems. Under this aspect we find additional support for the thesis already expressed in Section 5.1 that a clear qualitative distinction exists between the parametric linear complementarity problems considered on the one hand and even quadratic problems with parameter-dependent objective function matrices on the other. In our opinion one must proceed in new directions in analysing classes which are more general than linear parametric complementarity problems. We can no longer set the objective of determining a selection function $\hat{x}(\lambda) \in \psi(\lambda)$ on the entire solubility set.

Following E. S. LEVITIN [2], A. V. FIACCO [2], R. L. ARMACOST and A. V. FIACCO [2], [3], M. LIPS [1], and S. M. ROBINSON [9] we investigate in Section 6.3 an approximate solution of a selection function. After considering the approaches employed by E. S. LEVITIN and A. V. FIACCO we extend the theory developed in the papers cited, in particular the hypotheses of some theorems could be weakened. Under certain conditions, in particular differentiability assumptions, an approximation of a selection function in the neighbourhood of a given parameter point λ^0 is reduced to the calculation of a selection function for a particular multi-parametric linear optimization problem. This problem is obtained by linearizing the initial problem at a pair of points (x^0, λ^0), where x^0 is a solution of the initial problem for the parameter point λ^0. It is also possible in this manner to employ the simplex method effectively, we however only obtain an approximation locally.

An interesting global approach, namely a piece-wise linear approximation of a selection function on the entire solubility set may be found in the paper of M. KOJIMA [2] cited earlier, the computing effort involved is however correspondingly greater.

We have attempted to make it clear that the concept of a solution of a parametric optimization problem depends on the problem class considered, the computing facilities available and the practical applications envisaged. For this reason we do not speak globally of solution procedures in parametric optimization but rather more generally of procedures for analysing parametric optimization problems.

6.2. Two Procedures for Solving Parameter-Dependent Linear Complementarity Problems

The first of the two procedures described here, we call it *Procedure* (*A*), enables one to analyse the multiparametric problem (5.4.3) dealt with theoretically in Section 5.4. For convenience we write $B = (-K, E)$, $z = (x, y)$, so that (5.4.3) then has the form:

Determine for each $t \in \mathbb{E}_d$ the set

$$\psi_d(t) = \{z \in M_d(t) \mid z_i \cdot z_{n+i} = 0, i = 1, \ldots, n\} \tag{6.2.1}$$

where $M_d(t) = \{z \in \mathbb{E}_{2n} \mid Bz = q(t), z \geqq 0\}$.

Here $q(t)$ is given by (5.4.3) and we assume as in Section 5.4 that the condition (5.4.4) is satisfied. We also assume

$$\dim \mathfrak{A}_d = d \tag{6.2.2}$$

where $\mathfrak{A}_d$ is the solubility set of (6.2.1).

The procedure yields successive basis solutions of the convex polyhedron $M_d(t)$ which are affine-linearly dependent on the parameter vector t and satisfy the complementarity condition

$$z_i \cdot z_{n+i} = 0 \,, \qquad i = 1, \ldots, n \,, \tag{6.2.3}$$

If z_{BV} and z_{NBV} denote the vector of the basis variables and that of the non-basis variables respectively of a basis solution satisfying (6.2.3) then as is well-known there exist a vector function G and an (n, n)-matrix H such that

$$z_{\mathrm{BV}} = G(t) - Hz_{\mathrm{NBV}} \tag{6.2.4}$$

is equivalent to the system $Bz = q(t)$. Now we introduce the convention that the components of z_{BV} and z_{NBV} are numbered in such a manner that one of the two variables z_i and z_{n+i} is the i-th component of z_{BV} and the other is the i-th component of z_{NBV}. The elements of the matrix H are denoted by h_{ij}, $i, j = 1, \ldots, n$. This matrix is characterized by properties given in the following lemma.

Lemma 6.2.1

(1) *If* $K \in \mathbb{H}^P(n)$ *then* $-H \in \mathbb{H}^P(n)$ *and in particular* $h_{ii} < 0$, $i = 1, \ldots, n$.

(2) *If* $K \in \widetilde{\mathbb{H}}^+(n)$ *then* $-H \in \widetilde{\mathbb{H}}^+(n)$ *and in particular* $h_{ii} \leqq 0$, $i = 1, \ldots, n$; *for each* i *for which* $h_{ii} = 0$ *holds* $h_{ij} + h_{ji} = 0$, $j = 1, \ldots, n$, *follows.*

Proof. For each basis solution of $M_d(t)$ satisfying (6.2.3) the matrix $-H$ in (6.2.4) may be obtained by carrying out a block exchange step with the matrix K where the pivot block is a regular submatrix of K arising from K by deleting the same rows and columns. Since the matrix classes $\mathbb{H}^P(n)$ and $\widetilde{\mathbb{H}}^+(n)$ are closed with respect to such operations (see for example C. E. Lemke [2]), it follows that $-H$ is contained in the same class as K. The other properties given are simple consequences of this fact. □

The i-th component of the vector function $G(t)$ in (6.2.4) has the form

$$G_i(t) = g_{i0} + \sum_{j=1}^{d} g_{ij} t_j .$$

The relation (6.2.4) may then be represented as a tableau:

$$\begin{array}{cc|c|ccc|ccc} & & & & & & & Z_{\mathrm{NBV}} & \\ \cline{3-9} & & 1 & t_1 & \dots & t_d & 1 & \dots & n \\ \hline & 1 & g_{10} & g_{11} & \cdots & g_{1d} & h_{11} & \cdots & h_{1n} \\ Z_{\mathrm{BV}} & \vdots & \vdots & \vdots & & \vdots & \vdots & & \vdots \\ & n & g_{n0} & g_{n1} & \cdots & g_{nd} & h_{n1} & \cdots & h_{nn} \end{array} \tag{6.2.5}$$

If we wish to differentiate between different basis solutions to the same vertex of $M_d(t)$ with the property (6.2.3) then we provide the quantities in (6.2.4) with an additional upper index. We are interested in the parameter set

$$T^r = \{t \mid G^r(t) \geqq 0\} \tag{6.2.6}$$

corresponding to each such basis solution $z^r(t)$. This set is obviously a convex polyhedron in E_d. If $T^1, \dots, T^S$ are those sets (6.2.6) of dimension d then they may be characterized as follows.

Lemma 6.2.2

(1) $T^r = \{t \mid z^r(t) \in \psi_d(t)\}$.

(2) $\mathfrak{A}_d = \bigcup_{r=1}^{S} T^r$.

(3) *For each set* T^r, $r = 1, \dots, S$, *there exists exactly one partitioning polyhedron* $\mathfrak{B}_d^l$ *of problem* (6.2.1) (i.e. $\mathfrak{B}_d^l$ is the closure of a local stability set $\mathfrak{A}_d^l$ of maximal dimension d) *with the property* $T^r \subset \mathfrak{B}_d^l$.

(4) *If the set* $T_0{}^r = \{t \mid G^r(t) > 0\}$ *is non-empty then in addition* $\psi_d(t) = \{z^r(t)\}$ *holds for* $t \in T_0{}^r$ *and* $T^r = \mathfrak{B}_d^l$.

Proof. (1) By definition of $z^r(t)$ we have $z_{\mathrm{BV}}^r(t) = G^r(t)$ and $z_{\mathrm{NBV}}^r(t) = 0$ and the first proposition follows.

(2) This follows from Theorem A.3.3 and Theorem A.3.5 and the closedness of the set $\mathfrak{A}_d$ and the sets T^r which latter are finite in number.

(3) This follows from a proposition on $\mathfrak{A}_d$ which corresponds to Theorem 5.4.3.

(4) If $T_0{}^r \neq \emptyset$ then $T_0{}^r = \operatorname{int} T^r \subset \operatorname{int} \mathfrak{B}_d^l$ obviously holds. In analogy to Theorem 5.4.3 the dimension of the set $\psi_d(t)$ is however equal to a constant number k if $t \in \operatorname{int} \mathfrak{B}_d^l$. Because of the complementarity condition (6.2.3) each basis solution in $\psi_d(t)$ is for $t \in \operatorname{int} \mathfrak{B}_d^l$ at least k-fold degenerate.

The basis solution $z^r(t)$ is however not degenerate for $t \in T_0{}^r$. It follows that the dimension k of $\psi_d(t)$ is equal to zero on $\operatorname{int} \mathfrak{B}_d^l$. The only element of $\psi_d(t)$ in $\operatorname{int} \mathfrak{B}_d^l$ can however only be the point $z^r(t)$. The relation $\mathfrak{B}_d^l = T^r$ finally follows from a proposition on problem (6.2.1) which is analogous to Theorem 5.4.4. □

We make the following assumption with respect to the solution procedure which follows.

All $(d-1)$-dimensional faces of all convex polyhedra T^r, $r = 1, \dots, S$, in (6.2.6) are non-degenerate , (6.2.7) where a $(d-1)$-dimensional face S^{d-1} of a convex polyhedron (6.2.6) is said to be *non-degenerate* if there exists one and only one index $i_0 \in \{1, \dots, n\}$ such that the representation

$$S^{d-1} = \{t \mid G^r_{i_0}(t) = 0, G^r_i(t) > 0, i \neq i_0\}$$

holds. We then call the inequality $G^r_{i_0}(t) \geqq 0$ corresponding to the index i_0 a *binding inequality* of the convex polyhedron T^r.

It obviously follows from the preceding constraint qualification that the sets $T_0{}^r$ are necessarily non-empty for $r = 1, \dots, S$. Hence proposition (4) of Lemma 6.2.2 is available and the sets T^r determined in the course of the procedure are identical to the partitioning polyhedra of problem (6.2.1). The propositions of Theorem 5.4.4 are also available in the present sense with the exception of (1'').

To initiate the procedure we need a starting tableau of the form (6.2.5) with a first basis solution $z^1(t)$ for which the set T^1 in (6.2.6) has the dimension d. Possibilities of generating such a tableau using the second of the two procedures presented here are given in Note 6.2.2. The basic idea of the following first procedure is as follows: First all parameter sets T^l (numbered successively) with the property dim $(T^1 \cap T^l) = d - 1$ are determined. Such parameter sets T^l are called to be *adjacent* to the set T^1. The intersection $T^1 \cap T^l$ is then either a $(d-1)$-dimensional face of both convex polyhedra or a part of $(d-1)$-dimensional faces of T^1 and T^l. In order to determine these adjacent sets one needs the binding inequalities of T^1 which, as one easily confirms, may in principle be found by solving suitable linear optimization problems (see for example M. ZELENY [1]). The computing effort increases rapidly with increasing n and d, we will however not be concerned with this here. If some of the binding inequalities describing the set T^1 are at the same time binding inequalities for the solubility set $\mathfrak{A}_d$ then these are also determined. The system of basis variables corresponding to each already determined set T^l is stored in order to ensure that this set is not unneccessarily calculated a second time. In an index set J we store the numbers of those previously determined sets T^r for which with certainty all adjacent parameter sets T^l have been found. If this index set is not yet identical to the index set of all sets calculated (which is initially the case) then a next index $r \notin J$ is chosen and all sets T^l adjacent to T^r and further possible binding inequalities for the solubility set $\mathfrak{A}_d$ are determined. The procedure is terminated if all adjacent sets T^l of all previously calculated sets T^r have been determined. In this case one has found all partitioning polyhedra T^l as well as a complete description of the solubility set $\mathfrak{A}_d$.

In presenting the procedure we employ various additional quantities. Thus we define the matrices $G = (g_{ij})_{i=1,\dots,n;j=0,\dots,d}$ or $G^0 = (g^0_{ij})_{i=1,\dots,n;j=0,\dots,d}$ corresponding to a simplex tableau (6.2.5). By b_{li} we denote the index of the i-th basis variable of the l-th basis solution $z^l(t)$. We also use the vector B^0 formed by the elements b^0_i $i = 1, \dots, n$, for temporary storage of the numbers of the basis variables of the basis solutions generated in the course of the procedure.

Procedure (A) (multiparametric linear complementarity problems)

Step 0

Read n, d, B^0, G, H.

Print B^0, G.

Set $r := 1, l := 1, J = \{1\}, b_{1i} := b^0_i, i = 1, \dots, n$.

Step 1

Determine the index set $I_0 \subset \{1, \dots, n\}$ as the set of indices of the binding inequalities of the convex polyhedron given by $g_{i0} + \sum_{j=1}^{d} g_{ij}t_j \geqq 0,\ i = 1, \dots, n$.
Set $k := 1$.

Step 2

If $k > n$ then go to Step 10.
If $k \in I_0$ then go to Step 4.

Step 3

Set $k := k + 1$ and go to Step 2.

Step 4

If $h_{kk} = 0$ then go to Step 5.
Determine for $i = 1, \dots, n$

$$b_i^0 := \begin{cases} b_{ri} & \text{if} \quad i \neq k\,, \\ n + k & \text{if} \quad i = k \quad \text{and} \quad b_{rk} = k\,, \\ k & \text{if} \quad i = k \quad \text{and} \quad b_{rk} = n + k\,. \end{cases}$$

If $b_i^0 = b_{pi},\ i = 1, \dots, n$, for any $p = 1, \dots, l$, then set $k := k + 1$ and go to Step 2.
Calculate G^0 from (G, H) by pivotization with the pivot element h_{kk}.
Set $l := l + 1,\ b_{li} := b_i^0,\ i = 1, \dots, n$, and print B^0, G^0.
Set $k := k + 1$ and go to Step 2.

Step 5

If $h_{kj} \geqq 0,\ j = 1, \dots, n$, then print "$g_{k0}, g_{k1}, \dots, g_{kd}$ — binding inequality of the solubility set".
Set $k := k + 1$ and go to Step 2.

Step 6

Determine the set $I_1 = \{j \mid h_{kj} < 0\}$ and the index set $I_2 \subset I_1$ as the set of indices of the inequalities of type

$$g_{i0} + \sum_{j=1}^{d} g_{ij}t_j - h_{ik}t_0 \geqq 0\,, \qquad i \in I_1\,,$$

which are binding for the convex polyhedron

$$\begin{aligned} P = \Bigl\{(t_0, t) \in E_{d+1} \mid g_{k0} + \sum_{j=1}^{d} g_{kj}t_j = 0\,, \qquad & t_0 \geqq 0\,, \\ g_{i0} + \sum_{j=1}^{d} g_{ij}t_j - h_{ik}t_0 \geqq 0\,, \qquad & i \in I_1\,, \\ g_{i0} + \sum_{j=1}^{d} g_{ij}t_j \geqq 0\,, \qquad & i \notin I_1\,, \qquad i \neq k\Bigr\}\,. \end{aligned}$$

Set $q := 1$.

Step 7

If $q > n$ then set $k := k + 1$ and go to Step 2.
If $q \in I_2$ then go to Step 9.

Step 8

Set $q := q + 1$ and go to Step 7.

Step 9

Determine for $i = 1, \dots, n$

$$b_i^0 := \begin{cases} b_{ri} & \text{if} \quad i \neq k \quad \text{and} \quad i \neq q\,, \\ n+k & \text{if} \quad i = k \quad \text{and} \quad b_{rk} = k\,, \\ k & \text{if} \quad i = k \quad \text{and} \quad b_{rk} = n+k\,, \\ n+q & \text{if} \quad i = q \quad \text{and} \quad b_{rq} = q\,, \\ q & \text{if} \quad i = q \quad \text{and} \quad b_{rq} = n+q\,. \end{cases}$$

If $b_i^0 = b_{pi}$, $i = 1, \dots, n$, for any $p = 1, \dots, l$, then set $q := q + 1$ and go to Step 7.

Calculate G^0 from (G, H) by pivotization with the pivot block $\begin{pmatrix} h_{kk} & h_{kq} \\ h_{qk} & h_{qq} \end{pmatrix}$.
Set $l := l + 1$, $b_{li} := b_i^0$, $i = 1, \dots, n$, and print B^0, G^0.
Set $q := q + 1$ and got to Step 7.

Step 10

If $J = \{1, \dots, l\}$ then terminate the procedure. Otherwise choose an $r_0 \in \{1, \dots, l\} \setminus J$.
Determine (G, H) by pivotization from the previous matrix (G, H), the pivot rows and columns are these rows and columns respectively of the matrix H whose numbers are given by the indices i for which $b_{ri} \neq b_{r_0 i}$.
Set $J := J \cup \{r_0\}$, $r := r_0$ and go to Step 1.

The following lemma provides the basis for the method of determining all partitioning polyhedra adjacent to a given partitioning polyhedron.

Lemma 6.2.3. *Assume that condition* (6.2.7) *is fulfilled and* (6.2.5) *is the simplex tableau corresponding to a given basis solution* $z^r(t)$ *with* $\dim T^r = d$. *Further, let* S^{d-1} *be a* $(d-1)$*-dimensional face of the set* T^r *with the corresponding binding inequality* $g_{k0} + \sum_{j=1}^{d} g_{kj}t_j \geqq 0$, *the k-th inequality in* (6.2.6).

(1) *If* $h_{kj} \geqq 0$, $j = 1, \dots, n$, *then for all t satisfying* $g_{k0} + \sum_{j=1}^{d} g_{kj}t_j < 0$ $\psi_d(t)$ *is empty and the k-th inequality in* (6.2.6) *is a binding inequality in the description of the solubility set* $\mathfrak{A}_d$.

(2) *If* $h_{kk} < 0$ *then there exists a uniquely determined basis solution* $z^l(t)$ *with the property* (6.2.3) *such that* $\dim (T^r \cap T^l) = d - 1$ *and* $T^r \cap T^l \subset S^{d-1}$ *hold. This basis solution* $z^l(t)$ *and the corresponding simplex tableau may be determined from* (6.2.5) *by exactly one simplex step with the element* h_{kk} *as pivot element. In this case we even have* $T^r \cap T^l = S^{d-1}$.

(3) *If* $h_{kk} = 0$ *and* $I_1 = \{j \mid h_{kj} < 0\} \neq \emptyset$ *then there exists exactly the same number of basis solutions* $z^l(t)$ *with the property* (6.2.3), $T^r \cap T^l \subset S^{d-1}$ *and* $\dim (T^r \cap T^l) = d - 1$, *as there are binding inequalities of type*

$$g_{i0} + \sum_{j=1}^{d} g_{ij}t_j - h_{ik}t_0 \geqq 0\,, \qquad i \in I_1\,,$$

for the convex polyhedron P (see Step 6). *If* I_2 *is the index set of these binding inequalities then each of the corresponding basis solutions* $z^l(t)$ *and the corresponding simplex tableau may be obtained by carrying out a block exchange step with the pivot matrix* $\begin{pmatrix} h_{kk} & h_{kq} \\ h_{qk} & h_{qq} \end{pmatrix}$ *for a* $q \in I_2$ *and starting with* (6.2.5).

Proof. (1) The system of equations (6.2.4) is well-known to be equivalent to the system $Bz = q(t)$ in (6.2.1). If for some t the inequality $g_{k0} + \sum_{j=1}^{d} g_{kj}t_j < 0$ holds, then under the condition $h_{kj} \geqq 0, j = 1, \dots, n$, the equation

$$(z_{\mathrm{BV}})_k = g_{k0} + \sum_{j=1}^{d} g_{kj}t_j - \left(\sum_{j=1}^{n} h_{kj}(z_{\mathrm{NBV}})_j\right)$$

is obviously not soluble in non-negative variables $z_{\mathrm{BV}}, z_{\mathrm{NBV}}$, so that the set $M_d(t)$ in (6.2.1) is empty, i.e. $t \notin \mathfrak{A}_d$. Since $T^r \subset \mathfrak{A}_d$ and the k-th inequality in (6.2.6) is binding for T^r it follows that this inequality is also a binding condition for $\mathfrak{A}_d$.

(2) Because of condition (6.2.7) we have

$$S^{d-1} = \left\{t \mid g_{k0} + \sum_{j=1}^{d} g_{kj}t_j = 0, g_{i0} + \sum_{j=1}^{d} g_{ij}t_j < 0, i \neq k\right\}.$$

Thus in order to obtain a set T^l adjacent to T^r satisfying dim $(T^r \cap T^l) = d - 1$ and $T^r \cap T^l \subset S^{d-1}$ with a basis solution $z^l(t)$ which is feasible in T^l, the k-th row in (6.2.5) must be employed as pivot row. Since $h_{kk} < 0$ the k-th column of H can be used as pivot column, condition (6.2.3) is thus fulfilled for the new basis solution $z^l(t)$.

By the rules of the simplex method the new basis solution $z^l(t)$ has even the property $T^r \cap T^l = S^{d-1}$. Since by Lemma 6.2.2 (4) the sets T^r and T^l are partitioning polyhedra of problem (6.2.1) there can only be one set T^l satisfying $T^r \cap T^l \subset S^{d-1}$ and dim $(T^r \cap T^l) = d - 1$.

(3) As in the preceding part of the proof the k-th row in (6.2.5) must necessarily function as pivot row. However since $h_{kk} = 0$ this element can not be used as a pivot element, one must thus carry out at least two simplex steps or a block exchange step with a pivot matrix of order (2×2) to obtain a basis solution $z^l(t)$ that satisfies (6.2.3). One therefore first takes the k-th row of (6.2.5) and chooses a suitable column q for a simplex step, and likewise in the second simplex step the q-th row and the k-th column. From Lemma 6.2.1 we know that the condition $h_{kj} + h_{jk} = 0$ is fulfilled for all $j = 1, \dots, n$. In order to obtain a basis solution $z^l(t)$ which is feasible in a set T^l (with the given properties) after carrying out the above two simplex steps we find that by the rules of the simplex method only the elements of the index set I_1 come in question for q and moreover only those $q \in I_1$ for which in addition a $(d - 1)$-dimensional subset of S^{d-1} exists on which subset the quotient $\left(g_{q0} + \sum_{j=1}^{d} g_{qj}t_j\right)\Big/h_{qk}$ is minimal with respect to $\left(g_{i0} + \sum_{j=1}^{d} g_{ij}t_j\right)\Big/h_{ik}$, $i \in I_1$. This subset of S^{d-1} however is always just the set $T^r \cap T^l$. Consequently $T^r \cap \bigcup_l T^l = S^{d-1}$ where the union is taken over all those basis solutions $z^l(t)$ generated by the rules given with a $q \in I_2$. As in the second part of the proof it follows from condition (6.2.7) that these sets T^l are the only sets with the properties stipulated. □

Note 6.2.1

1. Some essential simplifications follow for single-parametric problems ($d = 1$). Since there are only two directions leading from T^1 on which the parameter can vary no set T^l can be calculated twice, the operations involving the index set J are there-

fore superfluous. The index set I_0 in Step 1 can obviously be determined in a very simple manner (if $d = 2$ then I_0 may still be determined with much less difficulty than in the general case, see K. TAMMER [4]). One also notes that this set contains not more than two elements. The relatively complicated sixth step may be simplified in as much as S^{d-1} contains only one element and the index set I_2 can only have exactly one element.

2. If $K \in \mathbb{H}^P(n)$ then we have the second case in Lemma 6.2.3 so that Steps 5 to 9 may be deleted (and in particular the complicated sixth step).

3. If $K \in \mathbb{H}^A(n)$ then this also holds for the matrices H in (6.2.4), which implies a reduction of storage capacity needed. Particularly extensive simplifications are possible in the case of linear optimization problems with parameters in the objective function and the right-hand sides of the constraints. We then have

$$K = \begin{pmatrix} 0 & -A^{\mathrm{T}} \\ A & 0 \end{pmatrix}$$

so that the matrices H in (6.2.4) and (6.2.5) have the same structure

$$H = \begin{pmatrix} 0 & -D^{\mathrm{T}} \\ D & 0 \end{pmatrix}.$$

If we partition the matrix G in (6.2.5) accordingly, $G = \begin{pmatrix} G_1 \\ G_2 \end{pmatrix}$, then instead of the entire matrix (G, H) we need only use $\begin{array}{|c:c|} G_1 & B \\ \hline & G_2 \end{array}$ if the calculation rules are adapted to the condensed notation. We then obtain the procedure of solving linear parametric optimization problems, which has been known for some time.

4. The complicated third case of Lemma 6.2.3 does not arise in quadratic optimization problems with parameters in the right-hand sides of the constraints alone, Steps 6 to 9 may therefore be omitted.

5. Apart from the antisymmetric case in which the equation $h_{kk} = 0$ is automatically satisfied the zero test in Step 4 will lead to numerical difficulties in machine calculations because of unavoidable rounding errors. One can therefore only consider a condition of the form $|h_{kk}| < \varepsilon$ with some given sufficiently small $\varepsilon > 0$. Since the actual value of ε can only be found by trial and error we cannot disregard the possibility that errors arise at this point which then considerably affect the calculations that follow.

6. In the present form the procedure is designed to calculate all partitioning polyhedra. However small modifications may be introduced to prematurely terminate the procedure if enough information has already been obtained at some point of operation. If enough storage capacity is available then one may store not only the system of basis variables but also the entire simplex tableau, thus dispensing with the need of repeated calculation as in Step 10.

7. In stating the method many authors apparently assume that the constraint qualification (6.2.7) is always fulfilled. This is however not the case as simple examples show, and this fact must be taken into account. One also needs to know what phenomena are to be expected if degeneracy occurs. In this case certain parameter sets T^r may be merely proper subsets of partitioning polyhedra and the solution set $\psi_d(t)$ may contain more than one element on the set int T^r. Above all however the rules for determining adjacent sets can not be established in the simple manner given by Lemma 6.2.3. Theoretically the constraint qualification may always be satisfied by

subjecting the data of problem (6.2.1) to certain arbitrarily small variations (cf. B. SCHMIDT [1]). In practice however one should use the given procedure as far as possible and only undertake further investigations in parameter regions in which difficulties arise because of degeneracy. This is reasonable because of the fact that degeneracy as a rule only occurs in a limited number of parameters sets T^r and on the other hand because the procedure need not necessarily fail in the degenerate case. One should also consider that additional investigations require a much more effort and should therefore be restricted to cases in which they are absolutely necessary.

Now we turn to *Procedure* (*B*) which describes a method for solving single-parametric linear complementarity problems. We again write the problem to be solved (5.4.28) in the form (6.2.1) where now of course $q(t) = q^0 + q^1 t$ because of $d = 1$. This problem was dealt with theoretically in much detail in Section 5.4. We use the notation of that section and in particular represent N and G_1 of (5.4.33) and (5.4.34) in the form

$$N = \{(z, t) \in E_{2n+1} \mid Bz - q^1 t = q^0, z \geqq 0\} , \qquad (6.2.8)$$

$$G_1 = \{(z, t) \in N \mid z_i \cdot z_{n+i} = 0, i = 1, \dots, n\} . \qquad (6.2.9)$$

The basic idea of the procedure is as follows. As starting point we take an arbitrary feasible basis solution (z^0, t^0) of the convex polyhedron N which is contained in the set G_1 and a corresponding simplex tableau. Using the simplex method (and taking the fact into account that the sign of the variable t is not restricted) we solve in arbitrary order the linear optimization problems

$$\min \{t \mid (z, t) \in N\} , \qquad (6.2.10)$$

$$\max \{t \mid (z, t) \in N\} \qquad (6.2.11)$$

under the additional condition that the edge polygon generated in this way is contained in the set G_1. In solving these linear problems we are not only interested in the solutions of (6.2.10) and (6.2.11) or in the result that t is not upper or lower bounded on N, but also in obtaining all feasible basis solutions found in the course of the procedure. In the manner to be described we generate an edge polygon of N as characterized by Corollary 5.4.4 and which is contained in G_1, the graph of the solution set mapping ψ_1 of the problem considered.

Before describing the procedure we first need the following notation. Consider an arbitrary basis solution of N which is contained in G_1, the corresponding simplex tableau is then assumed to have the form

		z_{NBV}	
	d_{10}	d_{11}	$\dots\ d_{1,n+1}$
z_{BV}	$\vdots$	$\vdots$	$\vdots$
	$d_{n-1,0}$	$d_{n-1,1}$	$\dots\ d_{n-1,n+1}$
t	d_{n0}	d_{n1}	$\dots\ d_{n,n+1}$

(6.2.12)

Since the variable t is always one of the basis variables the vector z_{BV} has exactly $n - 1$ elements and the vector z_{NBV} exactly $n + 1$, and there exists one and only one index $i_0 \in \{1, \dots, n\}$ such that z_{i_0} and z_{n+i_0} are both non-basis variables (see Lemma 5.4.3).

$D = (d_{ij})_{i=1,\dots,n;j=0,\dots,n+1}$ denotes the matrix of the elements of (6.2.12), b_i $i = 1, \dots, n-1$, the indices of the elements of the vector z_{BV} and $c_j, j = 1, \dots, n+1$ the indices of the elements of the vector z_{NBV}.

As initial information we need the simplex tableau D^0 corresponding to the initial solution (z^0, t^0), the corresponding vector B^0 of the indices of the basis variables z_i, the vector C^0 of the indices of the non-basis variables z_i, and finally the uniquely determined column numbers $j_1, j_2 \in \{1, \dots, n+1\}$ with $c_{j1}^0 = i_0$ and $c_{j2}^0 = n + i_0$, where i_0 is given above.

Procedure (B) (single-parametric linear complementarity problems)

Step 0

Read D^0, B^0, C^0, j_1, j_2.
Print $b_1^0, b_2^0, \dots, b_{n-1}^0$,
$d_{10}^0, d_{20}^0, \dots, d_{n-1,0}^0, d_{n0}^0$.
Set $k := 1$.

Step 1

Set $D := D^0, B := B^0, C := C^0, j^* := j_k$.

Step 2

If $d_{ij} \leqq 0, j = 1, \dots, n-1$, then go to Step 4.
Determine an index $i^* \in \{1, \dots, n-1\}$ satisfying

$$\frac{d_{i^*0}}{d_{i^*j^*}} = \min_{d_{ij^*}>0} \frac{d_{i0}}{d_{ij^*}}.$$

(If i^* is not unique and one expects a cycle, then use a subsidiary rule given in F. Nožička et al. [1].)

Step 3

Determine the new matrix D by pivotization of the previous matrix D with the pivot element $d_{i^*j^*}$.
Set $g := b_{i^*}, b_{i^*} := c_{j^*}, c_{j^*} := g$.
Print $b_1^0, b_2^0, \dots, b_{n-1}^0$,
$d_{10}^0, d_{20}^0, \dots, d_{n-1,0}^0, d_{n0}^0$.
If $g \leqq n$ then determine j^* from the condition $c_{j^*} = g + n$.
If $g > n$ then determine j^* from the condition $c_{j^*} = g - n$.
Go to Step 2.

Step 4

Print "$c_{j^*}, d_{1j^*}, \dots, d_{nj^*}$ — an unbounded edge".
If $k = 1$ then set $k := 2$ and go to Step 1.
If $k = 2$ then terminate the procedure.

Note 6.2.2

1. If the convex polyhedron N has only regular vertices then Procedure (B) is a direct consequence of Corollary 5.4.5. According to F. Nožička et al. [1] the regularity of all vertices may be achieved by suitably chosen, arbitrarily small variations of the vector q^0, it also follows from the same that the procedure also arrives at a solution even if an additional ε-variation is not carried out but instead only a subsidiary rule for determining the index i^* in Step 2 is employed. However since a cycle hardly ever occurs in practical applications of the simplex method this subsidiary rule was only referred to in brackets in Step 2.

2. B. C. EAVES [2], R. CHANDRASEKARAN [1], and R. SAIGAL [1] have shown that the procedure given here not only leads to a solution for arbitrary initial solutions in the case $K \in \tilde{\mathbb{H}}^+(n) \cup \mathbb{H}^P(n)$ but also even in the case $K \in \mathbb{H}^L(n) \cup \mathbb{H}^Z(n)$. However simple examples show that the graph G_1 of the mapping ψ_1 is not connected for these classes of matrices. The solution set $\psi_1(t)$ is also not necessarily connected or moreover convex. The fact that the procedure functions however indicates that the set G_1 must consist of a finite number of subsets for which in a corresponding sense Corollary 5.4.4.1 holds. Even if all vertices of the convex polyhedron N are regular one of course cannot expect that the procedure will enable one to determine the entire set $\psi_1(t)$ for each t as is the case for the matrix class $\tilde{\mathbb{H}}^+(n) \cup \mathbb{H}^P(n)$.

3. One needs a starting solution including the associated starting tableau for each of the procedures described here. Such a simplex tableau can be generated by Procedure (B) if problem (6.2.1) is solved for some fixed $t^0 \in \mathfrak{A}_d$ or $t^0 \in \mathfrak{A}_1$ where additional columns expressing the dependence of the right-hand sides on the parameter t must be included in the tableau. Apart from including these additional columns this way of constructing a starting tableau corresponds to the application of parametric optimization to the solution of fixed linear complementarity problems originally proposed by C. E. LEMKE [1].

We define the vector 1_q by

$$(1_q)_i = \begin{cases} 0 & \text{if} \quad q_i(t^0) \geqq 0 , \\ 1 & \text{if} \quad q_i(t^0) < 0 \end{cases}$$

and consider the single-parametric problem:

Determine for all $s \geqq 0$ the set

$$\psi_1(s) = \{(x, y) \in M_1(s) \mid x^\mathsf{T} y = 0\}$$

$$\text{where} \quad M_1(s) = \{(x, y) \in \mathbb{E}_{2n} \mid -Kx + y = q(t^0) + 1_q s, x \geqq 0, y \geqq 0\} . \tag{6.2.13}$$

The advantage of the thus constructed auxiliary problem consists of the fact that one immediately has a starting solution given by $x = 0$ and $y = q(t^0) + 1_q s$ which is feasible for $s \geqq -\min_i q_i(t^0)$. We now take s to be an additional variable and construct the tableau

		s	x	1	$t_1 \dots t_d$
y	$q(t^0)$	-1_q	$-K$	q^0	$q^1 \dots q^d$

(6.2.14)

and from this obtain a starting tableau for Procedure (B) to solve (6.2.13) by replacing the variable s with a simplex step by one of the variables $y_{i\bullet}$ for which $q_{i\bullet}(t_0) = \min q_i(t^0)$ holds. Starting with this initial tableau we minimize the value of s subject to the additional condition $s \geqq 0$ in accordance with problem (6.2.10). If $t^0 \in \mathfrak{A}_d$ was chosen then the minimal value of s will be equal to zero, and the end tableau (the variable s must be a non-basis variable in the last step) without the column corresponding to s is then an initial tableau for Procedure (A) to solve (6.2.1) providing the corresponding parameter set T^1 has the dimension d, which however is the case if condition (6.2.7) is satisfied and t^0 is not by chance a boundary point of a partitioning polyhedron.

In order to obtain a starting tableau for Procedure (B) from the above end tableau to solve (6.2.1) in the case $d = 1$ the parameter t must be considered as an additional variable, the column corresponding to t must be multiplied by -1 and the variable t likewise exchanged for one of the previous basis variables according to the rules of the simplex method.

6.3. On the Approximative Calculation of Selection Functions for Non-Linear Parametric Optimization Problems with Differentiability Properties

The results presented here have been sufficiently motivated in Section 6.1. We consider the following non-linear parametric optimization problem

$$(\mathrm{P}_\lambda) \qquad \min \{f(x, \lambda) \mid x \in M(\lambda)\}, \qquad \lambda \in \Lambda, \tag{6.3.1}$$

where

$$M(\lambda) = \{x \in \mathbb{E}_n \mid g_i(x, \lambda) \leqq 0, i = 1, \dots, m, h_j(x, \lambda) = 0, j = 1, \dots, p\}, \tag{6.3.2}$$

$f, g_i, i = 1, \dots, m, h_j, j = 1, \dots, p$, are mappings of the space $\mathbb{E}_n \times \mathbb{E}_k$ into the set of real numbers and Λ is a subset of the solubility set $\mathfrak{A}$. Our aim is to approximate a selection function $\bar{x}(\lambda) \in \psi(\lambda)$ at a point $\lambda^0 \in \Lambda$. In the following we give two definitions of an approximate selection function, the first of which is based on work in A. V. Fiacco [3], R. L. Armacost and A. V. Fiacco [2], S. M. Robinson [9]. The second is related to work by E. S. Levitin [2].

Definition of an approximate selection function. A function x which is defined on a neighbourhood $V\{\lambda^0\} \subset \Lambda$ is called an *approximate selection function in the sense of Fiacco at* λ^0 (or briefly *$\mathbb{F}$-function at* λ^0) if

$$d(x(\lambda), \psi(\lambda)) = o(\|\lambda - \lambda^0\|) \tag{6.3.3}$$

holds, i.e.

$$\lim_{\lambda^k \to \lambda^0} \frac{d(x(\lambda^k), \psi(\lambda^k))}{\|\lambda^k - \lambda^0\|} = 0 \qquad \forall \{\lambda^k\} \subset V\{\lambda^0\}, \qquad \lambda^k \to \lambda^0.$$

A mapping $x\colon V\{\lambda^0\} \to \mathbb{E}_n$ is called an *approximate selection function in the sense of Levitin at* λ^0 (or briefly *$\mathbb{L}$-function at* λ^0) if

$$d(x(\lambda), M(\lambda)) = o(\|\lambda - \lambda^0\|) \tag{6.3.4}$$

and

$$|\varphi(\lambda) - f(x(\lambda), \lambda)| = o(\|\lambda - \lambda^0\|) \tag{6.3.5}$$

hold. Condition (6.3.3) characterizes an $\mathbb{F}$-function at λ_0 in an equivalent manner as a mapping which approximates a selection function $\bar{x}(\lambda) \in \psi(\lambda)$ with an error of at most the order of $o(\|\lambda - \lambda^0\|)$. The existence of $\bar{x}$ is assured.

In order to determine whether or under what conditions an $\mathbb{L}$-function also approximates a selection function of the optimal set mapping one must examine the relation of these two concepts to one another. It then appears that the following concepts may also be introduced: A mapping $x\colon V\{\lambda^0\} \to \mathbb{E}_n$ is a *weak $\mathbb{F}$-function* at λ^0 if for each sequence

$$\{\lambda^k\} \subset V\{\lambda^0\}, \qquad \lambda^k \to \lambda^0,$$

$$\lim_{\lambda^k \to \lambda_0} d(x(\lambda^k), \psi(\lambda^k)) = 0 \tag{6.3.6}$$

holds. If instead of (6.3.6) the conditions

$$\lim_{\lambda^k \to \lambda_0} d(x(\lambda^k), M(\lambda^k)) = 0 \qquad \forall \{\lambda^k\} \subset V\{\lambda^0\}, \qquad \lambda^k \to \lambda^0, \tag{6.3.7}$$

and

$$\lim_{\lambda^k \to \lambda^0} |\varphi(\lambda^k) - f(x(\lambda^k), \lambda^k)| = 0 \tag{6.3.8}$$

are fulfilled then we say that the corresponding function is a *weak* $\mathbb{L}$*-function at* λ^0. In the first part of this section we establish properties of these approximate selection functions and in particular demonstrate various relations existing between the different concepts. We then consider the question of how one may calculate approximate selection functions.

Lipschitz continuity plays a certain role and we first recall the definition of the locally Lipschitz continuity at $x^0 \in \mathbb{E}_n$ as opposed to the usual Lipschitz continuity on a open subset of $\mathbb{E}_n$. A function $h\colon \mathbb{E}_n \to \mathbb{E}_m$ is said to be *locally Lipschitzian at* $x^0 \in \mathbb{E}_n$ if a constant K, $K > 0$, and a neighbourhood $U\{x^0\}$ of the point x^0 exist such that

$$||h(x) - h(x^0)|| \leqq K||x - x^0|| \tag{6.3.9}$$

holds for all $x \in U\{x^0\}$.

In the following theorem a relation between $\mathbb{F}$-function and $\mathbb{L}$-function is included.

Theorem 6.3.1. *If there is an* $\varepsilon > 0$ *such that* $f(\cdot, \lambda)$ *is Lipschitz continuous on* $\bigcup_{\lambda \in V\{\lambda^0\}} U_\varepsilon \psi(\lambda)$ *with a uniform Lipschitz constant* $K > 0$ *then each* $\mathbb{F}$*-function at* λ^0 *is also an* $\mathbb{L}$*-function at* λ^0.

Proof. The existence of a function $\bar{x}\colon V\{\lambda^0\} \to \mathbb{E}_n$, $\bar{x}(\lambda) \in \psi(\lambda)$, with the property $||x(\lambda) - \bar{x}(\lambda)|| = o(||\lambda - \lambda^0||)$ is given by supposition, we thus have

$$d(x(\lambda), M(\lambda)) \leqq ||x(\lambda) - \bar{x}(\lambda)|| .$$

For each sequence $\{\lambda^k\} \subset V\{\lambda^0)$, $\lambda^k \to \lambda^0$, we then obtain

$$0 \leqq \lim_{\lambda^k \to \lambda^0} \frac{d(x(\lambda^k), M(\lambda^k))}{||\lambda^k - \lambda^0||} \leqq \lim_{\lambda^k \to \lambda^0} \frac{||x(\lambda^k) - \bar{x}(\lambda^k)||}{||\lambda^k - \lambda^0||} = 0$$

which means that condition (6.3.4) for any $\mathbb{F}$-function is satisfied. The assumption of Theorem 6.3.1 yields

$$|f(x(\lambda), \lambda) - \varphi(\lambda)| = |f(x(\lambda), \lambda) - f(\bar{x}(\lambda), \lambda)| \leqq K||x(\lambda) - \bar{x}(\lambda)|| .$$

Since $x(\lambda)$ is an $\mathbb{F}$-function we have $||x(\lambda) - \bar{x}(\lambda)|| = o(||\lambda - \lambda^0||)$ and then

$$|f(x(\lambda), \lambda) - \varphi(\lambda)| = o(||\lambda - \lambda^0||)$$

follows, thus (6.3.5) also holds. □

We first give an example which is also employed in what follows but which here illustrates the preceding theorem. The second example demonstrates that the assumption of Theorem 6.3.1 cannot be relaxed.

Example 6.3.1

$$(\mathrm{P}_\lambda) \qquad \min \{-x_1^2 - x_2 \mid x_1^2 + x_2^2 - \lambda \leqq 0\} .$$

We have

$$\psi(\lambda) = \begin{cases} \{(0, \sqrt{\lambda})\} & \text{if } \lambda \in \langle 0, \tfrac{1}{4}\rangle , \\ \{(\sqrt{\lambda - \tfrac{1}{4}}, \tfrac{1}{2}), (-\sqrt{\lambda - \tfrac{1}{4}}, \tfrac{1}{2})\} & \text{if } \lambda \in \langle \tfrac{1}{4}, +\infty) . \end{cases}$$

and

$$\varphi(\lambda) = \begin{cases} -\sqrt{\lambda} & \text{if } \lambda \in \langle 0, \frac{1}{4}\rangle , \\ -\lambda - \frac{1}{4} & \text{if } \lambda \in \langle \frac{1}{4}, +\infty) . \end{cases}$$

The functions $x^{(n)}(\lambda) = ((\lambda - \frac{1}{4})^n, \sqrt{\lambda})$, $n > 1$, are then $\mathbb{F}$-functions at $\lambda^0 = \frac{1}{4}$. Since the objective function fulfils the assumption of Theorem 6.3.1 e.g. for choosing $V\{\lambda^0\} = (\frac{1}{3}, 1)$ it follows that the $x^{(n)}(\lambda)$ are also $\mathbb{L}$-functions at λ^0 for each $n > 1$.

Example 6.3.2

$$(\mathrm{P}_\lambda) \qquad \min \{\sqrt{x_1 + x_2} \mid x_1^2 + x_2^2 \leqq \lambda^2, x_1 + x_2 \geqq 0\} .$$

For all $\lambda \in (-\infty, +\infty)$ one finds

$$\psi(\lambda) = \{(x_1, x_2) \mid x_1 + x_2 = 0, x_1^2 + x_2^2 \leqq \lambda^2\} ,$$

$$\varphi(\lambda) = 0$$

and for $\lambda^0 = 0$ in particular $\psi(\lambda^0) = M(\lambda^0) = \{(0, 0)\}$. We then have $\bar{x}(\lambda) = (0, 0) \in \psi(\lambda)$ for all $\lambda \in (-\infty, +\infty)$ and $x(\lambda) = (0, \lambda^2)$ is consequently an $\mathbb{F}$-function at $\lambda^0 = 0$. Condition (6.3.4) is fulfilled for all $\lambda \in (-1, 1)$ but this is not the case for (6.3.5): we have

$$\lim_{\lambda^k \to \lambda^0} \frac{|\varphi(\lambda^k) - f(x(\lambda^k), \lambda^k)|}{||\lambda^k - \lambda^0||} = \lim_{\lambda^k \to 0} \frac{|\lambda^k|}{|\lambda^k|} = 1 .$$

$x(\lambda) = (0, \lambda^2)$ is thus not an $\mathbb{L}$-function at λ^0. The objective function is continuous but however not Lipschitz continuous on neighbourhoods of (0, 0).

Theorem 6.3.1 opens up the possibility of employing the conception developed by A. V. Fiacco [2], [3] and R. L. Armacost [2], [3] for calculating $\mathbb{F}$-functions to determine $\mathbb{L}$-functions. These authors use certain constraints qualifications and convexity conditions and construct $\mathbb{F}$-functions with the help of a linearization of a differentiable selection function $\bar{x}(\lambda) \in \psi(\lambda)$ in the following manner:

$$x(\lambda) = \bar{x}(\lambda^0) + \nabla_\lambda \bar{x}(\lambda^0)^T (\lambda - \lambda^0);$$

formulas for calculating $\nabla_\lambda \bar{x}(\lambda^0)$ are also derived, see also M. Lips [1].

In the following we consider under what circumstances an $\mathbb{L}$-function is also an $\mathbb{F}$-function or at least a weak $\mathbb{F}$-function; not every $\mathbb{L}$-function yields an $\mathbb{F}$-function or even a weak $\mathbb{F}$-function as the next example demonstrates.

Example 6.3.3

$$(\mathrm{P}_\lambda) \qquad \min \{(\lambda - \lambda^0)^2 x_2 \mid x_1 = \lambda, 0 \leqq x_2 \leqq 1\}$$

where $\lambda^0 \in \mathbb{R}$ is fixed.

$$\psi(\lambda) = \begin{cases} M(\lambda^0) & \text{if } \lambda = \lambda^0 , \\ \{(\lambda, 0)\} & \text{if } \lambda \neq \lambda^0 \end{cases}$$

and

$$\varphi(\lambda) = 0$$

hold for all $\lambda \in (-\infty, +\infty)$. If one chooses the selection function $x(\lambda) = (\lambda, 0) \in \psi(\lambda)$ for all $\lambda \in (-\infty, +\infty)$ then $x(\lambda) = (\lambda, k)$, $k \in (0, 1]$ fixed, is indeed an $\mathbb{L}$-function at $\lambda^0 \in (-\infty, +\infty)$ but not a weak $\mathbb{F}$-function: one finds on the one hand

$$x(\lambda) \in M(\lambda) , \qquad \lambda \in (-\infty, +\infty) , \qquad \lim_{\lambda \to \lambda^0} \frac{|\varphi(\lambda) - f(x(\lambda), \lambda)|}{||\lambda - \lambda^0||}$$

$$= \lim_{\lambda \to \lambda_0} |\lambda - \lambda^0| \, k = 0$$

but on the other hand $||\bar{x}(\lambda) - x(\lambda)|| = k$.

Theorem 6.3.2. *Any weak $\mathbb{L}$-function at λ^0 which is continuous at λ^0 is also a weak $\mathbb{F}$-function at λ^0 if each of the following conditions is satisfied:*

(1) *f is continuous on $\mathbb{E}_n \times \{\lambda^0\}$,*

(2) *M is closed at λ^0, and*

(3) *ψ is l.s.c.-B at λ^0.*

Proof. Assume that $x\colon V\{\lambda^0\} \to \mathbb{E}_n$ is a continuous weak $\mathbb{L}$-function at λ^0. We first demonstrate that

$$x(\lambda^0) \in \psi(\lambda^0) \tag{6.3.10}$$

holds. Let $\{\lambda^k\} \subset V\{\lambda^0\}$ be a sequence with the property $\lambda^k \to \lambda^0$. For sufficiently large k one can then find points $y(\lambda^k) \in M(\lambda^k)$ satisfying

$$||x(\lambda^k) - y(\lambda^k)|| \leqq ||\lambda^k - \lambda^0|| + d\big(x(\lambda^k), M(\lambda^k)\big) .$$

Since x is continuous at λ^0 and because of (6.3.7) we obtain $\lim\limits_{\lambda^k \to \lambda^0} y(\lambda^k) = x(\lambda^0)$. The constraint set mapping is closed at λ^0 and thus

$$x(\lambda^0) \in M(\lambda^0) \tag{6.3.11}$$

holds. By Theorem 4.2.4 the extreme value function is continuous at the point λ^0 and taking (6.3.8) into account one then obtains

$$0 = \lim_{\lambda^k \to \lambda^0} |\varphi(\lambda^k) - f\big(x(\lambda^k), \lambda^k\big)| = |f\big(x(\lambda^0), \lambda^0\big) - \varphi(\lambda^0)| \,. \tag{6.3.12}$$

This last together with (6.3.11) proves (6.3.10).

Since ψ is l.s.c.-B at λ^0 there exists a function $\bar{x}$ with the properties $\bar{x}(\lambda^k) \to x(\lambda^0)$ and $\bar{x}(\lambda^k) \in \psi(\lambda^k)$ for all k from which $\lim\limits_{\lambda^k \to \lambda^0} ||x(\lambda^k) - \bar{x}(\lambda^k)|| = 0$ follows. □

We note that condition (3) ensures that a selection function $\bar{x}(\lambda) \in \psi(\lambda)$ with $\bar{x}(\lambda^0) = x(\lambda^0)$ and which is continuous at λ^0 exists (see also the remarks on condition (A.3) in Section 2.3). This selection function is approximated at the point λ^0 by the weak $\mathbb{L}$-function x.

The following example shows that Theorem 6.3.2 is no longer valid if condition (1) alone is relaxed.

Example 6.3.4

(P_λ) $\quad \min \{f(x, \lambda) \mid x_1 = \lambda,\ 0 \leqq x_2 \leqq 1\}$

where

$$f(x, \lambda) = \begin{cases} (\lambda - \lambda^0)^2\, x_2 & \text{if } \lambda \neq \lambda^0 \,, \\ x_2 & \text{if } \lambda = \lambda^0 \end{cases}$$

whereby λ^0 is fixed. We then have for all $\lambda \in (-\infty, +\infty)$

$$\psi(\lambda) = \{(\lambda, 0)\} \,, \qquad \varphi(\lambda) = 0 \,.$$

One easily confirms that $x(\lambda) = (\lambda, \tfrac{1}{2})$ is an $\mathbb{L}$-function at λ^0, it is however not a weak $\mathbb{F}$-function at λ^0 since its distance to the only possible selection function $\bar{x}(\lambda) = (\lambda, 0)$ has the value $\tfrac{1}{2}$ and thus

$$\lim_{\lambda^k \to \lambda^0} ||x(\lambda^k) - \bar{x}(\lambda^k)|| = \tfrac{1}{2} \,.$$

Theorem 6.3.2 also indicates why the $\mathbb{L}$-functions $x^k(\lambda) = (\lambda, k)$ considered in Example 6.3.3. are not $\mathbb{F}$-functions or weak $\mathbb{F}$-functions. The lower semicontinuity of the optimal set-mapping required in the theorem, is not fulfilled at the point λ^0. As the assumptions (1), (2) of Theorem 6.3.2 are satisfied, this implies that we cannot dispense with the lower semicontinuity of ψ. Even if we consider an $\mathbb{L}$-function instead of a weak $\mathbb{L}$-function the assumptions of Theorem 6.3.2 are not sufficient to guarantee that an $\mathbb{L}$-function is also an $\mathbb{F}$-function.

This is shown by example 6.3.1

$$(\mathrm{P}_\lambda) \qquad \min \{-x_1^2 - x_2 \mid x_1^2 + x_2^2 - \lambda \leqq 0\}, \lambda \in \Lambda,$$

where we chose $\Lambda = \langle 0, \frac{1}{4}\rangle$. Thus $\bar{x}(\lambda) = (0, \sqrt{\lambda})$ is the only selection function from the optimal set-mapping in case $\lambda \in \Lambda$. It is easy to see that the assumptions (1), (2) and (3) of Theorem 6.3.2 are satisfied. We take also into consideration that

$$x(\lambda) = (\lambda - \tfrac{1}{4}, \lambda + \tfrac{1}{4})$$

represents an $\mathbb{L}$-function in λ^0 which, moreover, is continuous at λ^0. Due to Theorem 6.3.2 $x(\lambda) = (\lambda - \frac{1}{4}, \lambda + \frac{1}{4})$ is a weak $\mathbb{F}$-function:

$$\lim_{\lambda^k \to \lambda^0} \| x(\lambda^k) - \bar{x}(\lambda^k) \| = \lim_{\lambda^k \to 1/4^-} \sqrt{(\lambda^k - \tfrac{1}{4})^2 + (\lambda^k + \tfrac{1}{4} - \sqrt{\lambda^k})^2} = 0 .$$

But, this function does not represent an $\mathbb{F}$-function since

$$\lim_{\lambda^k \to \lambda^0} \frac{\|x(\lambda^k) - \bar{x}(\lambda^k)\|}{\|\lambda^k - \lambda^0\|} = \lim_{\lambda^k \to 1/4^-} \frac{\sqrt{(\lambda^k - \frac{1}{4})^2 + (\lambda^k + \frac{1}{4} - \sqrt{\lambda^k})^2}}{|\lambda^k - \frac{1}{4}|} \neq 0 .$$

On the other hand, this example shows that there is an $\mathbb{L}$-function which is an $\mathbb{F}$-function at the same time.

Let us consider the function $x(\lambda) = (0, \lambda + \frac{1}{4})$. Obviously this function fulfils the conditions (6.3.4) and (6.3.5) in $\lambda^0 = \frac{1}{4}$ and so it is an $\mathbb{L}$-function in $\lambda^0 = \frac{1}{4}$.

Furthermore it holds that

$$\lim_{\lambda^k \to \lambda^0} \frac{\|x(\lambda^k) - \bar{x}(\lambda^k)\|}{\|\lambda^k - \lambda^0\|} = \lim_{\lambda^k \to 1/4^-} \frac{|\lambda^k + \frac{1}{4} - \sqrt{\lambda^k}|}{|\lambda^k - \frac{1}{4}|} = 0$$

which implies that $x(\lambda)$ is also an $\mathbb{F}$-function at the point $\lambda^0 = \frac{1}{4}$.

In the following we shall deal with the $\mathbb{L}$-functions in some more detail and present a theoretic basis for their computation. Under certain assumptions a piecewise linear continuous vector function, which can be obtained as solution of a linear multi-parametric optimization problem and of a simplex technique, will provide an $\mathbb{L}$-function.

We now consider a linear parametric substitute problem corresponding to (P_λ) and which is obtained by linearizing the functions involved at a given point $(x^0, \lambda^0) \in \in \mathbb{E}_n \times \mathbb{E}_k$ under the following condition:

$$f, g_i, \quad i = 1, \dots, m, \qquad h_j, \quad j = 1, \dots, p, \quad \text{are}$$

$$\text{continuously differentiable on } \mathbb{E}_n \times \mathbb{E}_k . \tag{6.3.13}$$

The substitute problem is of the form

$$(\mathrm{P}_\lambda^{\mathfrak{L}}) \qquad \min \{\nabla_x f(x^0, \lambda^0)^\mathsf{T} y \mid y \in M^{\mathfrak{L}}(\lambda)\} \tag{6.3.14}$$

where

$$M^{\mathfrak{L}}(\lambda) = \{y \in \mathbb{E}_n \mid \nabla_x g_i(x^0, \lambda^0)^\mathsf{T} y \leqq -\nabla_\lambda g_i(x^0, \lambda^0)^\mathsf{T} (\lambda - \lambda^0), i \in I_0,$$

$$\nabla_x h_j(x^0, \lambda^0)^\mathsf{T} y = -\nabla_\lambda h_j(x^0, y\lambda^0)^\mathsf{T} (\lambda - \lambda^0), j = 1, \dots, p\}. \tag{6.3.15}$$

I_0 is the set of indices of the active constraints of $M(\lambda^0)$ at x^0. $\mathfrak{B}^{\mathfrak{L}}$ denotes the feasible parameter set, $\mathfrak{A}^{\mathfrak{L}}$ the solubility set, $\psi^{\mathfrak{L}}$ the optimal set mapping and $\varphi^{\mathfrak{L}}$ is the extreme value function of the parametric optimization problem $(\mathrm{P}_\lambda^{\mathfrak{L}})$.

Problem $(\mathrm{P}_\lambda^{\mathfrak{L}})$ is a multiparametric linear optimization problem with parameters in the right-hand sides of the constraints and as such a special case of problem (5.4.42). From the results of Section 5.5 there exists a continuous, piece-wise linear selection function $x^{\mathfrak{L}}(\lambda) \in \psi^{\mathfrak{L}}(\lambda)$ on $\mathfrak{A}^{\mathfrak{L}}$. A solution procedure for calculating $x^{\mathfrak{L}}(\lambda)$ based on the simplex technique was discussed in the previous section (Section 6.2). The aim of the following investigations is to find the conditions under which $x^{\mathfrak{L}}(\lambda)$ restricted to $V\{\lambda^0\}$ is an $\mathbb{L}$-function at λ^0. We first present a lemma which establishes a relation between the rate of convergence and the locally Lipschitz continuity.

Lemma 6.3.1. *Let c be a real-valued function defined on $\mathbb{E}_n \times \mathbb{E}_k$ and satisfying $c(x, \lambda) = o(\|x - x^0\|, \|\lambda - \lambda^0\|)$ and let x be a vector function which is locally Lipschitzian at λ^0, is defined on $V\{\lambda^0\}$ and which satisfies $x(\lambda^0) = x^0$.*
Then

$$c(x(\lambda), \lambda) = o(\|\lambda - \lambda^0\|)$$

holds.

The proof is obvious.

Theorem 6.3.3 (E. S. Levitin). *Assume*

(1) *g_i, $i = 1, \dots, m$, h_j, $j = 1, \dots, p$, are continuously differentiable on $\mathbb{E}_n \times \mathbb{E}_k$;*
(2) *$\lambda^0 \in \mathfrak{B}$ and $x^0 \in M(\lambda^0)$ hold;*
(3) *there exists a vector $a^0 \in \mathbb{E}_n$ such that*

$$\nabla_x g_i(x^0, \lambda^0)^\mathsf{T} a^0 < 0, \qquad i \in I_0,$$

$$\nabla_x h_j(x^0, \lambda^0)^\mathsf{T} a^0 = 0, \qquad j = 1, \dots, p,$$

and the vectors $\nabla_x h_j(x^0, \lambda^0)$, $j = 1, \dots, p$, are linearly independent (I_0 as in (6.3.15));
(4)[1] *let y be a selection function of the constraint set mapping $M^{\mathfrak{L}}$ which is locally Lipschitzian at λ^0, is defined on $V\{\lambda^0\}$ and satisfies $y(\lambda^0) = 0$.*
Then

$$d(x^0 + y(\lambda), M(\lambda)) = o(\|\lambda - \lambda^0\|) \tag{6.3.16}$$

holds.

Proof. Relation (6.3.16) is obviously equivalent to the existence of a function $w: V\{\lambda^0\} \to \mathbb{E}_n$ with the properties

$$x^0 + y(\lambda) + w(\lambda) \in M(\lambda) \tag{6.3.17}$$

[1]) see Note 6.3.1

and

$$||w(\lambda)|| = o(||\lambda - \lambda^0||) , \tag{6.3.18}$$

We construct such a function below, however first we determine a function which approximates the set

$$\{x \in E_n \mid h_j(x, \lambda) = 0, j = 1, \dots, p\}$$

with an error $o(||\lambda - \lambda^0||)$. Since the vectors $\nabla_x h_j(x^0, \lambda^0), j = 1, \dots, p$, are by supposition (3) linearly independent the system of equations

$$\nabla_x h_j(x^0, \lambda^0)^T r = \delta_{ij} , \qquad i = 1, \dots, p , \tag{6.3.19}$$

has a solution r^i for each $i \in \{1, \dots, p\}$. We then define the functions

$$T_j(y, t, \lambda) =_{Df} h_j\left(x^0 + y + \sum_{i=1}^{p} t_i r^i, \lambda\right), \qquad j = 1, \dots, p , \tag{6.3.20}$$

and the system of equations

$$T_j(y, t, \lambda) = 0 , \qquad j = 1, \dots, p . \tag{6.3.21}$$

The latter has for $\lambda = \lambda^0$ a regular Jacobian matrix $J_t(0, 0, \lambda^0)$ since $J_t(0, 0, \lambda^0)$ is by (6.3.19) the unit matrix. Using the implicit function theorem, i.e. there exists a uniquely defined function $\bar{t}: U\{(0, \lambda^0)\} \to E_p$[1]) satisfying $\bar{t}(0, \lambda^0) = 0$ and $T_j(y, \bar{t}(y, \lambda), \lambda) = 0$, $j = 1, \dots, p$, such that the functions $\bar{t}_i$, $i = 1, \dots, p$, the components of $\bar{t}$, are continuously differentiable at $(0, \lambda^0)$, it follows that

$$\begin{aligned}\bar{t}_j(y, \lambda) &= \bar{t}_j(0, \lambda^0) + \nabla_y \bar{t}_j(0, \lambda^0)^T y + \nabla_\lambda \bar{t}_j(0, \lambda^0)^T (\lambda - \lambda^0) \\ &\quad + o(||y||, ||\lambda - \lambda^0||) , \qquad j = 1, \dots, p ,\end{aligned}$$

holds. Because of $\bar{t}_j(0, \lambda^0) = 0, j = 1, \dots, p$, one obtains

$$\begin{aligned}\bar{t}_j(y, \lambda) &= \nabla_y \bar{t}_j(0, \lambda^0)^T y + \nabla_\lambda \bar{t}_j(0, \lambda^0)^T (\lambda - \lambda^0) \\ &\quad + o(||y||, ||\lambda - \lambda^0||) , \qquad j = 1, \dots, p .\end{aligned} \tag{6.3.22}$$

By replacing y in (6.3.21) by $y(\lambda)$ of condition (4) and exploiting the differentiability properties of the functions h_j, $j = 1, \dots, p$, we get

$$\begin{aligned}0 &= T_j\left(y(\lambda), \bar{t}(y(\lambda), \lambda), \lambda\right) = h_j(x^0, \lambda^0) + \nabla_x h_j(x^0, \lambda^0)^T (y(\lambda) \\ &\quad + \sum_{i=1}^{p} r^i (\nabla_y \bar{t}_i(y(\lambda^0), \lambda^0)^T y(\lambda) + \nabla_\lambda \bar{t}_i(y(\lambda^0)^T (\lambda - \lambda^0))) \\ &\quad + \nabla_\lambda h_j(x^0, \lambda^0)^T (\lambda - \lambda^0) + o(||y(\lambda)||, ||\lambda - \lambda^0||) \\ &= \nabla_y \bar{t}_j\left(y(\lambda^0), \lambda^0\right)^T y(\lambda) + \nabla_\lambda \bar{t}_j\left(y(\lambda^0), \lambda^0\right)^T (\lambda - \lambda^0) \\ &\quad + o(||y(\lambda)||, ||\lambda - \lambda^0||) , \qquad j = 1, \dots, p .\end{aligned} \tag{6.3.23}$$

This together with (6.3.22) and $y(\lambda^0) = 0$ (condition (4)) then gives

$$\bar{t}_j\left(y(\lambda), \lambda\right) = o(||y(\lambda)||, ||\lambda - \lambda^0||) , \qquad j = 1, \dots, p .$$

Lemma 6.3.2 may be applied since $y(\lambda)$ is locally Lipschitzian at λ^0 and we then obtain

$$\bar{t}_j\left(y(\lambda), \lambda\right) = o(||\lambda - \lambda^0||) . \tag{6.3.24}$$

[1]) $U\{0, \lambda^0\} \supset \{0\} \times V\{\lambda^0\}$

(6.3.23) and (6.3.20) yield

$$h_j(x^0 + y(\lambda) + w_1(\lambda), \lambda) = 0$$

where $w_1(\lambda) =_{\text{Df}} \sum_{i=1}^{p} \bar{t}_i(y(\lambda), \lambda)\, r^i$.

Because of (6.3.24) $||w_1(\lambda)|| = o(||\lambda - \lambda^0||)$ holds. The relation

$$d(x^0 + y(\lambda), \{x \in E_n \mid h_j(x, \lambda) = 0, j = 1, \dots, p\}) = o(||\lambda - \lambda^0||)$$

follows. We have at the same time indicated the idea underlying the next part of the proof in which we include the system of inequalities

$$g_i(x, \lambda) \leqq 0\,, \qquad i \in I_0\,.$$

Assume that $a^0 \in E_n$ is a vector as in condition (3). Consider again the solution r^i, $i \in \{1, \dots, p\}$, of the system of equations (6.3.21) and define

$$S_j(y, \lambda, c, s) =_{\text{Df}} h_j\left(x^0 + y + \sum_{i=1}^{p} \bar{t}_i(y, \lambda)\, r^i + ca^0 + \sum_{i=1}^{p} s_i r^i, \lambda\right). \tag{6.3.25}$$

In analogy to (6.3.21) we then examine the system of equations

$$S_j(y, \lambda, c, s) = 0\,, \qquad j = 1, \dots, p\,. \tag{6.3.26}$$

Because of $\bar{t}_i(0, \lambda^0) = 0$, $i = 1, \dots, p$, and (6.3.19) this system of equations has a Jacobian matrix with respect to s which is regular at $(0, \lambda^0, 0, 0) \in E_n \times E_k \times E_1 \times E_p$ so that we may again apply the implicit function theorem; there thus exists a function which is continuously differentiable at $(0, \lambda^0, 0) \in E_n \times E_k \times E_1$ and which is given by

$$\bar{s}\colon U\,\{(0, \lambda^0, 0)\} \times E_p$$

and which also satisfies

$$S_j(y, \lambda, c, \bar{s}(y, \lambda, c)) = 0\,, \qquad j = 1, \dots, p\,. \tag{6.3.27}$$

If $c = 0$ and $s = 0$ then we have (6.3.23). By the implicit function theorem

$$\bar{s}(y, \lambda, 0) = 0 \quad \text{for all} \quad (y, \lambda) \in U\{(0, \lambda^0)\} \tag{6.3.28}$$

follows and thus

$$\nabla_y \bar{s}_j(0, \lambda^0, 0) = 0\,, \qquad \nabla_\lambda \bar{s}_j(0, \lambda^0, 0) = 0\,, \qquad j = 1, \dots, p\,. \tag{6.3.29}$$

Since the functions h_j, $j = 1, \dots, p$ and $\bar{s}_j$, $j = 1, \dots, p$, are differentiable then from (6.3.27) and (6.3.25) it follows that

$$\begin{aligned} 0 = S_j\big(0, \lambda^0, c, \bar{s}(0, \lambda^0, c)\big) &= h_j\left(x^0 + ca^0 + \sum_{i=1}^{p} \bar{s}_i(0, \lambda^0, c)\, r^i, \lambda^0\right) \\ &= h_j(x^0, \lambda^0) + \nabla_x h_j(x^0, \lambda^0)^{\mathsf{T}} \left(a^0 + \sum_{i=1}^{p} r^i \frac{d\bar{s}_i(0, \lambda^0, 0)}{dc} \cdot c\right) + o(|c|) \\ &= \frac{d\bar{s}_j(0, \lambda^0, 0)}{dc} \cdot c + o(|c|)\,, \qquad j = 1, \dots, p\,. \end{aligned}$$

As in the first part of this proof

$$\frac{d\bar{s}_j(0, \lambda^0, 0)}{dc} = 0\,, \qquad j = 1, \dots, p\,, \tag{6.3.30}$$

and

$$\bar{s}_j(y, \lambda, c) = o(||y||, ||\lambda - \lambda^0||, |c|)\,, \qquad j = 1, \dots, p\,, \tag{6.3.31}$$

follow. Because of the differentiability of the functions g_i, $i = 1, \dots, m$, and (6.3.28), (6.3.29), and (6.3.30) we have

$$g_i\left(x^0 + y + \sum_{j=1}^{p} \bar{t}_j(y, \lambda)\, r^j + c a^0 + \sum_{j=1}^{p} \bar{s}_j(y, \lambda, c)\, r^j, \lambda\right) =$$
$$= g_i\left(x^0 + y + \sum_{j=1}^{p} \bar{t}_j(y, \lambda)\, r^j, \lambda\right)$$
$$+ c\nabla_x g_i\left(x^0 + y + \sum_{j=1}^{p} \bar{t}_j(y, \lambda)\, r^j, \lambda\right)^{\mathsf T} a^0 + o(|c|)\,, \qquad i = 1, \dots, m. \tag{6.3.32}$$

We define

$$c(\lambda) =_{\mathrm{Df}} \frac{2 \max\limits_{i \in I_0} \{\max \{g_i(x^0 + y(\lambda) + w_1(\lambda), \lambda), 0\}\}}{-\max\limits_{i \in I_0} \{\nabla_x g_i(x^0 + y(\lambda) + w_1(\lambda), \lambda)^{\mathsf T} a^0\}} \tag{6.3.33}$$

and

$$w(\lambda) =_{\mathrm{Df}} w_1(\lambda) + c(\lambda)\, a^0 + \sum_{j=1}^{p} \bar{s}_j(y(\lambda), \lambda, c(\lambda))\, r^j\,. \tag{6.3.34}$$

Because of (6.3.17) and (6.3.18) it suffices to show that $x^0 + y(\lambda) + w(\lambda) \in M(\lambda)$ for all $\lambda \in V\{\lambda^0\}$, and $\|w(\lambda)\| = o(\|\lambda - \lambda^0\|)$ hold. Since the functions g_i, $i = 1, \dots, m$, are continuously differentiable and

$$\nabla_x g_i(x^0, \lambda^0)^{\mathsf T} a^0 < 0\,, \qquad i \in I_0\,,$$

holds the validity of

$$\nabla_x g_i(x^0 + y(\lambda) + w_1(\lambda), \lambda)^{\mathsf T} a^0 \leqq v^0 < 0\,, \qquad i \in I_0\,, \tag{6.3.35}$$

is ensured. We first demonstrate that $c(\lambda)$ given by (6.3.33) fulfils

$$c(\lambda) = o(\|\lambda - \lambda^0\|)\,. \tag{6.3.36}$$

From (6.3.24), condition (3), the definition of w_1, and Lemma 6.3.1 (note condition (4) of the present theorem) we obtain for $i \in I_0$

$$g_i\left(x^0 + y(\lambda) + \sum_{j=1}^{p} \bar{t}_j(y(\lambda), \lambda)\, r^j, \lambda\right) = g_i(x^0, \lambda^0) + \nabla_x g_i(x^0, \lambda^0)^{\mathsf T} (y(\lambda) + w_1(\lambda))$$
$$+ \nabla_\lambda g_i(x^0, \lambda^0)^{\mathsf T} (\lambda - \lambda^0) + o(\|y(\lambda) + w_1(\lambda)\|, \|\lambda - \lambda^0\|)$$
$$\leqq \nabla_x g_i(x^0, \lambda^0)^{\mathsf T} w_1(\lambda) + o(\|\lambda - \lambda^0\|) = o(\|\lambda - \lambda^0\|)$$

rom which

$$\max_{i \in I^0} \{g_i(x^0 + y(\lambda) + w_1(\lambda), \lambda), 0\} = o(\|\lambda - \lambda^0\|)$$

and thus (6.3.36) follow. Taking (6.3.33), (6.3.34), and (6.3.35) into account we then obtain from (6.3.32) for $i \in I_0$:

$$g_i(x^0 + y(\lambda) + w(\lambda), \lambda) \leqq 0\,. \tag{6.3.37}$$

The continuity of the functions $y(\cdot)$, $\bar{t}_j(\cdot, \cdot)$, $\bar{s}_j(\cdot, \cdot, \cdot)$ and $c(\cdot)$ at the points λ^0, $(0, \lambda^0)$, $(0, \lambda^0, 0)$ and λ^0 respectively guarantees that (6.3.37) also holds for all $i \in \{1, \dots, m\} \setminus I_0$. Thus $g_i(x^0 + y(\lambda) + w(\lambda), \lambda) \leqq 0$, $i = 1, \dots, m$, and $\|w(\lambda)\| = o(\|\lambda - \lambda^0\|)$. From (6.3.27), (6.3.34), and (6.3.25) we then have

$$h_j(x^0 + y(\lambda) + w(\lambda), \lambda) = 0\,, \qquad j = 1, \dots, p\,.$$

Thus $d(x^0 + y(\lambda), M(\lambda)) = o(\|\lambda - \lambda^0\|)$ and the proof is complete. □

We now discuss the assumptions of the last theorem. We first consider whether the constraint qualification (3), which in Section A.5 of the Appendix is denoted by (R1) may be replaced by a weaker condition. We refer the reader to Section A.5 of the Appendix and in particular to Lemma A.5.1 for the relations existing between the various constraint qualifications considered here.

Example 6.3.6. Consider the set

$$M(\lambda) = \{x \in E_2 \mid -x_1 + \lambda \leqq 0, -x_2 \leqq 0, -x_1^3 + x_2 \leqq 0\}$$

and assume $\lambda^0 = 0$ and $x^0 = (0, 0)$. The linearization $M^x(\lambda)$ of $M(\lambda)$ at the point (x^0, λ^0) is then given by

$$M^x(\lambda) = \{x \in E_2 \mid -x_1 + \lambda \leqq 0, x_2 = 0\} .$$

For $\lambda \leqq 0$ we choose the function $y(\lambda) = (\lambda, 0) \in M^x(\lambda)$ which is locally Lipschitzian at λ^0 and obviously satisfies $y(\lambda^0) = 0$. Since $y(\lambda) \in M^x(\lambda)$ and $d(y(\lambda), M(\lambda)) = \lambda$ we obtain $\lim\limits_{\lambda \to -0} \dfrac{d(y(\lambda), M(\lambda))}{|\lambda|} = -1$. The constraint qualifications (R2) and (R3) given by (A.5.5) and (A.5.6) are however satisfied.

Example 6.3.7. Consider the set

$$M(\lambda) = \{x \in E_2 \mid -x_1^2 + x_2 \leqq 0, -x_1 + \lambda = 0, 2\lambda x_1 - x_2 \leqq 0\}$$

and assume $\lambda^0 = 0$ and $x^0 = (0, 0)$. The linearization $M^x(\lambda)$ of $M(\lambda)$ at the point (x^0, λ^0) is then given by

$$M^x(\lambda) = \{x \in E_2 \mid -x_1 + \lambda \leqq 0, x_2 = 0\} .$$

We choose $y(\lambda) = (\lambda, 0) \in M^x(\lambda)$ for $\lambda > 0$. Because of $x_1^2 \geqq x_2 \geqq 2\lambda x_1$ and $x_1 \geqq \lambda$ we have for each $x \in M(\lambda)$ the inequality $x_1 \geqq 2\lambda$, i.e. $d(y(\lambda), M(\lambda)) \geqq \lambda$. Although the constraint qualification (R4) given by (A.5.7) is satisfied the proposition of Theorem 6.3.3 is not valid.

The question arises of whether condition (R1) in (A.5.4) may be replaced by (R2) as given by (A.5.5) if additional convexity conditions are imposed on the functions g_i, $i \in I_0$, the next example demonstrates that this is however not the case.

Example 6.3.8. Consider the set

$$M(\lambda) = \{x \in E_2 \mid -x_1 \leqq 0, x_1 + \lambda \leqq 0, x_1^2 - x_2 \leqq 0, x_1^2 + x_2 \leqq 0\}$$

and assume $\lambda^0 = 0$ and $x^0 = (0, 0)$. One then has

$$M^x(\lambda) = \{x \in E_2 \mid x_2 = 0, -x_1 \leqq 0, x_1 + \lambda \leqq 0\} .$$

For $\lambda \leqq 0$ and $y(\lambda) = (-\lambda, 0) \in M^x(\lambda)$

$$\lim_{\lambda \to 0^-} \frac{d(y(\lambda), M(\lambda))}{|\lambda|} \neq 0$$

holds.

The next two examples indicate the significance of assumption (4) of Theorem 6.3.3.

Example 6.3.9. Assume

$$M(\lambda) = \{x \in E_2 \mid x_1^2 + x_2^2 - \lambda^2 \leqq 0\} , \quad \lambda^0 = 1, \quad x^0 = (1, 0) .$$

For $\lambda \geqq 1$ and $y(\lambda) = (\lambda - 1, \sqrt{\lambda - 1}) \in M^{\mathfrak{L}}(\lambda) = \{x \in E_2 \mid 2x_1 - 2(\lambda - 1) \leqq 0\}$ we obtain

$$\lim_{\lambda \to \lambda^{0+}} \frac{d(x^0 + y(\lambda), M(\lambda))}{|\lambda - \lambda^0|} = \lim_{\lambda \to \lambda^{0+}} \frac{\sqrt{\lambda^2 + \lambda - 1}}{|\lambda - \lambda^0|} \neq 0 .$$

The function $y(\lambda)$ is not locally Lipschitzian at λ^0.

Example 6.3.10. Assume

$$M(\lambda) = \{x \in E_2 \mid x_1^2 + x_2^2 - \lambda \leqq 0\} , \qquad \lambda^0 = \tfrac{1}{4} , \qquad x^0 = (0, \tfrac{1}{2}) \in M(\lambda^0) .$$

We then have $M^{\mathfrak{L}}(\lambda) = \{x \in E_2 \mid x_2 \leqq \lambda - \frac{1}{4}\}$. Choosing for example

$$y(\lambda) = (0, \lambda - \tfrac{5}{4}) \in M^{\mathfrak{L}}(\lambda)$$

we obtain $y(\lambda^0) \neq 0$ and

$$\lim_{\lambda \to \lambda^0} \frac{d(x^0 + y(\lambda), M(\lambda))}{|\lambda - \lambda^0|} = 2 .$$

We therefore can not dispense with either locally Lipschitz continuity of $y(\lambda)$ at λ^0 or the condition $y^{\mathfrak{L}}(\lambda^0) = 0$. One easily confirms that (4) does not restrict the form of the constraint set $M(\lambda)$: one may show that a selection function $y(\lambda) \in M(\lambda)$ which is locally Lipschitzian at λ^0 and satisfies $y(\lambda^0) = 0$ always exists. The constraint set in (6.3.15) may be seen as a special case of the solution set of the parametric optimization problem given in (5.5.42). The existence of a piece-wise linear selection function $y(\lambda) \in M^{\mathfrak{L}}(\lambda)$ on $\mathfrak{B}$ is thus established. This function is, as one easily confirms, locally Lipschitzian at $\lambda^0 \in \mathfrak{B}$. Finally, one must demonstrate that a function $y(\lambda) \in M^{\mathfrak{L}}(\lambda)$ which is locally Lipschitzian at λ^0 and moreover satisfies $y(\lambda^0) = 0$ also exists. From the concrete representation (6.3.15) of $M^{\mathfrak{L}}(\lambda)$ the point $y(\lambda^0) = 0$ is contained in $M^{\mathfrak{L}}(\lambda^0)$ and satisfies all constraints in $M^{\mathfrak{L}}(\lambda^0)$ as equations. With the help of the simplex technique the possibility is therefore given of calculating a piece-wise linear and continuous function $y(\lambda) \in M^{\mathfrak{L}}(\lambda)$ on $V\{\lambda^0\}$ satisfying $y(\lambda^0) = 0$ (see also Section 6.2: the present considerations are in principle a specialization of the procedure given in that section).

With regard to assumptions (2) and (4) of Theorem 6.3.3 we consider the question of for which $\lambda \in E_k$ the set $M^{\mathfrak{L}}(\lambda)$ is non-empty, i.e. of how the set $V\{\lambda^0\}$ may in fact be chosen as a domain of $x(\lambda) = x^0 + y(\lambda)$. With other words we would like to characterize the feasible parameter set $\mathfrak{B}^{\mathfrak{L}} = \{\lambda \in E_k \mid M^{\mathfrak{L}}(\lambda) \neq \emptyset\}$. Besides the fact that λ^0 is contained in $\mathfrak{B}_y^{\mathfrak{L}}$ we also have the following lemma.

Lemma 6.3.2. *Let condition* (3) *of Theorem* 6.3.3 *be fulfilled then* $\mathfrak{B}^{\mathfrak{L}} = E_k$.

Proof. Consider the linear system of equations

$$\nabla_x h_j(x^0, \lambda^0)^{\mathsf{T}} y = -\nabla_\lambda h_j(x^0, \lambda^0)^{\mathsf{T}} (\lambda - \lambda^0) , \qquad j = 1, \dots, p ,$$

and denote the coefficient matrix by A. We partition $A = [A_1 \mid A_2]$ in such a manner that A_1 is regular, and analogously $x = \begin{pmatrix} x^1 \\ x^2 \end{pmatrix}$. The vector a^0, which by assumption, exists, likewise has the form $a^0 = \begin{pmatrix} a^{01} \\ a^{02} \end{pmatrix}$. Because of (R1) $\bar{x}(\lambda) \in M^{\mathfrak{L}}(\lambda)$ for all $\lambda \in V_{t_0}\{\lambda^0\}$ $t_0 > 0$, holds for the point $\bar{x}(\lambda) = \begin{pmatrix} \bar{x}^1(\lambda) \\ \bar{x}^2(\lambda) \end{pmatrix}$ where

$$\bar{x}^1 = -A_1^{-1} A_2 \bar{x}^2(\lambda) - A_1^{-1} \begin{pmatrix} \nabla_\lambda h_1(x^0, \lambda^0)^{\mathsf{T}} (\lambda - \lambda^0) \\ \cdots\cdots\cdots\cdots \\ \nabla_\lambda h_p(x^0, \lambda^0)^{\mathsf{T}} (\lambda - \lambda^0) \end{pmatrix}$$

and $\bar{x}^2(\lambda) = a^{02}$. Consequently, $M^{\mathcal{L}}(\lambda) \neq \emptyset$ for all $\lambda \in V_{t_0}\{\lambda^0\}$. If $\|\lambda - \lambda^0\| \geqq t_0$ then consider the point $\lambda^1 =_{\mathrm{Df}} \lambda^0 + \frac{t_0}{2\,\|\lambda - \lambda^0\|}(\lambda - \lambda^0)$; we then have $\frac{2\,\|\lambda - \lambda^0\|}{t_0}\,\bar{x}(\lambda^1) \in$ $\in M^{\mathcal{L}}(\lambda)$, and $M^{\mathcal{L}}(\lambda) \neq \emptyset$ for all $\lambda \in \mathbb{E}_k$ follows. □

We summarize the above considerations concerning the assumptions of Theorem 6.3.3 as follows.

Note 6.3.1. The set $M^{\mathcal{L}}(\lambda)$ (6.3.15) obtained by linearizing $M(\lambda)$ (6.3.2) at $\lambda^0 \in \mathfrak{B}$ and $x^0 \in M(\lambda^0)$ is non-empty for all $\lambda \in \mathbb{E}_k$. Using the simplex technique one may find a selection function $y(\lambda) \in M^{\mathcal{L}}(\lambda)$ on a neighbourhood $V\{\lambda^0\}$ which satisfies $y(\lambda^0) = 0$ and is continuous and piece-wise linear on $V\{\lambda^0\}$. The existence of a selection function of the constraint set mapping $M^{\mathcal{L}}(\lambda)$ which satisfies $y(\lambda^0) = 0$ and is locally Lipschitzian at λ^0 is thus demonstrated. The only restrictive assumptions are therefore the differentiability conditions imposed on the functions $g_i(x, \lambda)$, $i = 1, \ldots, m$, and the Mangasarian-Fromovitz condition (R1) (see A.5.4). These conditions can not be replaced by weaker ones if we follow this approach.

In the following theorems we present sufficient conditions for a selection function of the optimal set mapping of the linear substitute problem $(\mathrm{P}_\lambda^{\mathcal{L}})$ (6.3.20) to be an $\mathbb{L}$-function. We however first need two preliminary lemmas.

Lemma 6.3.3. *Let $V\{\lambda^0\} \subset \Lambda$, $\Lambda \subset \mathfrak{A}$, be given. Furthermore, assume that the following conditions are satisfied:*

(1) *f, g_i, $i = 1, \ldots, m$, h_j, $j = 1, \ldots, p$, are continuously differentiable on $\mathbb{E}_n \times \mathbb{E}_k$;*

(2) $x^0 \in \psi(\lambda^0)$;

(3) *there exists a vector $a^0 \in \mathbb{E}_n$ satisfying*

$$\nabla_x g_i(x^0, \lambda^0)^{\mathrm{T}} a^0 < 0\,, \qquad i \in I_0\,, \qquad \nabla_x h_j(x^0, \lambda^0)^{\mathrm{T}} a^0 = 0\,, \qquad j = 1, \ldots, p\,;$$

the vectors $\nabla_x h_j(x^0, \lambda^0)$, $j = 1, \ldots, p$, are linearly independent;

(4)[1] *there exists a selection function $y^{\mathcal{L}}(\lambda)$ of the optimal set mapping $\psi^{\mathcal{L}}(\lambda)$ of the problem $(\mathrm{P}_\lambda^{\mathcal{L}})$ which is defined on $V\{\lambda^0\}$, is locally Lipschitzian at $\lambda^0 \in V\{\lambda^0\}$ and which satisfies* $y(\lambda^0) = 0$.

Then the following is true:

$$\varphi(\lambda) \leqq \varphi(\lambda^0) + \max_{(u,v)\in N_0} \nabla_\lambda L(x^0, u, v, \lambda^0)^{\mathrm{T}}(\lambda - \lambda^0) + o(\|\lambda - \lambda^0\|) \tag{6.3.38}$$

where

$$L(x, u, v, \lambda) = f(x, \lambda) + \sum_{i=1}^{m} u_i g_i(x, \lambda) + \sum_{j=1}^{p} v_j h_j(x, \lambda) \tag{6.3.39}$$

is the Lagrangian corresponding to (P_λ) (6.3.1), *and the set of Lagrange multipliers of the problem* (P_{λ^0}) *is denoted by* $N(x^0, \lambda^0) = N_0$, *i.e.*

$$N_0 = \{(u, v) \in \mathbb{E}_m \times \mathbb{E}_p \mid \nabla_x L(x^0, u, v, \lambda^0) = 0,\ u_i g_i(x^0, \lambda^0) = 0,\ u_i \geqq 0,\ i = 1, \ldots, m\}\,. \tag{6.3.40}$$

Proof. The hypotheses of Theorem 6.3.3 are obviously satisfied, there thus exists a function w: $V\{\lambda^0\} \to \mathbb{E}_n$ with the properties $\|w(\lambda)\| = o(\|\lambda - \lambda^0\|)$ and $x^0 + y^{\mathcal{L}}(\lambda) + w(\lambda) \in M(\lambda)$. Exploiting the differentiability of f, the properties of w, hypotheses (2)

[1]) see Note 6.3.2

and (4), and Lemma 6.3.1 we obtain for all $\lambda \in V\{\lambda^0\}$:

$$\begin{aligned}\varphi(\lambda) &\leqq f(x^0 + y^{\mathscr{L}}(\lambda) + w(\lambda), \lambda) = f(x^0, \lambda^0) + \nabla_x f(x^0, \lambda^0)^{\mathrm{T}} \left(y^{\mathscr{L}}(\lambda) + w(\lambda)\right) \\ &\quad + \nabla_\lambda f(x^0, \lambda^0)^{\mathrm{T}} (\lambda - \lambda^0) + o(\|\lambda - \lambda^0\|, \|y^{\mathscr{L}}(\lambda) - y^{\mathscr{L}}(\lambda^0)\|) \\ &= \varphi(\lambda^0) + \nabla_x f(x^0, \lambda^0)^{\mathrm{T}} y^{\mathscr{L}}(\lambda) + \nabla_\lambda f(x^0, \lambda^0)^{\mathrm{T}} (\lambda - \lambda^0) + o(\|\lambda - \lambda^0\|) \,. \end{aligned} \tag{6.3.41}$$

By assumption (4) $y^{\mathscr{L}}(\lambda)$ is a selection function, defined on $V\{\lambda^0\}$, of the optimal set mapping of the linear parametric problem $(\mathrm{P}_\lambda^{\mathscr{L}})$. It is well-known that the dual problem corresponding to $(\mathrm{P}_\lambda^{\mathscr{L}})$ has the form

$$(\mathrm{D}_\lambda^{\mathscr{L}}) \qquad \min \left\{ \sum_{i \in I_0} u_i \nabla_\lambda g_i(x^0, \lambda^0)^{\mathrm{T}} (\lambda - \lambda^0) + \sum_{j=1}^{p} v_j \nabla_x h_j(x^0, \lambda^0)^{\mathrm{T}} (\lambda - \lambda^0) \mid (u, v) \in N_0 \right\}.$$

By the duality theorem of linear optimization and from $u_i = 0$, $i \in \{1, \dots, m\} \setminus I_0$, we have

$$\begin{aligned}\nabla_x f(x^0, \lambda^0)^{\mathrm{T}} y^{\mathscr{L}}(\lambda) = \max_{(u,v) \in N_0} \Bigg\{ & \sum_{i=1}^{m} u_i \nabla_\lambda g_i(x^0, \lambda^0)^{\mathrm{T}} (\lambda - \lambda^0) \\ & + \sum_{j=1}^{p} v_j \nabla_\lambda h_j(x^0, \lambda^0)^{\mathrm{T}} (\lambda - \lambda^0) \Bigg\}. \end{aligned} \tag{6.3.42}$$

Substituting this into (6.3.41) we obtain

$$\varphi(\lambda) \leqq \varphi(\lambda^0) + \max_{(u,v) \in N_0} \{\nabla_\lambda L(x^0, u, v, \lambda^0)^{\mathrm{T}} (\lambda - \lambda^0)\} + o(\|\lambda - \lambda^0\|)$$

where $L(x, u, v, \lambda)$ is given by (6.3.39). The proof is complete. □

Lemma 6.3.4. *Let $V\{\lambda^0\} \subset \Lambda$, $\Lambda \subset \mathfrak{A}$, be given. Furthermore, let the following conditions be fulfilled:*

(1) *$f: \mathbb{E}_n \times \mathbb{E}_k \to \mathbb{R}$ is differentiable at (x^0, λ^0);*
(2) *$x^0 \in \psi(\lambda^0)$;*
(3) *the function $x: V\{\lambda^0\} \to \mathbb{E}_n$ has the properties*

$$\begin{aligned}&\varphi(\lambda) = \varphi(\lambda^0) + \nabla_x f(x^0, \lambda^0)^{\mathrm{T}} \left(x(\lambda) - x^0\right) + \nabla_\lambda f(x^0, \lambda^0)^{\mathrm{T}} (\lambda - \lambda^0) + o(\|\lambda - \lambda^0\|)\,, \\ &x(\lambda^0) = x^0 \end{aligned} \tag{6.3.43}$$

and is locally Lipschitzian at λ^0.

Then

$$|\varphi(\lambda) - f(x(\lambda), \lambda)| = o(\|\lambda - \lambda^0\|)\,.$$

Proof. Assumptions (1) and (3) give

$$\begin{aligned} f(x(\lambda), \lambda) &= f(x^0, \lambda^0) + \nabla_x f(x^0, \lambda^0)^{\mathrm{T}} \left(x(\lambda) - x^0\right) + \nabla_\lambda f(x^0, \lambda^0) (\lambda - \lambda^0) \\ &\quad + o(\|x(\lambda) - x^0\|, \|\lambda - \lambda^0\|) \\ &= f(x^0, \lambda^0) + \varphi(\lambda) - \varphi(\lambda^0) + o\left(\|x(\lambda) - x^0\|, \|\lambda - \lambda^0\|\right). \end{aligned} \tag{6.3.44}$$

By assumption (2) $f(x^0, \lambda^0) = \varphi(\lambda^0)$ holds and since x is locally Lipschitzian at λ^0 Lemma 6.3.1 implies

$$f(x(\lambda), \lambda) = \varphi(\lambda) + o(\|\lambda - \lambda^0\|)$$

thus completing the proof. □

We remark that in his concept of a linear correction E. S. Levitin [2] uses the property (6.3.43) of the extreme value function φ instead of condition (6.3.5).

Theorem 6.3.4.[1]) *Let conditions* (1) *to* (4) *of Lemma* 6.3.4 *be fulfilled and in addition,*

(5) *there exists a selection function* $x(\lambda) \in M(\lambda)$ *defined on* $V\{\lambda^0\}$ *which is locally Lipschitzian at* λ^0 *and has the following properties*:

(a) $f(x(\lambda), \lambda) \leqq \varphi(\lambda) + o(\|\lambda - \lambda^0\|)$,

(b) $x(\lambda^0) = x^0$.

The function $x^0 + y^{\mathscr{Y}}(\lambda)$ *is then an* $\mathbb{L}$*-function at* λ^0.

Proof. It suffices to demonstrate that

$$\varphi(\lambda) \geqq \varphi(\lambda^0) + \max_{(u,v)\in N_0} \{\nabla_\lambda L(x^0, u, v, \lambda^0)^{\mathsf{T}} (\lambda - \lambda^0)\} + o(\|\lambda - \lambda^0\|) \tag{6.3.45}$$

holds, where L and N_0 are given by (6.3.39) and (6.3.40) respectively. Together with (6.3.38) we obtain

$$\varphi(\lambda) = \varphi(\lambda^0) + \max_{(u,v)\in N_0} \{\nabla_\lambda L(x^0, u, v, \lambda^0)^{\mathsf{T}} (\lambda - \lambda^0)\} + o(\|\lambda - \lambda^0\|)$$

and thus the conditions (6.3.43) are fulfilled for $x^0 + y^{\mathscr{Y}}$ because of $(u, v) \in N_0$, (6.3.39), and (6.3.42). Moreover $x^0 + y^{\mathscr{Y}}$ is locally Lipschitzian at λ^0 by assumption (4) so that Lemma 6.3.4 may be invoked:

$$|\varphi(\lambda) - f(x^0 + y^{\mathscr{Y}}(\lambda), \lambda)| = o(\|\lambda - \lambda^0\|)\,.$$

Condition (6.3.5) is also satisfied and (6.3.4) follows from Theorem 6.3.3. We then find that $x^0 + y^{\mathscr{Y}}$ is an $\mathbb{L}$-function if (6.3.45) can be verified. To this purpose we consider $x(\lambda)$ of assumption (5) which function is defined on $V\{\lambda^0\}$. For arbitrary $(u, v) \in N_0$ we obtain

$$\begin{aligned}\varphi(\lambda) &\geqq f(x(\lambda), \lambda) + o(\|\lambda - \lambda^0\|) \geqq L(x(\lambda), u, v, \lambda) + o(\|\lambda - \lambda^0\|)\\ &= L(x^0, u, v, \lambda^0) + \nabla_x L(x^0, u, v, \lambda^0)^{\mathsf{T}} (x(\lambda) - x^0)\\ &\quad + \nabla_\lambda L(x^0, u, v, \lambda^0)^{\mathsf{T}} (\lambda - \lambda^0) + o(\|x(\lambda) - x^0\|, \|\lambda - \lambda^0\|)\,.\end{aligned}$$

Since $x^0 \in \psi(\lambda^0)$ we have $\nabla_x L(x^0, u, v, \lambda^0) = 0$ and $L(x^0, u, v, \lambda^0) = f(x^0, \lambda^0) = \varphi(\lambda^0)$. Using Lemma 6.3.1, which is possible since x is locally Lipschitzian at λ^0, and $x(\lambda^0) = x^0$ the inequality (6.3.45) may then be infered, thus completing the proof. □

We first remark that no convexity conditions imposed on the problem class (P_λ) (6.3.1) were necessary in the preceding theorem. The essential difficulty however in implementing this theorem is confirming that condition (5) is fulfilled for a given nonlinear parametric optimization problem. Lipschitzian selection functions $x(\lambda)$ of the optimal set mapping of problem (P_λ) having the property $x(\lambda^0) = x^0$ obviously also satisfy the condition (5) given in Theorem 6.3.4

$$f(x(\lambda), \lambda) \leqq \varphi(\lambda) + o(\|\lambda - \lambda^0\|)\,.$$

One thus looks for those classes of parametric optimization problem for which the existence of locally Lipschitzian selection functions of the optimal set mapping is assured. We will return later to the question of the existence of such locally Lipschitzian selection functions.

---

[1]) This theorem, as well as the following, may be found in a weaker version in E. S. Levitin [2].

Theorem 6.3.5. *Let conditions* (1) *to* (4) *of Lemma* 6.3.3 *be fulfilled and in addition,*

(6) *there exists a selection function* $x(\lambda) \in M(\lambda)$ *defined on* $V\{\lambda^0\}$ *which is continuous at* λ^0 *and satisfies*

(a) $f(x(\lambda), \lambda) \leqq \varphi(\lambda) + o(\|\lambda - \lambda^0\|)$,

(b) $x(\lambda^0) = x^0$,

and

(7) *the functions* $f(\cdot, \lambda^0)$, $g_i(\cdot, \lambda^0)$, $i \in I_0$, *are convex on* $U(x^0)$ *and the functions* $h_j(\cdot, \lambda^0)$, $j = 1, \dots, p$, *are affine-linear.*

The function $x^0 + y^{\mathcal{I}}(\lambda)$ *is then an* $\mathbb{L}$*-function at* λ^0.

Proof. By Lemmas 6.3.3 and 6.3.4 it suffices to verify the inequality (6.3.45) (see also the proof of Theorem 6.3.4). By assumption (6)

$$\varphi(\lambda) \geqq f(x(\lambda), \lambda) + o(\|\lambda - \lambda^0\|) \tag{6.3.46}$$

holds for the function $x(\lambda) \in M(\lambda)$ considered there. Since $x(\lambda) \in M(\lambda)$ we find that

$$f(x(\lambda), \lambda) \geqq L(x(\lambda), u, v, \lambda)$$

follows for arbitrary $(u, v) \in N_0$, We then obtain

$$\varphi(\lambda) \geqq L(x(\lambda), u, v, \lambda) + o(\|\lambda - \lambda^0\|) \quad \forall (u, v) \in N_0$$

from (6.3.46) and by assumption (1)

$$\varphi(\lambda) \geqq L(x(\lambda), u, v, \lambda^0) + \nabla_\lambda L(x(\lambda), u, v, \lambda^0)^{\mathsf{T}} (\lambda - \lambda^0) + o(\|\lambda - \lambda^0\|) \ \forall (u, v) \in N_0 \tag{6.3.47}$$

then follows. Since x^0 is a stationary point of $L(\cdot, u, v, \lambda^0)$ assumption (7) implies

$$L(x(\lambda), u, v, \lambda^0) \geqq L(x^0, u, v, \lambda^0) = \varphi(\lambda^0) \quad \text{for all} \quad (u, v) \in N_0 . \tag{6.3.48}$$

Because of (1) and (6) (b) the continuity of x gives for each sequence $\{\lambda^k\} \subset V\{\lambda^0\}$, $\lambda^k \to \lambda^0$,

$$\lim_{\lambda^k \to \lambda^0} \frac{|(\nabla_\lambda L(x(\lambda^k), u, v, \lambda^0) - \nabla_\lambda L(x^0, u, v, \lambda^0))^{\mathsf{T}} (\lambda^k - \lambda^0)|}{\|\lambda^k - \lambda^0\|} = 0 ,$$

i.e. $\nabla_\lambda L(x(\lambda), u, v, \lambda^0)^{\mathsf{T}} (\lambda - \lambda^0) = \nabla_\lambda L(x^0, u, v, \lambda^0)^{\mathsf{T}} (\lambda - \lambda^0) + o(\|\lambda - \lambda^0\|)$. Taking (6.3.48) into account we get from (6.3.47)

$$\varphi(\lambda) \geqq \varphi(\lambda^0) + \nabla_\lambda L(x^0, u, v, \lambda^0)^{\mathsf{T}} (\lambda - \lambda^0) + o(\|\lambda - \lambda^0\|) \quad \forall (u, v) \in N_0$$

which then gives (6.3.45). The proof is complete. □

As noted earlier lower semicontinuity (B) of ψ at λ^0 is a sufficient condition for the existence of a selection function $x(\lambda) \in \psi(\lambda)$ which is continuous at λ^0. We refer in particular to Theorem 4.3.6 in this context.

In the next theorem we again assume that a continuous selection function of the optimal set mapping exists, we however replace convexity properties by other conditions. In this connection we recall the definition of a pseudo-face (cf. (5.2.16)). Assume $I \subset \{1, \dots, m\}$ and

$$\Sigma^I(\lambda) = \{x \in E_n \mid g_i(x, \lambda) = 0, i \in I, g_i(x, \lambda) < 0, i \in \{1, \dots, m\} \setminus I, h_j(x, \lambda) = 0, j = 1, \dots, p\} . \tag{6.3.49}$$

Theorem 6.3.6.[1]) *Let the conditions* (1), (2), (4) *of Lemma* 6.3.3 *be fulfilled. Assume further that either*

(i)
- (8) *there exists a selection function* $x(\lambda) \in \psi(\lambda)$ *defined on* $V\{\lambda^0\}$ *which is continuous at* λ^0 *and satisfies* $x(\lambda^0) = x^0$;
- (9) (x^0, λ^0) *satisfies* (R5), *i.e.* (A.5.8);
- (10) $V\{\lambda^0\} \subset \{\lambda \in \mathfrak{A} \mid \psi(\lambda) \subset \Sigma^{I_0}(\lambda)\}$ *where* $\Sigma^{I_0}(\lambda)$ *is given by* (6.3.49);
- (11) $a^{\mathrm{T}} \nabla_x^2 L(x^0, u^0, v^0, \lambda^0)\, a > 0$ *for each* $a \neq 0$ *satisfying*

$$\nabla_x g_i(x^0, \lambda^0)^{\mathrm{T}} a = 0\,, \qquad i \in I_0\,, \qquad \nabla_x h_j(x^0, \lambda^0)^{\mathrm{T}} a = 0\,, \qquad j = 1, \dots, p;$$

or

(ii)
- (8) *and*
- (12) *for each* $(u, v) \in N_0$ (N_0 is given by (6.3.40)) $\nabla_x^2 L(x^0, u, v, \lambda^0)$ *is positive definite, or*

(iii)
- (8), (9) *and*
- (13) (x^0, λ^0) *satisfies* (R8), *i.e.* (A.5.11),
- (14) *the set* I_0 *of indices of the active constraints is equal to the set* $I_1(u^0)$ *of the positive Lagrange multipliers in* (A.5.10) (*strict complementary sladeness*)

Then $x^0 + y^{\mathfrak{L}}(\lambda)$ *is an* $\mathbb{L}$*-function at* λ^0.

Proof. We first show that there exists a selection function of the optimal set mapping of (P_λ) which is locally Lipschitzian under the conditions (1), (2), (4), (8), (9), (10), (11). Then we apply Theorem 6.3.4 by using Lemma A.5.1. Consider the function x of hypothesis (8). The point $x(\lambda)$ satisfies the local Kuhn-Tucker conditions, i.e. the set $N(x(\lambda), \lambda)$ of the corresponding Lagrange multipliers (cf. (6.3.40)) is non-empty. It follows that $(x(\lambda), u(\lambda), v(\lambda), \lambda)$, where $(u(\lambda), v(\lambda)) \in N(x(\lambda), \lambda)$, solves the system of equations

$$\begin{aligned} &\nabla_x L(x, u, v, \lambda) = 0\,, \\ &h_j(x, \lambda) = 0\,, \qquad j = 1, \dots, p\,, \\ &u_i g_i(x, \lambda) = 0\,, \qquad i = 1, \dots, m\,. \end{aligned} \tag{6.3.50}$$

The particular choice of the set $V\{\lambda^0\}$ (hypothesis (10)) implies $g_i(x(\lambda), \lambda) = 0$, $i \in I_0$. It follows that $(x(\lambda), u(\lambda), v(\lambda), \lambda)$ solves the system of equations

$$\begin{aligned} &\nabla_x L(x, u, v, \lambda) = 0\,, \\ &g_i(x, \lambda) = 0\,, \qquad i \in I_0\,, \\ &h_j(x, \lambda) = 0\,, \qquad j = 1, \dots, p\,. \end{aligned} \tag{6.3.51}$$

The regularity of the corresponding Jacobian matrix $J_1(x^0, u^0, v^0, \lambda^0)$ follows from Lemma A.5.2 and hypotheses (9) and (11). Because of the implicit function theorem $(x(\lambda), u(\lambda), v(\lambda))$ is the function uniquely determined by (6.3.51), which is moreover continuously differentiable and thus locally Lipschitzian at λ^0. The first proposition is proved.

Consider the case that (8) and (12) are fulfilled. From (12) and $\nabla_x L(x^0, u, v, \lambda^0) = 0$ for all $(u, v) \in N_0$ it follows that $L(x, u, v, \lambda^0) \geqq L(x^0, u, v, \lambda^0)$ holds for all $x \in E_n$. Using (6.3.47) one may repeat the corresponding part of the proof of Theorem 6.3.5 in analogous fashion to verify inequality (6.3.45). We now show that there exists a selection function of the optimal set mapping of (P_λ) which is locally Lipschitzian at λ^0 under the conditions (1), (2), (4), (9), (13), (14). Consider $(x^0, u^0, v^0, \lambda^0)$ where

[1]) The study of the papers of A. V. Fiacco [3] and R. L. Armacost and A. V. Fiacco [2] was helpfull to prove this theorem.

$x^0 \in \psi(\lambda^0)$ and $(u^0, v^0) \in N_0$. Then $(x^0, u^0, v^0, \lambda^0)$ satisfies the system (6.3.50). The regularity of the corresponding Jacobian matrix $J_1(x^0, u^0, v^0, \lambda^0)$ follows from Lemma A.5.2 and hypotheses (9), (13), (14). Because of the implicit function theorem, there exists a uniquely determined continuously differentiable function $(x(\lambda), u(\lambda), v(\lambda))$ which is defined on a certain neighbourhood of λ^0 and has the following properties:

(i) $(x(\lambda^0), u(\lambda^0), v(\lambda^0)) = (x^0, u^0, v^0)$,

(ii) $(x(\lambda), u(\lambda), v(\lambda))$ satisfies (6.3.50).

Because of the continuity of $x(\lambda)$ and $u(\lambda)$ at λ^0 and (14) it holds that

$$g_i(x(\lambda), \lambda) \leqq 0\,, \qquad u_i(\lambda) \geqq 0\,, \qquad i = 1, \dots, m\,,$$

for all λ in a sufficiently small neighbourhood of λ^0. Then $(x(\lambda), u(\lambda), v(\lambda), \lambda)$ satisfies the local Kuhn-Tucker condition on a certain neighbourhood of λ^0. Now we show that $(x(\lambda), u(\lambda), v(\lambda))$ fulfils condition (R8). The proof follows by contradiction. We assume that there exists a sequence $\{\lambda^k\}$ where $\lambda^k \to \lambda^0$ and a sequence $\{y^k\}$ where

$$y^{k\mathrm{T}} \nabla_x^2 L(x(\lambda^k), u(\lambda^k), v(\lambda^k), \lambda^k)\, y^k \leqq 0\,,$$
$$\nabla_x g_i(x(\lambda^k), \lambda^k)^\mathrm{T} y^k = 0\,, \qquad i \in I(\lambda^k) = I_0 = I_1(u^0) \quad (\text{cf. } (14))\,,$$
$$\nabla_x h_j(x(\lambda^k), \lambda^k)^\mathrm{T} y_k = 0\,, \qquad j = 1, \dots, p\,,$$

Without loss of generality let $\|y^k\| = 1$. Without loss of generality * $\{y^k\}$ converges to $\bar{y}$ where $\bar{y} \neq 0$. Then we obtain for $k \to \infty$

$$\bar{y}^\mathrm{T} \nabla_x^2 L(x^0, u^0, v^0, \lambda^0)\, \bar{y} \leqq 0\,,$$
$$\nabla_x g_i(x^0, \lambda^0)^\mathrm{T}\, \bar{y} = 0\,, \qquad i \in I_0\,,$$
$$\nabla_x h_j(x^0, \lambda^0)^\mathrm{T}\, \bar{y} = 0\,, \qquad j = 1, \dots, p\,.$$

This is a contradiction to (13). Then $x(\lambda)$ is a locally unique local minimal point of (P_λ) (by using Lemma A.5.3) if λ lies in a sufficiently small neighbourhood of λ^0. Let $V\{\lambda^0\}$ be this neighbourhood what yields that $x(\lambda)$ is the function being assumed in (8). Then $x(\lambda)$ is a selection function of $\psi(\lambda)$ which is locally Lipschitzian at λ^0 and we apply Theorem 6.3.4. □

We now consider a class of parametric optimization problems in which the objective and constraint functions have a special form.

Theorem 6.3.7. *Let conditions* (1) *to* (4) *of Theorem* 6.3.5 *be fulfilled. Assume further that either* (7) (Theorem 6.3.5) *and in addition*

(15) (a) *there exist functions* $g_{i1}(x)$ *and* $g_{i2}(\lambda)$, $i \in I_0$, *such that* $g_i(x, \lambda)$, $i \in I_0$, *may be represented in the form*

$$g_i(x, \lambda) = g_{i1}(x) + g_{i2}(\lambda)\,, \qquad i \in I_0\,,$$

(b) *there exist functions* $h_{j1}(x)$ *and* $h_{j2}(\lambda)$, $j = 1, \dots, p$, *such that* $h_j(x, \lambda)$, $j = 1, \dots, p$, *may be represented in the form*

$$h_j(x, \lambda) = h_{j1}(x) + h_{j2}(\lambda)\,, \qquad j = 1, \dots, p\,,$$

(c) $f(x, \lambda) \equiv f(x) \qquad \forall \lambda \in \mathbb{E}_k$,

or (12) (Theorem (6.3.6) *and* (15) *are also satisfied.*

Then $x^0 + y^\mathcal{I}(\lambda)$ *is an* $\mathbb{L}$*-function at* λ^0.

Proof. In both cases we demonstrate the validity of inequality (6.3.45). First assume that (1) to (4) and (7) and (15) hold. Choosing $x(\lambda) \in \psi(\lambda)$ we then have

$$\varphi(\lambda) = f(x(\lambda), \lambda) \geqq L(x(\lambda), u, v, \lambda)$$

for arbitrary $(u, v) \in N_0$ (N_0 is given in (6.3.40)). By hypothesis (15) we have then $\nabla_\lambda L(x(\lambda), u, v, \lambda^0) = \nabla_\lambda L(x^0, u, v, \lambda^0)$ and as in the proof of Theorem 6.3.5 we can conclude that (6.3.45) is valid. Now assume (1) to (4) and (12) and (15). From $\nabla_x L(x^0, u, v, \lambda^0) = 0$, $(u, v) \in N_0$, and (12) it follows that $L(x, u, v, \lambda^0) \geqq L(x^0, u, v, \lambda^0)$ holds for all $x \in \mathbb{E}_n$. The proof for this case then follows as above. □

Convex optimization problems with parameter-dependent right-hand sides of the constraints obviously fulfil conditions (7) and (15), we therefore have established preliminaries for the calculation of an $\mathbb{L}$-function for an important class of optimization problems in many applications.

In the following we discuss the more important conditions of Theorems 6.3.4 to 6.3.7. We first give some properties of the multiparametric linear optimization problem $(P_\lambda^{\mathscr{Y}})$ in (6.3.14). In Lemma 6.3.2 we found that the feasible parameter set $\mathfrak{B}^{\mathscr{Y}}$ is equal to $\mathbb{E}_k$ if the constraints qualification (R1) is satisfied; this latter was included as hypothesis (3) in Theorems 6.3.4 to 6.3.7. After proving Theorem 6.3.3 we demonstrated with some examples that this constraints qualification cannot be replaced by weaker ones. Since Theorem 6.3.3 is necessary for the proofs of Theorems 6.3.4 to 6.3.7 (the former establishes property (6.3.4) of an $\mathbb{L}$-function) the condition (R1) is essential for the problem classes considered in Theorems 6.3.4 to 6.3.7. We now take up the question of what property the solubility set possesses under this condition.

Lemma 6.3.5. *Assume that* (R1) *as given in* (A.5.4) *is satisfied for* (x^0, λ^0) *where* $\lambda^0 \in \mathfrak{A}$ *and* x^0 *is a local optimal point of* (P_{λ^0}). *Then* $\mathfrak{A}^{\mathscr{Y}} = \mathbb{E}_k$.

Proof. By Lemma 6.3.2 we have $\mathfrak{B}^{\mathscr{Y}} = \mathbb{E}_k$. Since (R1) is fulfilled at (x^0, λ^0) and x^0 is a local optimal point for (P_{λ^0}) the set of Lagrange multipliers in (6.3.40) is non-empty (cf. Lemma A.5.1). As is well-known N_0 is the constraint set of the dual problem corresponding to (P_{λ^0}), the lemma then follows from the duality theorem of linear optimization. □

Note 6.3.2. The results of Section 5.5 imply that a continuous and piece-wise linear selection function $y^{\mathscr{Y}}(\lambda) \in \psi^{\mathscr{Y}}(\lambda)$ which is locally Lipschitzian at λ^0 exists on $\mathfrak{A}^{\mathscr{Y}}$.

Because of hypothesis (4) of Theorems 6.3.4 to 6.3.7 we need such a function $y^{\mathscr{Y}}(\lambda) \in \psi^{\mathscr{Y}}(\lambda)$ satisfying $y^{\mathscr{Y}}(\lambda^0) = 0$. We have the following lemma.

Lemma 6.3.6. *Assume that* $\lambda^0 \in \mathfrak{A}$, x^0 *is a local optimal point of* (P_{λ^0}) *and* (x^0, λ^0) *satisfies the constraints qualification* (R1).

Then

(1) $\psi^{\mathscr{Y}}(\lambda^0) = \{x \in M^{\mathscr{Y}}(\lambda^0) \mid \nabla_x g_i(x^0, \lambda^0)^{\mathsf{T}} x = 0, i \in I_0 \setminus I_2, \nabla x\, g_i(x^0, \lambda^0)^{\mathsf{T}} x \leqq 0, i \notin I_0 \setminus I_2\}$ *where* $I_2 = \{i \in I_0 \mid u_i = 0\ \forall (u, v) \in N_0\}$ *and* N_0 *is given by* (6.3.40),

(2) *if* $\widetilde{M}^{\mathscr{Y}}(\lambda^0)$ *is the polyhedron obtained from* $M^{\mathscr{Y}}(\lambda^0)$ *by introducing slack variables and non-negative variables it follows that an optimal vertex of* $\widetilde{M}^{\mathscr{Y}}(\lambda^0)$ *may be assigned to* $y(\lambda^0) = 0$.

Proof. (1) This easily follows by application of the duality theorem of linear optimization.

(2) All inequalities occurring in $M^{\mathscr{Y}}(\lambda^0)$ are fulfilled as equalities by $y(\lambda^0) = 0$. Then one may assign a vertex of the polyhedron $\widetilde{M}^{\mathscr{Y}}(\lambda^0)$ to $y(\lambda^0) = 0$. It follows from (1) that $y(\lambda^0) = 0$ is optimal for $(P_{\lambda^0}^{\mathscr{Y}})$ and the vertex of $\widetilde{M}^{\mathscr{Y}}(\lambda^0)$ assigned to $y(\lambda^0) = 0$ is thus also optimal for the problem $(\widetilde{P}_{\lambda^0}^{\mathscr{Y}})$ having the objective function $\nabla_x f(x^0, \lambda^0)^{\mathsf{T}}\, y$ and the constraint set $\widetilde{M}^{\mathscr{Y}}(\lambda^0)$. □

On the basis of this lemma and the results of Section 6.2 one may easily find a procedure for determining a piece-wise linear and continuous selection function $y^{\mathcal{L}}(\lambda) \in \psi(\lambda)$ with $y^{\mathcal{L}}(\lambda^0) = 0$ based on the simplex technique. In the following we outline this procedure. Assume that problem $(\mathrm{P}_\lambda^{\mathcal{L}})$ has been converted to a problem $(\tilde{\mathrm{P}}_\lambda^{\mathcal{L}})$ (equation form, sign restrictions).

We determine an optimal simplex tableau for $(\tilde{\mathrm{P}}_\lambda^{\mathcal{L}})$ with $y(\lambda^0) = 0$ from which we read off the associated parameter set (usually a subset of a local stability set) and the corresponding vector function. By computing all adjacent parameter sets and the corresponding vector function we can compose a piece wise linear continuous vector function defined on a neighbourhood $V\{\lambda^0\}$.

We now illustrate Theorem 6.3.4 with an example.

Example 6.3.11. We employ example 6.3.1:

$$(\mathrm{P}_\lambda) \qquad \min\{-x_1^2 - x_2 \mid x_1^2 + x_2^2 - \lambda \leqq 0\}, \qquad \lambda^0 = \tfrac{1}{4},$$

and choose $x^0 = (0, \frac{1}{2}) \in \psi(\lambda^0)$ (condition (2)). Conditions (1) and (3) are also fulfilled. The selection function $\bar{x}(\lambda) = (0, \sqrt{\lambda})$ is a function which is locally Lipschitzian at $\lambda^0 = \frac{1}{4}$ (condition (5)). The at $(x^0, \lambda^0) = (0, \frac{1}{2}, \frac{1}{4})$ linearized problem is given by

$$(\mathrm{P}_\lambda^{\mathcal{L}}) \qquad \min\{-x_2 \mid x \in M^{\mathcal{L}}(\lambda)\},$$

where $M^{\mathcal{L}}(\lambda) = \{x \in \mathbb{E}_2 \mid x_2 \leqq \lambda - \frac{1}{4}\}$. We have $\psi^{\mathcal{L}}(\lambda) = \{(x_1, \lambda - \frac{1}{4}) \mid x_1 \in \mathbb{R} \text{ arbitrary}\}$. Then, for instance the functions

$$y^{\mathcal{L}}(\lambda) = ((\lambda - \tfrac{1}{4})^n, \lambda - \tfrac{1}{4}), \qquad n = 1, 2, \dots,$$

$$y^{\mathcal{L}}(\lambda) = (0, \lambda - \tfrac{1}{4})$$

are locally Lipschitzian selection functions of $\psi^{\mathcal{L}}(\lambda)$ at λ^0 which moreover satisfy $y^{\mathcal{L}}(\lambda^0) = 0$. The only functions of these which are (piece-wise) linear are $y^{\mathcal{L}}(\lambda) = (\lambda - \frac{1}{4}, \lambda - \frac{1}{4})$ and $y^{\mathcal{L}}(\lambda) = (0, \lambda - \frac{1}{4})$. We then obtain $\mathbb{L}$-functions at λ^0 given by $x^0 + y^{\mathcal{L}}(\lambda) = ((\lambda - \frac{1}{4})^n, \lambda + \frac{1}{4})$, $n = 1, 2, \dots$, and $x^0 + y^{\mathcal{L}}(\lambda) = (0, \lambda + \frac{1}{4})$. We note that $(\lambda - \frac{1}{4}, \lambda + \frac{1}{4})$ is not an $\mathbb{F}$-function at λ, whereas $(0, \lambda + \frac{1}{4})$ has this property, thus one in general does not obtain $\mathbb{F}$-functions with the help of the linearized problem $(\mathrm{P}_\lambda^{\mathcal{L}})$.

If one considers the functions $x(\lambda) = (0, \lambda^2 + \frac{1}{2}\lambda + \frac{5}{16})$ and $x(\lambda) = (0, -\lambda^2 + \frac{3}{2}\lambda + \frac{3}{16})$ then one finds that they are also $\mathbb{L}$-functions at λ^0, they however may not be obtained from the linearized problem. It is thus apparent that although Theorem 6.3.4 provides a basis for a convenient method for calculating $\mathbb{L}$-functions we in general do not find all $\mathbb{L}$-functions by this approach.

The question of whether the class of locally Lipschitzian selection functions $y^{\mathcal{L}}(\lambda) \in \psi^{\mathcal{L}}(\lambda)$ with the property $y^{\mathcal{L}}(\lambda^0) = 0$ which yield $\mathbb{L}$-functions under conditions (1), (2), (3), and (5) of Theorem 6.3.4, can be extended has been answered in the negative by Examples 6.3.9 and 6.3.10.

In the following example we show that condition (5) cannot be replaced by the existence of a selection function that is continuous at λ^0.

Example 6.3.12

$$(\mathrm{P}_\lambda) \qquad \min\{x_2 \mid x_1^2 - x_2 \leqq 0, -x_1^2 - x_2 + \lambda \leqq 0\}, \qquad \lambda \geqq 0.$$

We obtain

$$\psi(\lambda) = \{(\sqrt{\tfrac{1}{2}\lambda}, \tfrac{1}{2}\lambda), (-\sqrt{\tfrac{1}{2}\lambda}, \tfrac{1}{2}\lambda)\},$$
$$\varphi(\lambda) = \tfrac{1}{2}\lambda.$$

Taking $\lambda^0 = 0$ and $x^0 = (0, 0) \in \psi(\lambda^0)$ we then have the problem

$$(\mathrm{P}_\lambda^{\mathscr{I}}) \qquad \min\{x_2 \mid -x_2 \leqq 0, -x_2 \leqq -\lambda\}$$

where $\psi^{\mathscr{I}}(\lambda) = \{x \in E_2 \mid x_2 = \lambda\}$. Choosing $y^{\mathscr{I}}(\lambda) = (0, \lambda) \in \psi^{\mathscr{I}}(\lambda)$ we find that $x(\lambda) = x^0 + y^{\mathscr{I}}(\lambda) = (0, \lambda)$ is not an $\mathbb{L}$-function at $\lambda^0 = 0$ since

$$\lim_{\lambda\to\lambda_0} \frac{|\varphi(\lambda) - f(x(\lambda), \lambda)|}{\|\lambda - \lambda^0\|} = \frac{1}{2}$$

holds.

We note that although two selection functions of $\psi(\lambda)$ which are continuous at λ^0 exist neither of them is locally Lipschitzian at λ^0. The existence of locally Lipschitzian selection functions of the optimal set mapping has up to now not been examined to any extent in the literature; much more attention has been payed to the existence of continuous selection functions x satisfying $x(\lambda^0) = x^0$ as was postulated in Theorem 6.3.5. As already noted there is a close connection to lower semicontinuity of the optimal set mapping at λ^0 in this respect. The lower semicontinuity results presented in Chapters 4 and 5 with respect to the optimal set mapping as well as the constraint set mapping thus have a special significance for the investigations presented in this section.

Appendix

In this volume we frequently use certain concepts and theorems, mostly more than once, which however belong to other areas of optimization theory and which are not further explained in the main body of the text in order to keep our presentation as concise as possible. This subject matter is therefore relegated to the Appendix which is accordingly divided into five sections: A.1 is concerned with convex polyhedra cones and their polar cones, A.2 with various classes of square matrices, A.3 is devoted to the solution set of linear complementarity problems, A.4 deals with convex quadratic optimization problems and finally A.5 contains material on constraint qualifications and optimality conditions for non-linear optimization problems.

A.1. Convex Polyhedral Cones and Associated Polar Cones

(This and other related material may for instance be found in F. Nožička et al. [1], S. N. Černikov [1], and R. T. Rockafellar [2].)

A set $M \subset E_n$ is a convex polyhedron if it may be represented as the intersection of a finite number of closed halfspaces, i.e. in the form

$$M = \{x \in E_n \mid Ax \leqq b\}$$

where A is an (m, n)-matrix. If $I \subset \{1, \dots, m\}$ and a_i is the i-th row of the matrix A, then a non-empty set

$$S^I = \{x \in M \mid a_i x = b_i, i \in I, a_j x < b_j, j \in \{1, \dots, m\} \backslash I$$

which is not equal to ri M is called (open) *face* of M and I is the *characteristic index set of the face* S^I. The corresponding closed face is then

$$\bar{S}^I = \{x \in M \mid a_i x = b_i, i \in I, a_j x \leqq b_j, j \in \{1, \dots, m\} \backslash I .$$

If the relation

$$\text{ri } M = \{x \in M \mid a_i x = b_i, i \in I, a_j x < b_j, j \in \{1, \dots, m\} \backslash I$$

holds for an index set I then $\dim M = \dim \{y \mid a_i y = 0, i \in I\}$. A set $Z \subset E_n$ is called (affine) *convex polyhedral cone* if Z is both a convex polyhedron and a (affine) cone. The definition of a (affine) cone and its vertex set is given in Section 2.1. A convex polyhedron is an affine cone if and only if it possesses not more than one face of minimal dimension. If it has one and only one face of smallest dimension then this face constitutes the vertex set, if it has no face of minimal dimension then it does not possess any faces, it is therefore an affine subspace and identical to its vertex set.

In the following we restrict our considerations to convex polyhedral cones. In this case the vertex set is a linear subspace.

A set of the form

$$Z = \{x \mid Ax \leqq 0\}$$

is a convex polyhedral cone with the vertex set

$$S_0 = \{x \mid Ax = 0\} .$$

Every convex polyhedral cone may be represented in this form.

Assume that Z is a convex polyhedral cone with a vertex set S_0 whose dimension is equal to d. If $\dim Z \geqq d + 2$ and $S_1^{d+1}, \dots, S_r^{d+1}$ are all $(d + 1)$-dimensional faces of Z then each collection of vectors $h^1, \dots, h^r$ satisfying $h^i \in S_i^{d+1}$, $i = 1, \dots, r$, constitutes a fundamental system of Z by definition. If $\dim Z = d + 1$ then each vector $h^i \in \operatorname{ri} Z$ is a fundamental system of Z and in the case $\dim Z = d$ the fundamental system of Z is the empty set.

Theorem A.1.1

(1) *If $h^1, \dots, h^r$ is a fundamental system of a convex polyhedral cone Z and if $k^1, \dots, k^d$ is a basis of its vertex set S_0 then Z and* ri *Z may be represented as*

$$Z = \left\{x \,\middle|\, x = \sum_{i=1}^{r} \alpha_i h^i + \sum_{j=1}^{d} \beta_j k^j, \alpha_i \geqq 0 \ \forall i, \beta_j \in R \ \forall j\right\}, \tag{A.1.1}$$

and

$$\operatorname{ri} Z = \left\{x \,\middle|\, x = \sum_{i=1} \alpha_i h^i + \sum_{j=1}^{d} \beta_j k^j, \alpha_i > 0 \ \forall i, \beta_j \in R \ \forall j\right\} \tag{A.1.2}$$

respectively.

(2) *Every set Z which may be represented in the form* (A.1.1) *is a convex polyhedral cone and* ri *Z then has the representation* (A.1.2) (*the vectors h^i and k^j then need not necessarily be a fundamental system Z and a basis of S_0 respectively*).

Let Z be a cone. The set

$$Z^p = \{y \mid x^\mathsf{T} y \leqq 0 \ \forall x \in Z\}$$

is then the polar cone of Z. Since the relation $Z^{pp} = \operatorname{cl} \operatorname{conv} Z$ always holds, $Z^{pp} = Z$ follows if Z is convex and closed (for instance if Z is a convex polyhedral cone).

Theorem A.1.2. *If Z is a convex polyhedral cone, $h^1, \dots, h^r$ a fundamental system of Z and $k^1, \dots, k^d$ a basis of its vertex set S_0, then Z^p and* ri *Z^p may be represented as*

$$Z^p = \{y \mid h^{i\mathsf{T}} y \leqq 0, i = 1, \dots, r, k^{j\mathsf{T}} y = 0, j = 1, \dots, d\}$$

and

$$\operatorname{ri} Z^p = \{y \mid h^{i\mathsf{T}} y < 0, i = 1, \dots, r, k^{j\mathsf{T}} y = 0, j = 1, \dots, d\}$$

respectively. The set Z^p is thus also a convex polyhedral cone.

Theorem A.1.3. *If $Z \subset E_n$ is a convex polyhedral cone with the vertex set S_0 then* $\dim Z^p + \dim S_0 = n$.

If $Z = \{x \mid Ax \leqq 0\}$ then $Z^p = \{y \mid y = A^\mathsf{T} u, u \geqq 0\}$ follows. The set $\{x \mid Ax = b, x \geqq 0\}$ is non-empty if and only if $b \in \{u \mid A^\mathsf{T} u \leqq 0\}^p$ holds (Theorem of Farkas).

A.2. Certain Classes of Square Matrices and Their Interrelations

(We refer the reader to, for instance, C. E. LEMKE [2] and H.-J. LÜTHI [1].)

The following is a list of various classes of square matrices of order n and the symbols used to denote them.

$\tilde{\mathbb{H}}^+(n)$ — Class of all positive semidefinite matrices K defined by $x^\mathsf{T}Kx \geqq 0 \ \forall x \in \mathbb{E}_n$;

$\tilde{\mathbb{H}}^{++}(n)$ — Class of all positive definite matrices K defined by $x^\mathsf{T}Kx > 0 \ \forall x \neq 0$;

$\mathbb{H}^c(n)$ — Class of all copositive plus matrices K defined by $x^\mathsf{T}Kx \geqq 0 \ \forall x \geqq 0$ and $(x^\mathsf{T}Kx = 0 \wedge x \geqq 0) \Rightarrow (K + K^\mathsf{T})\, x = 0$;

$\mathbb{H}^{P_0}(n)$ — Class of all matrices whose main minors are all non-negative;

$\mathbb{H}^P(n)$ — Class of all matrices whose main minors are all positive;

$\mathbb{H}^{E_0}(n)$ — Class of all matrices K with the property that for all $x \in \mathbb{E}_n$ satisfying $0 \neq x \geqq 0$ there exists an index i such that $x_i > 0$ and $(Kx)_i \geqq 0$ hold;

$\mathbb{H}^E(n)$ — Class of all matrices K with the property that for all $x \in \mathbb{E}_n$ satisfying $0 \neq x \geqq 0$ there exists an index i such that $x_i > 0$ and $(Kx)_i > 0$ hold;

$\mathbb{H}^{E_1}(n)$ — Class of all matrices K with the property that for all $x \in \mathbb{E}_n$ satisfying $0 \neq x \geqq 0$, $Kx \geqq 0$ and $x^\mathsf{T}Kx = 0$ there exist non-negative diagonal matrices Λ and Ω such that $(\Lambda K + K^\mathsf{T}\Omega)\, x = 0$ and $\Omega x \neq 0$ hold;

$\mathbb{H}^L(n)$ — Class defined by $\mathbb{H}^{E_0}(n) \cap \mathbb{H}^{E_1}(n)$;

$\mathbb{H}^T(n)$ — Class of all matrices K with the property that for all $x \geqq 0$ satisfying $K^\mathsf{T}x \leqq 0$ there exists a non-negative diagonal matrix T such that $(TK^\mathsf{T} + K)\, x = 0$ holds;

$\mathbb{H}^A(n)$ — Class of all antisymmetric matrices K defined by $K = -K^\mathsf{T}$;

$\mathbb{H}^S(n)$ — Class of all symmetric matrices K defined by $K = K^\mathsf{T}$;

$\mathbb{H}^Z(n)$ — Class of all matrices K whose elements are all non-positive except for those in the main diagonal.

The following relations hold between the above classes:

$$\mathbb{H}^A(n) \subset \tilde{\mathbb{H}}^+(n) \subset \mathbb{H}^C(n) \subset \left\{ \begin{matrix} \mathbb{H}^L(n) \\ \mathbb{H}^{P_0}(n) \end{matrix} \right\} \subset \mathbb{H}^{E_0}(n)\,,$$

$$\tilde{\mathbb{H}}^{++}(n) \subset \mathbb{H}^P(n) \subset \mathbb{H}^E(n) \subset \mathbb{H}^L(n) \subset \mathbb{H}^{E_1}(n)\,,$$

$$\tilde{\mathbb{H}}^+(n) \subset \mathbb{H}^T(n)\,, \quad \tilde{\mathbb{H}}^{++}(n) \subset \tilde{\mathbb{H}}^+(n)\,, \quad \mathbb{H}^P(n) \subset \mathbb{H}^{P_0}(n)\,, \quad \mathbb{H}^E(n) \subset \mathbb{H}^{E_0}(n)\,,$$

$$\mathbb{H}^S(n) \cap \mathbb{H}^{P_0}(n) = \mathbb{H}^S(n) \cap \tilde{\mathbb{H}}^+(n)\,, \quad \mathbb{H}^S(n) \cap \mathbb{H}^P(n) = \mathbb{H}^S(n) \cap \tilde{\mathbb{H}}^{++}(n)\,.$$

Class $\mathbb{H}^Z(n)$ is neither contained in nor contains any other of the above classes.

A.3. Various Results on the Solution Set of a Linear Complementarity Problem

(Some references are B. C. EAVES [2], C. E. LEMKE [1], H.-J. LÜTHI [1], W. S. DORN [1], R. W. COTTLE [3] and S. R. MOHAN [1].)

We consider the following linear complementarity problem:

$$\begin{aligned} &\text{Determine the set } N = \{(x, y) \in M \mid x^\mathsf{T}y = 0\} \\ &\text{where} \quad M = \{(x, y) \in \mathbb{E}_{2n} \mid -Kx + y = q,\, x \geqq 0,\, y \geqq 0\}\,. \end{aligned} \tag{A.3.1}$$

Theorem A.3.1. *If $K \in H^L(n) \cup H^Z(n)$ then $M \neq \emptyset$ implies $N \neq \emptyset$.*

Theorem A.3.2. *If $K \in H^L(n) \cup H^Z(n)$ then (A.3.1) is equivalent to the quadratic optimization problem*

$$\min \{x^\mathrm{T} Kx + q^\mathrm{T} x \mid Kx + q \geqq 0, x \geqq 0\} \tag{A.3.2}$$

in the sense that

(1) if (x^0, y^0) is a solution of (A.3.1) then x^0 is an optimal solution of (A.3.2) and $x^{0\mathrm{T}} Kx^0 + q^\mathrm{T} x^0 = 0$,

(2) if x^0 is an optimal solution of (A.3.2) then (x^0, y^0) is a solution of (A.3.1) where $y^0 = Kx^0 + q$.

Theorem A.3.3

(1) If $K \in \widetilde{H}^+(n) \cup H^P(n)$ and $M \neq \emptyset$ then the set N is a closed face of M or is equal to M.

(2) If $K \in H^P(n)$ then M is always non-empty and N contains only one element, the latter is thus a uniquely determined vertex of M.

Proof. The case $K \in H^P(n)$ may be found in H. Samelson et al. [1] and we thus restrict ourselves to the case $K \in \widetilde{H}^+(n)$ with respect to (1). Clearly, if K is an arbitrary (n, n)-matrix then one always has

$$N = \bigcup_{(I,J)\in S} \{(x, y) \in M \mid x_i = 0, i \in I, y_j = 0, j \in J\}$$

where

$$S = \{(I, J) \mid I \cup J = \{1, \dots, n\}\}\,.$$

The set N is thus the union of a finite number of closed faces of M or is equal to M itself. However (A.3.2) is a convex problem in the case $K \in \widetilde{H}^+(n)$. By Theorem A.3.2 N is then necessarily convex and it can only be equal to a closed face of M or to M itself. □

In the case $K \in \widetilde{H}^+(n) \cup H^P(n)$ assuming $M \neq \emptyset$ let the pair $(\hat{I}, \hat{J})$ be the uniquely determined characteristic index set pair corresponding to N so that

$$N = \{(x, y) \in M \mid x_i = 0, i \in \hat{I}, y_j = 0, j \in \hat{J}\}$$

and

$$\operatorname{ri} N = \{(x, y) \in N \mid x_i > 0, i \notin \hat{I}, y_j > 0, j \notin \hat{J}\}$$

hold. We always have $\hat{I} \cup \hat{J} = \{1, \dots, n\}$. If in addition $K \in H^A(n)$ then $\hat{I} \cap \hat{J} = \emptyset$ is also true.

Theorem A.3.4

(1) Assume $K \in H^{E_1}(n) \cup [H^Z(n) \cap (H^{P_0}(n) \cup H^S(n))]$.
If the set

$$M^0 = \{(x, y) \in E_{2n} \mid -Kx + y = q, x > 0, y > 0\}$$

is non-empty then the set N is bounded.

(2) Assume $K \in H^T(n)$.
If $M^0 = \emptyset$ and $N \neq \emptyset$ then the set N is unbounded.

Proof. (1) The proof has been given by S. R. Mohan [1] for the case $K \in H^Z(n) \cap (H^{P_0}(n) \cup H^S(n))$. Here we generalize the proof of R. W. Cottle [3] for $H^C(n)$ to the case $K \in H^{E_1}(n)$.

According to R. W. COTTLE [3] the set N is unbounded if and only if a pair $(\bar{x}, \bar{v})$ with $\bar{v} \neq 0$ exists such that the conditions

$$K\bar{x} + q \geqq 0\,, \qquad \bar{x} \geqq 0\,, \qquad \bar{x}^{\mathsf{T}}K\bar{x} + q^{\mathsf{T}}\bar{x} = 0\,, \tag{A.3.3}$$

$$K\bar{v} \geqq 0\,, \qquad \bar{v} \geqq 0\,, \qquad \bar{v}^{\mathsf{T}}K\bar{v} = 0\,, \tag{A.3.4}$$

$$\bar{v}^{\mathsf{T}}(K\bar{x} + q) = 0\,, \tag{A.3.5}$$

$$\bar{x}^{\mathsf{T}}K\bar{v} = 0 \tag{A.3.6}$$

are satisfied.

We must therefore demonstrate that $\bar{v} = 0$ holds for each pair $(\bar{x}, \bar{v})$ satisfying (A.3.3) to (A.3.6).

Under the assumption $K \in \mathbb{H}^{E_1}(n)$ condition (A.3.4) implies the existence of diagonal matrices Λ, $\Omega \geqq 0$ satisfying

$$K^{\mathsf{T}}\Omega\bar{v} + \Lambda K\bar{v} = 0$$

and $\Omega\bar{v} \neq 0$ if $\bar{v} \neq 0$. For each $x \in E_n$ we then have

$$x^{\mathsf{T}}K^{\mathsf{T}}\Omega\bar{v} + x^{\mathsf{T}}\Lambda K\bar{v} = 0\,.$$

Because of $\bar{x} \geqq 0$, $K\bar{v} \geqq 0$ and $\bar{x}^{\mathsf{T}}K\bar{v} = 0$ it obviously follows that $\bar{x}^{\mathsf{T}}\Lambda K\bar{v} = 0$ and hence

$$\bar{x}^{\mathsf{T}}K^{\mathsf{T}}\Omega\bar{v} = 0$$

holds. Analogously

$$(K\bar{x} + q)^{\mathsf{T}}\,\Omega v = 0$$

because of $K\bar{x} + q \geqq 0$, $\bar{v} \geqq 0$ and $(K\bar{x} + q)^{\mathsf{T}}\,\bar{v} = 0$. By hypothesis one can find an $x^0 > 0$ satisfying $Kx^0 + q > 0$. If $\bar{v} \neq 0$ were true then $\Omega\bar{v} \neq 0$ and consequently

$$0 = (K\bar{x} + q)^{\mathsf{T}}\,\Omega v > -x^{0\mathsf{T}}K^{\mathsf{T}}\Omega\bar{v} + \bar{x}^{\mathsf{T}}K^{\mathsf{T}}\Omega\bar{v}$$
$$= -x^{0\mathsf{T}}K^{\mathsf{T}}\Omega\bar{v} = x^{0\mathsf{T}}\Lambda K\bar{v} \geqq 0$$

would follow, which however leads to a contradiction. It follows that $\bar{v} = 0$ must hold, i.e. N is bounded.

(2) Assume $\mathfrak{B} = \{\lambda \mid \lambda = -Kx + y, x \geqq 0, y \geqq 0\}$. This set is obviously a convex polyhedral cone and

$$\operatorname{int}\mathfrak{B} = \{\lambda \mid \lambda = -Kx + y, x > 0, y > 0\}$$

holds. Since $M^0 = \emptyset$ we have $q \in \partial\mathfrak{B}$. By a separation theorem a supporting hyperplane thus exists for $\mathfrak{B}$ at the point $\lambda = q$, i.e. there exists a vector $\bar{v} \neq 0$ satisfying $q^{\mathsf{T}}\bar{v} = 0$ and $\bar{v}^{\mathsf{T}}\lambda \geqq 0$ for all $\lambda \in \mathfrak{B}$. This implies $\bar{v}^{\mathsf{T}}(-Kx + y) = -\bar{v}^{\mathsf{T}}Kx + \bar{v}^{\mathsf{T}}y \geqq 0$ for all $x \geqq 0$, $y \geqq 0$ and it necessarily follows that $K^{\mathsf{T}}\bar{v} \leqq 0$ and $\bar{v} \geqq 0$ hold. Because of $K \in \mathbb{H}^{\mathsf{T}}(n)$ we have $K\bar{v} = -TK^{\mathsf{T}}\bar{v} \geqq 0$, and $K^{\mathsf{T}}\bar{v} \leqq 0$, $K\bar{v} \geqq 0$ and $\bar{v} \geqq 0$ yield

$$\bar{v}^{\mathsf{T}}K\bar{v} = 0\,.$$

Assume now that $\bar{x} \in N$ is arbitrary. Since $\bar{v} \geqq 0$ and $K\bar{x} + q \geqq 0$ we have

$$\bar{v}^{\mathsf{T}}(K\bar{x} + q) \geqq 0\,.$$

However on the other hand we also have $\bar{v}^{\mathsf{T}}(K\bar{x} + q) \leqq 0$ because of $\bar{v}^{\mathsf{T}}K \leqq 0$, $\bar{x} \geqq 0$, and $\bar{v}^{\mathsf{T}}q = 0$. This and the previous inequality then imply

$$\bar{v}^{\mathsf{T}}(K\bar{x}) + q) = 0\,.$$

Finally,

$$0 = -\bar{x}^{\mathsf{T}} T K^{\mathsf{T}} \bar{v} = \bar{x}^{\mathsf{T}} K \bar{v}$$

follows from $\bar{x} \geqq 0$, $K^{\mathsf{T}}\bar{v} \leqq 0$ and $\bar{x}^{\mathsf{T}} K^{\mathsf{T}} \bar{v} = \bar{v}^{\mathsf{T}} K \bar{x} = 0$ since $K \in \mathbb{H}^{\mathsf{T}}(n)$. The pair $(\bar{x}, \bar{v})$ therefore fulfills the conditions (A.3.3) to (A.3.6) for arbitrary $\bar{x} \in N$, the set N is thus unbounded. □

Theorem A.3.5. *Assume* $K \in \widetilde{\mathbb{H}}^{+}(n) \cup \mathbb{H}^{S}(n) \cup \mathbb{H}^{A}(n) \cup \mathbb{H}^{P}(n)$.

To each vertex (x^0, y^0) *of* M *there corresponds a basis with the property that for each index* $i \in \{1, \dots, n\}$ *one and only one of the variables* x_i *or* y_i *is a basis variable.*

This theorem is trivial if (x^0, y^0) is non-degenerate. The proof for the degenerate case is however relatively long and for space reasons we therefore refer the reader to K. Tammer [9] or K. Tammer [6] for an important special case.

A.4. Convex Quadratic Optimization Problems

(See for instance H. P. Künzi and W. Krelle [1] and K.-H. Elster et al. [1]).

We consider a convex quadratic optimization problem of the form

$$\min \{\tfrac{1}{2} x^{\mathsf{T}} C x + p^{\mathsf{T}} x \mid Ax \leqq b, x \geqq 0\}, \tag{A.4.1}$$

where $C \in \widetilde{\mathbb{H}}^{+}(n) \cap \mathbb{H}^{S}(n)$ is assumed.

The associated Lagrange dual problem has the form

$$\max \{h(u) \mid u \geqq 0, u \in Df(h)\} \tag{A.4.2}$$

where

$$h(u) = \inf \{\tfrac{1}{2} x^{\mathsf{T}} C x + p^{\mathsf{T}} x + u^{\mathsf{T}}(Ax - b) \mid x \geqq 0\}$$

and

$$Df(h) = \{u \mid h(u) > -\infty\}.$$

Problem (A.4.2) may be put in the equivalent form

$$\max \{-\tfrac{1}{2} x^{\mathsf{T}} C x - b^{\mathsf{T}} u \mid Cx + A^{\mathsf{T}} u \geqq -p, x \geqq 0, u \geqq 0\}. \tag{A.4.3}$$

The following linear complementarity problem is associated with problem (A.4.1):

$$\begin{aligned} &\text{Determine the set } N = \{(x, u, y, z) \in M \mid x^{\mathsf{T}} y + u^{\mathsf{T}} z = 0\} \\ &\text{where} \\ &M = \{(x, u, y, z) \in \mathbb{E}_{2n+2m} \mid -Cx - A^{\mathsf{T}} u + y = p, x \geqq 0, u \geqq 0, \\ &\qquad\qquad Ax + z = b, y \geqq 0, z \geqq 0\}. \end{aligned} \tag{A.4.4}$$

Theorem A.4.1. *The following statements are equivalent:*

(1) (A.4.1) *is soluble;*
(2) (A.4.2) *is soluble;*
(3) (A.4.3) *is soluble;*
(4) (A.4.1) *and* (A.4.2) *possess feasible points;*
(5) (A.4.1) *and* (A.4.3) *possess feasible points;*
(6) *the set* N *is non-empty;*
(7) *the set* M *is non-empty.*

Theorem A.4.2

(1) *The point* (x^0, u^0, y^0, z^0) *is contained in* N *if and only if* x^0 *is an optimal solution of* (A.4.1) u^0 *is an optimal solution of* (A.4.2) *and* $y^0 = p + Cx^0 + A^\mathsf{T}u^0$ *and* $z^0 = b - Ax^0$ *hold.*

(2) *If problems* (A.4.1), (A.4.2), *and* (A.4.3) *are soluble then the optimal values of their objective functions are equal to each other.*

Theorem A.4.3

(1) *If* M_{opt} *denotes the set of all optimal points of problem* (A.4.1) *and* $x^0 \in \operatorname{aff} M_{\text{opt}}$ *is arbitrary then the representation*

$$M_{\text{opt}} = \{x \mid Ax \leqq b, x \geqq 0, Cx = Cx^0, p^\mathsf{T}x = p^\mathsf{T}x^0\}$$

is valid.

(2) *If* $(\hat{I}_1, \hat{I}_2, \hat{J}_1, \hat{J}_2)$ *is the characteristic index set four-tuple for the set* N, *i.e. if the relations*

$$N = \{(x, u, y, z) \in M \mid x_i = 0, i \in \hat{I}_1, u_j = 0, j \in \hat{J}_1,$$
$$y_i = 0, i \in \hat{I}_2, z_j = 0, j \in \hat{J}_2\}$$

and

$$\operatorname{ri} N = \{(x, y, u, z) \in N \mid x_i > 0, i \notin \hat{I}_1, u_j > 0, j \notin \hat{J}_1,$$
$$y_i > 0, i \notin \hat{I}_2, z_j > 0, j \notin \hat{J}_2\}$$

hold, then with the notation of (1)

$$M_{\text{opt}} = \{x \mid a_ix = b_i, i \in \hat{I}_2, a_ix \leqq b, i \notin \hat{I}_2, x_i = 0, i \in \hat{I}_1,$$
$$x_i \geqq 0, i \notin \hat{I}_1, Cx = Cx^0\}$$

and

$$\operatorname{ri} M_{\text{opt}} = \{x \mid a_ix = b_i, i \in \hat{I}_2, a_ix < b_i, i \notin \hat{I}_2, x_i = 0, i \in \hat{I}_1,$$
$$x_i > 0, i \notin \hat{I}_1, Cx = Cx^0\} .$$

follow.

A.5. Constraint Qualifications and Optimality Conditions for Non-Linear Optimization Problems: Some Results

In this section of the Appendix we present some results relating to constraint qualifications and optimality conditions which are needed in Chapter 6.

Consider the non-linear optimization problem (P_{λ^0}) given by (6.3.1) for a given parameter point $\lambda^0 \in E_k$, i.e.

$$(P_{\lambda^0}) \qquad \min \{f(x, \lambda^0) \mid g_i(x, \lambda^0) \leqq 0, i = 1, \dots, m, h_j(x, \lambda^0) = 0, j = 1, \dots, p\} .$$

Assume further that $x^0 \in E_n$ is given and fixed, $I_0 =_{\text{Df}} I(x^0, \lambda^0) =_{\text{Df}} \{i \mid g_i(x^0, \lambda^0) = 0\}$ is *the set of indices of the active constraints at* x^0 and that the functions g_i, $i = 1, \dots, m$, and h_j, $j = 1, \dots, p$, are continuously differentiable on $E_n \times E_k$. We denote the constraint set of (P_{λ^0}) by $M(\lambda^0)$ and introduce the following notation:

$$M^{\mathfrak{L}}(\lambda^0) = \{y \in E_n \mid \nabla_x g_i(x^0, \lambda^0)^\mathsf{T}\, y \leqq 0, i \in I_0, \nabla_x h_j(x^0, \lambda^0)^\mathsf{T}\, y = 0, j = 1, \dots, p\} \tag{A.5.1}$$

is the *linearization* of $M(\lambda^0)$ at x^0,

$$F(x^0, \lambda^0) = \{y \in E_n \mid \exists t_0 > 0 \colon x^0 + ty \in M(\lambda^0)\ \forall t \in (0, t_0)\} \tag{A.5.2}$$

is the *cone of feasible directions to the set* $M(\lambda^0)$ *at the point* x^0, and

$$T(x^0, \lambda^0) = \{y \in E_n \mid y = \lim_{k\to\infty} t_k(x^k - x^0),\ t_k > 0,\ x^k \in M(\lambda^0),\ x^k \to x^0\} \quad (A.5.3)$$

is the *tangent cone to the set* $M(\lambda^0)$ *at the point* x^0.

We say that $(x^0, \lambda^0) \in E_n \times E_k$ *fulfils the constraints qualification*

(R1) if there exists a vector $a^0 \in E_n$ satisfying

$$\nabla_x g_i(x^0, \lambda^0)^T a^0 < 0\,, \quad i \in I_0\,, \quad \nabla_x h_j(x^0, \lambda^0)^T a^0 = 0\,, \quad j = 1, \dots, p\,, \quad (A.5.4)$$

and the vectors $\nabla_x h_j(x^0, \lambda^0)$, $j = 1, \dots, p$, are linearly independent ,

(R2) if

$$M^{\mathscr{L}}(\lambda^0) \subset F(x^0, \lambda^0) \quad ((A.5.1), (A.5.2))\,, \quad (A.5.5)$$

(R3) if

$$M^{\mathscr{L}}(\lambda^0) \subset T(x^0, \lambda^0) \quad \text{and} \quad (A.5.6)$$

(R4) if

$g_i(\cdot, \lambda^0)$, $i \in I_0$, are pseudo-convex at x^0 and

$h_j(\cdot, \lambda^0)$, $j = 1, \dots, p$, are pseudo-convex and pseudo-concave

at x^0 . (A.5.7)

(A real-valued function defined on E_n is said to be *pseudo-convex* at x^0 if for each $x \in E_n$ the relation $f(x) \geqq f(x^0)$ follows from $\nabla f(x^0)^T (x - x^0) \geqq 0$, likewise a function f is *pseudo-concave* at x^0 if $-f$ is pseudo-convex at that point).

Following M. S. Bazaraa and C. M. Shetty [1] (R1) to (R4) are called Mangasarian-Fromovitz *constraints qualification, Zangwill constraints qualification, Abadie constraints qualification* and *reverse constraints qualification* respectively.

Furthermore, the condition (R5) holds if

the vectors $\nabla_x g_i(x^0, \lambda^0)$, $i \in I_0$, $\nabla_x h_j(x^0, \lambda^0)$, $j = 1, \dots, p$,

are linearly independent , (A.5.8)

likewise $x^0 \in E_n$ *fulfils the optimality condition* (R6) if

$$x^0 \in M(\lambda^0)\,,$$

$$N_0 =_{\mathrm{Df}} N(x^0, \lambda^0) =_{\mathrm{Df}} \{(u, v) \in E_m \times E_p \mid \nabla_x L(x^0, u, v, \lambda^0) = 0,$$
$$u_i \geqq 0,\ i = 1, \dots, m,\ u_i g_i(x^0, \lambda^0) = 0,$$
$$i = 1, \dots, m\} \neq \emptyset$$

$$\left(L(x, u, v, \lambda) =_{\mathrm{Df}} f(x, \lambda) + \sum_{i=1}^{m} u_i g_i(x, \lambda) + \sum_{j=1}^{p} v_j h_j(x, \lambda) \text{ is the } \textit{Lagrangian}\right), \quad (A.5.9)$$

(R7) if

the functions f, g_i, $i = 1, \dots, m$, h_j, $j = 1, \dots, p$, are twice continuously differentiable on $E_n \times E_k$, and there exist elements $(u^0, v^0) \in N_0$ (N_0 is defined in (A.5.9)) such that

$$y^T \nabla_x^2 L(x^0, u^0, v^0, \lambda^0)\, y \geqq 0$$

holds for each $y \neq 0$ satisfying (A.5.10)

(a) $\nabla_x h_j(x^0, \lambda^0)^\mathsf{T} y = 0\,, \qquad j = 1, \dots, p\,,$

and

(b) $\nabla_x g_i(x^0, \lambda^0)^\mathsf{T} y = 0\,, \qquad i \in I_1(u^0) =_{\mathrm{Df}} \{i \in I_0 \mid u_i^0 > 0\}\,,$

(c) $\nabla_x g_i(x^0, \lambda^0)^\mathsf{T} y \leqq 0\,, \qquad i \in I_0 \setminus I_1(u^0)\,.$

(R8) if

instead of $y^\mathsf{T} \nabla L(x^0, u^0, v^0, \lambda^0)\, y \geqq 0$ it holds in (A.5.10)

$$y^\mathsf{T} \nabla_x^2 L(x^0, u^0, v^0, \lambda^0)\, y > 0\,. \tag{A.5.11}$$

In (R7) and (R8) $\nabla_x^2 L$ is the *Hessian matrix* of the Lagrangian.

Finally, we have (R9) if

the functions $f, g_i,\ i = 1, \dots, m,\quad h_j,\ j = 1, \dots, p,$ are twice continuously differentiable on $E_n \times E_k$ and there exist elements $(u^0, v^0) \in N_0$ and $a^0 \in E_n$ such that

$$\begin{aligned} &\nabla_x h_j(x^0, \lambda^0)^\mathsf{T} a^0 = 0\,, && j = 1, \dots, p\,,\\ &\nabla_x g_i(x^0, \lambda^0)^\mathsf{T} a^0 = 0\,, && i \in I_1(u^0)\,,\\ &\nabla_x g_i(x^0, \lambda^0)^\mathsf{T} a^0 < 0\,, && i \in I^0 \setminus I_1(u^0)\,,\end{aligned} \tag{A.5.12}$$

hold and the vectors

$$\begin{aligned} &\nabla_x g_i(x^0, \lambda^0)\,, && i \in I_1(u^0)\,,\\ &\nabla_x h_j(x^0, \lambda^0)\,, && j = 1, \dots, p\,,\end{aligned}$$

are linearly independent.

(R6) is the well-known local Kuhn-Tucker condition and (R7) and (R8) are a necessary and a sufficient second order condition.

Proofs of the propositions collected in the following lemma may be found for instance in M. J. Bazaraa, C. M. Shetty [1], K. H. Elster et al. [1].

Lemma A.5.1

(1) (R1) *follows from* (R5).

(2) *Let x^0 be a local minimal point of* (P_{λ^0}), *then* (R6) *follows from* (R1).

(3) *If* $(u^0, v^0) \in N_0$ (N_0 is defined in (A5.9)) *then* (R9) *follows from* (R5).

(4) *Let x^0 be a local minimal point of* (P_{λ^0}), *then* (R7) *follows from* (R6) *and* (R9).

(5) (R1) *and* (R4) *are sufficient conditions for* (R2) *and* (R3).

One easily confirms the following proposition.

Lemma A.5.2. *Let x^0 be a local minimal point of* (P_{λ^0}), *satisfying condition* (R6) *and* $(u^0, v^0) \in N_0$ *where* (R6) *is defined in* (A.5.9). *The Jacobian*

$$J_1(x^0, u^0, v^0, \lambda^0) =_{\mathrm{Df}}$$

$$\begin{bmatrix} \nabla_x^2 L(x^0, u^0, v^0, \lambda^0) & \nabla_x g_1(x^0, \lambda^0) \dots \nabla_x g_m(x^0, \lambda^0) & \nabla_x h_1(x^0, \lambda^0) \dots \nabla_x h_p(x^0, \lambda^0) \\ u_1^1 \nabla_x g_1(x^0, \lambda^0)^\mathsf{T} & g_1(x^0, \lambda^0) \quad \dots \quad 0 & 0 \\ \vdots & \vdots \qquad\qquad \vdots & \vdots \\ u_m^0 \nabla_x g_m(x^0, \lambda^0)^\mathsf{T} & 0 \quad \dots \quad g_m(x^0, \lambda^0) & 0 \\ \nabla_x h_1(x^0, \lambda^0) & 0 \qquad\qquad 0 & 0 \\ \vdots & \vdots \qquad\qquad \vdots & \vdots \\ \nabla_x h_p(x^0, \lambda^0) & 0 \qquad\qquad 0 & 0 \end{bmatrix} \tag{A.5.13}$$

of the function

$$K(x, u, v, \lambda^0) =_{\mathrm{Df}} \begin{pmatrix} \nabla_x L(x, u, v, \lambda^0) \\ u_1 g_1(x, \lambda^0) \\ \vdots \\ u_m g_m(x, \lambda^0) \\ h_1(x, \lambda^0) \\ \vdots \\ h_p(x, \lambda^0) \end{pmatrix} \tag{A.5.14}$$

at the point (x^0, u^0, λ^0) *is regular if*

(1) $I_0 = I_1(u^0)$ (I_0 is the set of the indices of constraints active at x^0, $I_1(u^0)$ is defined in (A.5.10)), ((1) is the so-called strict complementary slackness),

(2) (R5),

(3) (R8).

A proof of the following Lemma may be found in A. V. Fiacco and G. P. McCormick [1].

Lemma A.5.3. *Let* x^0 *fulfil the condition* (R6) *as is given in* (A.5.9). *Let condition* (R8) *for* x^0 *be fulfilled. The point* x^0 *is then a locally unique local minimal point of* (P_{λ^0}), *i.e. there exists a neighbourhood* $U\{x^0\}$ *with* $f(x, \lambda^0) > f(x^0, \lambda^0)$ *for all* $x \in U\{x^0\} \cap M(\lambda^0)$.

Bibliography

I denotes parametric optimization, II related literature, III basic literature

ABADIE, J.: Programmes linéaires: Paramétrisation du second membre. Note E.D.R., H.R. 3861. (I)

ABADIE, J., and R. C. WILLIAMS: Dual and parametric-methods in decomposition. *In*: WOLFE, PH., and R. L. GRAVES (eds.): Recent advances in mathematical programming. McGraw Hill, New York 1963, 149–158. (II)

ABRHAM, J.: Über die Stabilität von Lösungen im Transportproblem der linearen Programmierung, Czechosl. Math. J. 8 (83) (1958) 1, 131–138. (I)

ADBY, P. R., and M. A. H. DEMPSTER: Introduction to optimization methods. Chapman and Hall, London/Halsted Press, New York 1974. (III)

AGGARWAL, S. P.:

[1] Stability of the solution to a linear fractional functional programming problem. Z. angew. Math. Mech. 46 (1966), 343–349. (I)

[2] Parametric linear fractional functionals programming. Metrika 12 (1968), 107 to 114. (I)

AGGARWAL, S. P., and K. SWARUP: Fractional functionals programming with a quadratic constraint. Oper. Res. 14 (1966), 950–956. (I)

ALLENDE, S.: Globale Stabilitätsuntersuchungen rein- und gemischt-ganzzahliger parametrischer Optimierungsprobleme. Diss. (A), Humboldt-Univ. Berlin 1980. (I)

ALT, W.: Stabilität mengenwertiger Abbildungen mit Anwendungen auf nichtlineare Optimierungsprobleme. Diss., Bayreuther Math. Schriften (1979) 3, Selbstverlag der Univ. Bayreuth. (I, II)

ALVAREZ, R.: Sensitivity of optimal inventory policies. Oper. Res. Suppl. (1966), B 212 to B 213. (II)

ARANA, R. M.: Programming with parametric elements of the matrix coefficients. RAIRO Ser. Recherche Operationelle 11 (1977) 2, 233–238. (I)

ARBUZOVA, N. I., and V. L. DANILOV (Арбузова, Н. И., и В. Л. Данилов): К расширению понятия устойчивости задачи линейного программирования. Кибернетика 4 (1970), 139–140. (II)

ARMACOST, R. L., and A. V. FIACCO:

[1] Computational experience in sensitivity analysis for nonlinear programming. Math. Programming 6 (1974), 301–326. (I, II)

[2] Second-order parametric sensitivity analysis in nonlinear programming and estimates by penalty function methods. Technical Paper Serial T-324, The George Washington Univ., Inst. of Management, Sci., and Engin., Seattle 1975. (I, II)

[3] Nonlinear programming sensitivity for R.R.S. perturbation: A brief survey and recent second-order-extensions. Technical Paper Serial T-334, The George Washington Univ., Inst. of Management, Sci., and Engin., Seattle 1976. (I)

ARMACOST, R. L., and W. CH. MYLANDER: A guide to a SUMT-Version 4 computer subroutine for implementing sensitivity analysis in nonlinear programming. Working Paper Serial T-287, The George Washington Univ., Seattle 1973. (II)

ARNOFF, E. L., and S. S. SENGUPTA: Mathematical programming. Progress in Operations Research, vol 1. John Wiley, New York 1961, 105–210. (I, III)

Arutjunjan, Ju. V. (Арутюнян, Ю. В.): Об одной параметрической задаче дробно-линейного программирования. В. сб.: Применение новейш. мат. методов (кибернет.) в экон. энергетики. Ереван 1969, 129—140. (I)

Astaf'ev, N. N. (Астафьев, Н. Н.):

[1] О прямой и обратной теоремах двойственности в выпуклом программировании. Оптимальн. планирование 4 (1969), 137—149. (II)

[2] О маргинальных значениях задачи выпуклого программирования. Труды инст. мат. и мех. Ураль. научн. центр акад. наук СССР 18 (1975), 62—70. (I)

[3] Устойчивость и маргинальные значения задачи выпуклого программирования. Сибирск. мат. ж. 19 (1978), 491—503. (I)

[4] Двойственная регуляризация задачи линейного программирования, заданной последовательностью реализации. Ж. вычисл. мат. и мат. физ. 18 (1978), 1129—1138. (II)

Aubin, J. P.:

[1] Selection de solutions d'un problème d'optimization à criteres multiples. Cahiers mathematiques de la decision 7 (1971) 4. (II)

[2] Applied functional analysis. Wiley-Interscience, New York 1978. (III)

[3] Mathematical methods of game and economic theory. North-Holland, Amsterdam 1979. (III)

Aumann, R. J.: Integrals of set valued functions. J. Math. Anal. Appl. 12 (1965), 1—12. (II)

Auslender, A.:

[1] La méthode géométrique en programmation linéaire. Application à l'étude de la dégénérescence et des problèmes paramétriques. Thèse de 3e cycle, Fac. Sci. Grenoble 1964. (I, II)

[2] Algorithmes des programmes linéaires et paramétriques. I. Chiffres R.F.T.I. 8 (1965), 135—155. (I)

[3] Algorithmes et programmes linéaires et paramétriques. II. Algorithme géométrique des programmes paramétriques. Chiffres R.F.T.I. 8 (1965), 225—239. (II)

[4] Differentiable stability in nonconvex and nondifferentiable programming. Math. Programming Stud. 10 (1979), 29—42. (I, II)

Avriel, M.: Nonlinear programming. Analysis and methods. Prentice-Hall, Englewood Cliffs (N.J.) 1976. (III)

Balachandran, V., V. Srinivasan, and G. L. Thompson: A survey of the applications of the operator theory of parametric programming for the transportation and generalized transportation problem. Discussion Paper No. 300, North Western Univ., Center for Math. Studies in Econ. and Management Sci., Evanston 1977. (II)

Balachandran, V., and G. L. Thompson:

[1] An operator theory of parametric programming for the generalized transportation problem. I. Basic theory. Naval Res. Logist. Quart. 22 (1975), 79—100. (I)

[2] II. Cost, rim and bound operators. Naval Res. Logist. Quart. 22 (1975), 101—125. (I)

[3] III. Weight operators. Naval Res. Logist. Quart. 22 (1975), 297—315. (I)

[4] IV. Global operators. Naval Res. Logist. Quart. 22 (1975), 317—339.

Balakrishnan, A. V.: Introduction to optimization theory in a Hilbert space. Lecture Notes in Operation Research and Mathematical Systems, vol. 42. Springer-Verlag, Berlin-Heidelberg-New York 1971. (III)

Balaș, E., and P. Ivănescu (Hammer):

[1] In legătură cu problema transporturilor. I: Problema transporturilor on centre legate. Studii și cercetări mat. Acad. RPR 11 (1960), 439—450. (II)

[2] Transportation problems with variable data. Revue Math. pures appl. 6 (1961), 713—734. (I)

[3] In legătură cu problema transporturilor. II: Probleme de transporturilor cu parametri. Studii și cercetări mat. Acad. RPR 12 (1961), 413—427. (I)

[4] In legătură cu problema transporturilor. VI: Stabilitatea soluțiilor optime în raport cu variațiile costurilor. Comun. Acad. RPR 13 (1963), 249—251. (I)

Balder, E. J.: An extension on duality-stability relation to nonconvex optimization problems. SIAM J. Control Optim. 15 (1977), 329—343. (II)

Bank, B.:

[1] Qualitative Stabilitätsuntersuchungen rein- und gemischtganzzahliger linearer parametrischer Optimierungsprobleme. Seminarbericht Nr. 6, Humboldt Univ. Berlin, Sektion Mathematik 1978. (I)

[2] Stability analysis in pure and mixed-integer linear programming. In: Proc. 9th IFIP Conf. Optimization Techniques Warsaw, 1979. Lecture Notes in Control Inform. Sci., vol. 23. Springer-Verlag, Berlin—Heidelberg—New York 1980, 149—153. (I)

Bank, B., R. Mandel, and K. Tammer: Parametrische Optimierung und Aufteilungsverfahren. *In*: Lommatzsch, K. (ed.): Anwendungen der linearen parametrischen Optimierung. Akademie-Verlag, Berlin 1979, 107—123. (I)

Barabaš, S. B. (Барабаш, С. Б.): Об устойчивости одной экстремальной задачи. В сб.: Багриновский, К. А., и Е. Л. Берлянд: Модели и методы анализа экономических целенаправлённых систем. Серия: Математический анализ экономических моделей. Наука, Сибирское отделение, Новосибирск, 139—150. (I)

Barankin, E. W., and R. Dorfman: On quadratic programming. Univ. California Publ. Statist. **2** (1958), 285—317. (I)

Barnett, S.:

[1] Stability of the solution to a linear programming problem. Operat. Res. Quart. **13** (1962), 219—228. (I)

[2] A simple class of parametric linear programming problems. Oper. Res. **16** (1968), 1160—1165. (I)

Bazaraa, M. S., and C. M. Shetty: Foundations of optimization. Springer-Verlag, Berlin—Heidelberg—New York 1976. (III)

Beale, E. M. L.: Numerical methods. *In*: Abadie, J. (ed.): Nonlinear programming. North-Holland, Amsterdam/Wiley-Interscience, New York 1967, 133—205. (III)

Bector, C. R.: Indefinite cubic programming with standard errors in the objective function. Unternehmensforsch. **12** (1968), 113—120. (I)

Bellmore, M., H. J. Greenberg, and J. J. Jarvis: Generalized penalty functions concepts in mathematical optimization. Oper. Res. **18** (1970), 229—252. (II)

Belousov, E. G. (Белоусов, Е. Г.): Введение в выпуклый анализ и целочисленное программирование. Издат. Московск. унив., Москва 1977. (III)

Beluhin, V. P. (Белухин, В. П.): Метод решения одной задачи параметрического линейного программирования. Автомат. и телемех. **10** (1973), 136—145. (I)

Ben-Israel, A., and A. Ben-Tal: On a characterization of optimality in convex programming. Math. Programming **11** (1976), 81—88. (II)

Ben-Israel, A., A. Ben-Tal, and S. Zlobec: Optimality conditions in convex programming. *In*: Proceedings of the IX International Symposium on Mathematical Programming. North-Holland, Amsterdam—Oxford—New York 1979, 153—170. (II)

Benson, H. P., and L. P. Morin: The vector maximization problem: proper efficiency and stability. SIAM J. Appl. Math. **32** (1977), 64—72. (II)

Ben-Tal, A., A. Ben-Israel, and S. Zlobec: Characterization of optimality in convex programming without a constraint qualification. J. Optim. Theory Appl. **20** (1976), 417—437. (II)

Ben-Tal, A., and S. Zlobec: A new class of feasible direction methods. Report No. CCS-216, Univ. of Texas, Center for Cybernetics Studies, Austin 1977. (II)

Berdyšev, V. I. (Бердышев, В. И.):

[1] Непрерывная зависимость элемента, реализующего минимум выпуклого функционала, от множества допустимых элементов. Мат. заметки **19** (1976), 501 до 512. (I)

[2] Устойчивость проблем минимизации при возмущении множества допустимых элементов. Мат. сб. **103** (1977), 467—489. (I)

[3] Непрерывность многозначных отображений связанных с минимизацией выпуклых функционалов. Докл. Акад. Наук **243** (1978), 361—364. (II)

Bereanu, B.:

[1] On stochastic linear programming I: Distribution problems: A single random variable. Rev. Roumaine Math. Pures Appl. **8** (1963), 683—697. (II)

[2] On stochastic linear programming II: Distribution problems: Nonstochastic technological matrix. Rev. Roumaine Math. Pures Appl. **11** (1966), 713—725. (II)

[3] On stochastic linear programming. Distribution problems. Stochastic technological matrix. Z. Wahrsch. Verw. Gebiete **8** (1967), 148—152. (II)

[4] The distribution problem in stochastic linear programming. The cartesian integration method. Reprint No. 7103, Center of Math. Stat. of the Acad. RSR, Bucharest 1971. (II)

[5] The continuity of the optimum in parametric programming and applications to stochastic programming. J. Optim. Theory Appl. **18** (1976), 319—333. (II)

[6] Stochastic-parametric linear programs I. Rev. Roumaine Math. Pures Appl. **22** (1977), 1367—1380. (II)

Berežinskiĭ, T. A., and G. M. Ostrovskiĭ (Бережинский, Т. А., и Г. М. Островский): Об оптимальном управлении процессами с параметрами при ограниченных фазовых координатах. Автомат. и телемех. **2** (1970), 7—13. (II)

Berge, C.:

[1] Théorie général des jeux à n personnes. Gauthier-Villars, Paris 1957. (I, III)

[2] Topological spaces. Oliver and Boyd, London 1963. (I, III)

Berger, A., and K.-H. Elster:

[1] Zum Begriff der halbstetigen Punkt-Menge-Abbildungen in topologischen Strukturen. Wiss. Z. Techn. Hochsch. Ilmenau **21** (1975) 1, 39—56. (II)

[2] Untersuchungen über halbstetige Punkt-Menge-Abbildungen. Wiss. Z. Tech. Hochsch. Ilmenau **22** (1976) 4, 11—29. (II)

Bertsekas, D. P.: Necessary and sufficient conditions for a penalty method to be exact. Math. Programming **9** (1975), 87—99. (II)

Bigelow, J. H., and N. Z. Shapiro: Optimization problems with large parameters. SIAM J. Appl. Math. **24** (1973), 152—163. (I)

Billera, L. J.: Topologies for 2^X; set-valued functions and their graphs. Trans. Amer. Math. Soc. **155** (1971), 137—147. (II)

Bitram, G. R., and T. L. Magnati: Duality and sensitivity analysis for fractional programs. Oper. Res. **24** (1976), 675—699. (II)

Blair, C. E., and R. G. Jeroslow: The value function of a mixed integer program. Discrete Math. **19** (1977), 121—138. (I)

Bloech, J.: Zum Problem der nachträglichen Änderung industrieller Produktionsprogramme. Z. Betriebswirtschaft **36** (1966), 186—197. (I)

Bod, P.:

[1] Lineare Optimierung im Falle mehrerer gleichzeitig gegebener Zielfunktionen. Publ. math. Inst. Hungar. Acad. Sci. **8**, Ser. B (1963), 541—556. (Hungarian) (I)

[2] Lineare Optimierung mittels simultan gegebener Zielfunktionen. Colloqu. Appl. Math. Econ., Budapest 1963 (1965), 55—60. (I)

Böhm, V.: On the continuity of the optimal policy set for linear programs. SIAM J. Appl. Math. **28** (1975), 303—306. (I)

Boot, J. C. G.:

[1] On sensitivity analysis in convex quadratic programming problems. Oper. Res. **11** (1963), 771—786. (I)

[2] Quadratic programming, algorithms — anomalies — applications. Studies in Mathematical and Managerial Economics, vol. 2. North-Holland, Amsterdam 1964. (I)

Borwein, J.: Multivalued convexity and optimization: a unified approach to inequality and equality constraints. Math. Programming **13** (1977), 183—199. (II)

Bose, S. C.: Common fixed points of mappings in a uniformly convex Banach space. J. London Math. Soc. **18** (1978) 1, 151—156. (II)

Bouligand, G.:

[1] Sur la semi-continuité d'inclusion et quelques sujets connexes. Enseignement mathématique **30** (1931), 240; **31** (1932), 14—22. (II)

[2] Propriétés générales des correspondances multiformes. Comptes Rendus de l'Académie des Sciences, Paris, **196** (1933), 1767—1769. (II)

Bowman Jr., V. J.: Sensitivity analysis in linear integer programming. AIIE Techn. Paper, 1972. (I)

Brisac, R.: Sur le fonctions multiformes. Comptes Rendus de l'Académie des Sciences Paris **224** (1947), 92—94. (II)

Brosowski, B.:

[1] Zur stetigen Abhängigkeit der Menge der Minimalpunkte bei gewissen Minimierungsaufgaben. *In*: Lecture Notes in Mathematics, vol. 556. Springer-Verlag, Berlin—Heidelberg—New York 1976, 63—72. (II)

[2] Zur stetigen Parameterabhängigkeit der Menge der Minimalpunkte bei Minimax-Aufgaben. *In*: Operations Research Verfahren XXV, Teil 1, Verlag Anton Hain, Meisenheim/Glan 1976, 28—33. (I)

[3] On parametric linear optimization. *In*: Lecture Notes in Economics and Mathematical Systems, vol. 157. Springer-Verlag, Berlin—Heidelberg—New York 1978, 37—44. (I)

[4] On parametric optimization IV, differentiable parameterfunctions. *In*: Lecture Notes in Economics and Mathematical Systems, vol. 179. Springer-Verlag, Berlin—Heidelberg—New York 1980, 31—39. (I)

[5] On the continuity of the optimum set in parametric linear programming. Preprint, J. W. Goethe Univ. Frankfurt/Main 1980. (I)

Brosowski, B., F. Deutsch, and G. Nürnberger: Parametric approximation. J. Approximation Theory **29** (1980). (II)

Brosowski, B., and K. Schnatz: Parametric optimization, differentiable parameterfunctions. Methods of Oper. Res. **37** (1980), 99—118. (I)

Brucker, P.: Diskrete parametrische Optimierungsprobleme und wesentliche effiziente Punkte. Z. Oper. Res. **16** (1972), 189—197. (II)

Bruns, P.: Fehlerabschätzungen und Stetigkeitsuntersuchungen bei konvexen Optimierungsaufgaben. Diss., Univ. Hamburg 1972. (I, II)

Budak, B. M., and E. M. Berkovič (Будак, Б. М., и Е. М. Беркович):

[1] Об аппроксимации экстремальных задач I. Ж. вычисл. мат. и мат. физ. **11** (1971), 580—596. (II)

[2] Об аппроксимации экстремальных задач II. Ж. вычисл. мат. и мат. физ. **11** (1971), 870—884. (II)

Bulavskii, V. A., and G. S. Rubinšteĭn (Булавский, В. А., и Г. Ш. Рубинштейн): О решении задач выпуклого программирования с линейными ограничениями методом последовательного улучшения допустимого вектора. Докл. Акад. Наук СССР **150** (1963), 231—234. (II)

Candler, W.: A modified simplex solution for linear programming with variable capital restrictions. J. Farm Econ. **38** (1956), 940—955. (I)

Carpentier, J.: A method for solving linear programming problems in which the cost depends nonlinearly on a parameter. Électricité de Françe, Diréction des Études et Recherches, May 1959. (I)

Castaing, C., and M. Valadier: Convex analysis and measurable multifunctions. Springer-Verlag, Berlin—Heidelberg—New York 1977. (II)

Cea, J.: Optimization: Théorie et algorithmes, Dunod, Paris 1971. (III)

Černikov, S. N. (Черников, С. Н.): Линейиыс неравенства. Наука, Моска 1968.

Černov, Ju. P. (Чернов, Ю. П.): Об одной задаче параметрического дробно-линейного программирования. Кырг. ССР Илимдер Акад. кабарлары, Изв. АН Кирг. ССР **3** (1970), 20—27. (I)

Chadha, S. S.: A linear fractional program with a two parameter objective function. Z. Angew. Math. Mech. **51** (1971), 479—481. (I)

Chadha, S. S., and S. Shivpuri: A simple class of parametric linear fractional functionals programming. Z. Angew. Math. Mech. **53** (1973), 644—646. (I)

Chadha, S. S., and J. M. Gupta: Sensitivity analysis of the solution of a generalized linear and piece-wise linear program. Cahiers Centre Études Rech. Opér. **18** (1976) 3, 309 to 321. (I)

Haĭrutdinov, M. (Хайрутдинов, М.): Алгоритм для решения параметрической задачи линейного программирования. Итоги научн. конференции Казанского университета 1963, 54—57. (I)

Chandrasekaran, R.: A special case of the complementarity pivot problem. Opsearch **7** (1970), 263—268. (II)

Choquet, G.: Convergences. Annales de l'Université de Grenoble **23** (1947—1948), 55 to 112. (II, III)

CLARKE, F. H.:
[1] Generalized gradients and applications. Trans. Amer. Math. Soc. **205** (1975), 247 to 262. (III)
[2] A new approach to Lagrange multipliers. Math. Oper. Res. **1** (1976), 165—174. (II)
COLLATZ, L., and W. WETTERLING:
[1] Optimierungsaufgaben. Springer-Verlag, Berlin—Heidelberg—New York 1966. (III)
[2] (eds.): Numerische Methoden bei Optimierungsaufgaben. Birkhäuser-Verlag, Basel 1973. (III)
CONON, M. D., C. D. CULLUM, and E. POLAK: Theory of optimal control and mathematical programming. McGraw-Hill, New York 1975. (III)
COTTLE, R. W.:
[1] Monotone solutions of the parametric linear complementarity problem. Math. Programming **3** (1972), 210—224. (I)
[2] Solution of the "Problem 27-7, A Parametric Linear Complementarity Problem" by G. Maier. SIAM Rev. **15** (1973), 381—384. (I)
[3] Solution rays for a class of complementarity problems. Math. Programming Stud. **1** (1974), 59—70. (III)
COURTILLOT, M.:
[1] Programmation linéaire. Étude de la modification de tous les paramètres. Méthode de résolution séquentielle. Comptes Rendus de l'Académie des Sciences **247** (1958), 670—673. (I)
[2] Study of the variation of all parameters of a linear program and study of a linear program in the case of a linear relation between the terms of the second member. *In*: Proc. 2nd internat. Conf. Oper. Res., Aix-en-Provence 1960, Engl. Univ. Press, London 1961, 292. (I)
[3] On varying all the parameters in a linear programming problem. Oper. Res. **10** (1962), 471—475. (I)
[4] Paramétrisation de la matrice. Monographie de Recherche opérationelle **1** (1964), 93—98. (I)
COWLES, D. E.: Upper semi-continuity properties of variables sets in optimal control. J. Optim. Theory Appl. **10** (1972), 226—236. (II)
CROUZEIX, J. P.: Continuité des applications linéaires multivoques. Rev. Française d'Automatique, d'Informatique et de Recherche Opérationelle No. **R 1** (1973), 62—67. (II)
CRUZ, J. B., and W. R. PERKINS: A new approach to the sensitvity problem in multivariable system design. IEEE Trans. Aut. Contr., vol. AC 9, July 1964. (II)
CULLUM, J.:
[1] Discrete approximations to continuous optimal control problems. SIAM J. Control **7** (1969), 32—49. (II)
[2] An explicit procedure for discretizing continuous, optimal control problems. J. Optim. Theory Appl. 8 (1971), 15—34. (II)

DANIEL, J. W.:
[1] The approximate minimization of functionals. Prentice-Hall, Englewood Cliffs (N.J.) 1971. (II)
[2] On perturbations in systems of linear inequalities. SIAM J. Numer. Anal. **10** (1973), 299—307. (II)
[3] Stability of the solution of definite quadratic programs. Math. Programming **5** (1973), 41—53. (I)
[4] Remarks on perturbations in linear inequalities. SIAM J. Numer. Anal. **12** (1975), 770—773. (II)
DANØ, S.: Parametrische lineare Programmierung. Jahrb. Nationalökon. Statist. **182** (1968) 1, 49—58. (I)
DANTZIG, G. B.: Linear programming and extensions (A RAND Corp. Res. Study)/Lineare Programmierung und Erweiterungen (Deutsche Bearbeitung: A. JAEGER). Princeton Univ. Press, Princeton (N.J.) 1963/Springer-Verlag, Berlin—Heidelberg—New York 1966. (III)
DANTZIG, G. B., J. FOLKMAN, and N. SHAPIRO: On the continuity of the minimum set of a continuous function. J. Math. Anal. Appl. **17** (1967), 519—548. (I)
DEBREU, C.: Theory of value. John Wiley, New York 1959. (II, III)

DELAHAYE, J. P., and J. DENEL: Equivalences des continuités des applications multivoques dans des espaces topologiques. Laboratoire de Calcul, publication No. 111, Univ. of Lille 1978. (II)

DEM'JANOV, V. F., and V. N. MALOZJOMOV (Демьянов, В. Ф., и В. Н. Малозёмов): Введение в минимакс. Наука, Москва 1972. (III)

DEM'JANOV, V. F., and A. B. PEVNIĬ (Демьянов, В. Ф., и А. Б. Певний):
[1] Вычисление первых и вторых маргинальных значений в задачах математического программирования I. Оптимизация 5 (1972), 63—72. (I)
[2] Маргинальные значения задач математического программирования и минимаксных задач. Вестн. Ленинградск. унив. 19 (1973), 35—41. (I)
[3] Разложение экстремальных значений игровых задач по параметру. Ж. вычисл мат. и мат. физ. 14 (1974), 1118—1130. (I)

DEM'JANOV, V. F., and A. M. RUBINOV (Демьянов, В. Ф., и А. М. Рубинов): Приближённые методы решения экстремальных задач. Изд. Ленинградск. унив., Ленинград 1968. (III)

DEM'JANOV, V. F., and V. K. ŠOMESOVA (Демьянов, В. Ф., и В. К. Шомесова): Условие зубдифференциалы выпуклых функций. Докл. Акад. Наук СССР 242 (1978), 753—756. (II)

DENEL, J.:
[1] Propriétés de continuité des familles p-décroissantes d'applications multivoques. Laboratoire de Calcul, publication No. 87, Univ. of Lille 1977. (II)
[2] On the continuity of point-to-set maps with applications to optimization. *In*: Proc. 2nd symposium on operations research, Aachen 1977. (I, II)
[3] Extensions of the continuity of point-to-set maps: applications to fixed point algorithms. Math. Programming Stud. 10 (1979), 48—68. (II)

DENT, W., R. JAGANNATHAN, and M. R. RAO: Parametric linear programming: some special cases. Naval Res. Logist. Quart. 20 (1973) 4, 725—728. (I)

DEUTSCH, F., W. POLLUL, and I. SINGER: On set-valued metric projections, Hahn-Banach-extension maps, and spherical image maps. Duke Math. J. 40 (1973), 355—370. (II, III)

DIEUDONNÉ, J.:
[1] Grundzüge der modernen Analysis 1. VEB Deutscher Verlag der Wissenschaften, Berlin 1971 (transl. from Engl.). (III)
[2] Grundzüge der modernen Analysis 2. VEB Deutscher Verlag der Wissenschaften, Berlin 1975 (transl. from French). (III)

DINKELBACH, W.:
[1] Sensitivitätsanalysen und parametrische Programmierung. Ökonometrie und Unternehmensforschung, vol. 12. Springer-Verlag, Berlin—Heidelberg—New York 1969. (I, III)
[2] Anmerkungen zur Sensitivitätsanalyse. Management Inf. Syst. 14 (1971), 473—481. (I)

DINKELBACH, W., and P. HAGELSCHUER: On multiparametric linear programming. Oper. Res. Verf. 6 (1969), 86—92. (I)

DOLECKI, S.:
[1] Extremal measurable selections. Bull. Acad. Pol. Sci. 25 (1977), 355—360. (II)
[2] Remarks on semicontinuity. Bull. Acad. Pol. Sci. 25 (1977), 863—867. (II)
[3] Constraints stability and moduli of semicontinuity. Preprint of the 2nd IFAC Symposium, June 1977, Warwick. (I)
[4] Semicontinuity in constrained optimization. Part 1: Metric spaces. Control Cybernet. 7 (1978) 2, 5—16. (II)
[5] Semicontinuity in constrained optimization. Part 1b: Norm of spaces. Control Cybernet. 7 (1978) 3, 17—26. (II)
[6] Semicontinuity in constrained optimization. Part 2: Control Cybernet. 7 (1978) 4, 51—68. (II)

DOLECKI, S., and S. ROLEWICZ:
[1] A characterization of semicontinuity-preserving multifunctions. J. Math. Anal. Appl. 65 (1978), 26—31. (II)
[2] Metric characterizations of upper semicontinuity. J. Math. Anal. Appl. 69 (1979), 146—152. (II, III)

DOMBROWSKIJ, R.: Zadania transportowe z parametrycznymi ograniczeniami. Przeglad. Statyst. **15** (1968), 103—117. (I)

DONTCHEV, A.:
[1] Metody analizy wrazliwosci układów sterowania optymalnego. Diss., Inst. Autom. Politech. Warszawskiej, Warszawa 1974. (I, II)
[2] Sensitivity analysis of linear infinite-dimensional optimal control systems under changes of system order. Control Cybernet. **3** (1975) 3/4, 21—35. (I, II)
[3] Sensitivity analysis of optimal control systems with small time delay. Control Cybernet. **4** (1976). (I, II)

DORN, W. S.: Self-dual quadratic programs. J. Soc. Ind. and Appl. Math. **9** (1961), 51 to 54. (III)

DRĂGAN, I.:
[1] Asupra unei clase de probleme de programare parametrică. Studii şi cercetări mat. Acad RPR **17** (1965), 445—449. (I)
[2] Un algorithme pour la résolution des certains problèmes paramétriques, avec un seul paramètre contenu dans la fonction économique. Rev. Roumaine Math. Pures Appl. **11** (1966), 447—451. (I)
[3] Algoritmul lui Beale pentru programe liniare parametrice raţionale. Studii şi cercetări mat. Acad. RSR **18** (1966), 1525—1528. (I)
[4] L'algorithme primal du simplexe pour certains problèmes paramétriques. Rev. Roumaine Math. Pures Appl. **4** (1966). (I)
[5] Sur la structure des domaines d'optimalité du paramètre, pour les programmes lineaires uniparamétriques rationels. An. şti. Univ. Iaşi, Secţ. Ia. Mat. **13** (1967), 177—192. (I)

DRAGOMIRESCU, M.: Asupra sensitivităţiĭ în programarea liniară. Studii şi cercetări mat. Acad. RSR **20** (1968), 965—967. (I)

DUBOVICKIĬ, A. JA., and A. A. MILJUTIN (Дубовицкий, А. Я., и А. А. Милютин): Задачи на экстремум при наличин ограничений. Ж. вылисл. мат. и мат. физ. **5** (1965), 395—453. (II)

DUMITRU, V., and V. IONESCU: Notă asupra parametrizării generale. Studii şi cercetări calcul econ. şi cibern. econ. **5/6** (1966), 167—174. (I)

EAVES, B. C.:
[1] Nonlinear programming via Kakutani fixed points. Working Paper No. 294, Center for Res. in Management Sci., Univ. of California, Berkeley 1970. (I, II)
[2] The linear complementarity problem. Managament Sci. **17** (1971), 612—634. (II, III)
[3] On quadratic programming. Management Sci. **17** (1971), 698—711. (I, III)

EICHHORN, W., and W. OETTLI: Parameterabhängige Optimierung und die Geometrie des Prinzips von Le Châtelier-Samuelson. Z. Oper. Res. Ser. A **16** (1972), 233—244. (I, II)

EKELAND, I., and G. LEBOURG: Generic Fréchet-differentiability and perturbed optimization problems in Banach-spaces. Trans. Amer. Math. Soc. **224** (1976) 2, 193—216. (II)

EKELAND, I., and R. TEMAN: Convex analysis and variational problems. North-Holland, Amsterdam—Oxford 1978. (II, III)

ELLENBERG, S.: Über quantitative Konvergenzeigenschaften von Lösungsalgorithmen in der nichtlinearen Optimierung. Dipl.-Arbeit, Humboldt Univ. Berlin, Sektion Mathematik 1979. (II)

ELMAGHRABY, S. E.: Sensitivity analysis of multiterminal flow networks. Oper. Res. **12** (1964), 680—688. (I)

EL RAMLY, N.: On the application of parametric programming on vector optimization. Diss. A, Humboldt Univ. Berlin 1977. (II)

ELSTER, K.-H., et al.: Einführung in die nichtlineare Optimierung. BSB B. G. Teubner-Verlagsgesellschaft, Leipzig 1977. (III)

ENGL, H. W., and A. KIRSCH: On the continuity of the metric projection onto a convex set subject to an additional constraint. Institutsbericht Nr. 153, Johannes Kepler Univ. Linz, Institut für Mathematik, September 1979. (II)

EREMIN, I. I., and N. N. ASTAF'JEV: (Еремин, И. И., и Н. Н. Астафьев): Введение в теорию линейного и выпуклого программирования. Наука, Москва 1976. (I, III)

ERMOL'JEV, JU. M. (Ермольев, Ю. М.): Методы стохастического программирования. Наука, Москва 1976. (I, III)

Esser, H.: Über die Stetigkeit des Extremalwertes nichtkonvexer Optimierungsaufgaben mit einer Anwendung auf die Diskretisierung von Kontrollproblemen. Z. Angew. Math. Mech. **52** (1972), 535—542. (I, II)

Evans, J. P., and F. J. Gould: Stability in nonlinear programming. Oper. Res. **18** (1970), 107—118. (I)

Evans, J. P., F. J. Gould, and J. W. Tolle: Exact penalty functions in nonlinear programming. Math. Programming **4** (1973), 72—97. (II)

Faber, M. M.: Stochastisches Programmieren. Physica-Verlag, Würzburg—Wien 1970. (III)

Faselt, S.: Über Vorschläge zur Bestimmung der Lösungsfunktion eines einparametrischen linearen gemischt-ganzzahligen Optimierungsproblems mit konstanter Zielfunktion. Dipl.-Arbeit, Humboldt Univ. Berlin, Sektion Mathematik 1979. (I)

Fedorov, V. V. (Федоров, В. В.):

[1] К вопросу об устойчивости задачи линейного программирования. Ж. вылисл. мат. и мат. физ. **15** (1975), 1412—1423. (I, II)

[2] Численные методы максимина. Наука, Москва 1979. (III)

Fenchel, W.:

[1] On conjugate convex functions. Canadian J. Math. **1** (1949), 73—77. (II)

[2] Convex cones, sets and functions. Lecture Notes, Princeton Univ., Department of Math., Princeton 1953. (III)

Fiacco, A. V.:

[1] Penalty methods for mathematical programming in E^n with general constraint sets. J. Optim. Theory Appl. **6** (1970) 3, 252—268. (I, II)

[2] Convergence properties of local solutions of sequences of mathematical programming problems in general spaces. J. Optim. Theory Appl. **13** (1974) 1, 1—12. (I, II)

[3] Sensitivity analysis for nonlinear programming using penalty functions. Math. Programming **10** (1976), 287—311. (I, II)

Fiacco, A. V., and G. P. McCormick: Nonlinear programming: sequential unconstrained minimization techniques. John Wiley, New York—London—Sydney—Toronto 1968. (III)

Findeisen, W.:

[1] Parametric optimization by primal method in a two-level system. Arch. Automat. Telemech. **12** (1967), 391—404. (I)

[2] Parametric optimization by primal method in multilevel systems. IEEE Trans. Syst. Sci. Cybernet. **4** (1968), 155—164. (I)

Finkel'šteĭn, B. V. (Финкельштейн, Б. В.):

[1] О методе пересчета матрицы коеффициентов полных затрат в случае изменения технологии в нескольких отраслях производства. *В*: Труды научн. совещания о применении мат. методов в экон. исслед. и планировании. АН СССР, Москва 1961, 115—119. (I)

[2] Обобщение параметрической задачи линейного программирования. Экон. и мат. методы **1** (1965), 442—450. (I)

Finkel'šteĭn, B. V., and L. P. Gumenjuk (Финкельштейн, Б. В., и Л. П. Гуменюк): Алгоритмы решения задачи параметрического программирования в случае зависимости от параметрици условий. Экон. и мат. методы **13** (1977), 342—347. (I)

Fiorot, J. Ch., and P. Huard:

[1] Une approach théorique du problème de linéarisation en programmation mathématique convexe. Laboratoire de Calcul, publication No. 43, Univ. of Lille 1974. (II)

[2] Composition and union of général algorithms of optimization. Math. Programming Stud. **10** (1979), 69—85. (II)

Fischer, M.: Některé otázky citlivosti úlohy lineárniho programování, vznikahící při řešení metodou parametrického programování. Ekon.-mat. obzor **4** (1968), 337—350. (I)

Fleming, W. H.: The Cauchy problem for a non-linear first order partial differential equation. J. Differential Equations **5** (1969), 515—530. (II)

Focke, J.: Vektormaximumprobleme und parametrische Optimierung. Math. Operationsforsch. Stat. **4** (1973) 5, 365—369. (I)

Fong, C. O., and M. R. Rao: Parametric studies in transportationtype problems. Naval Res. Logist. Quart. **22** (1975), 355—364. (I)

Fort Jr., M. K.: A unified theory of semicontinuity. Duke Math. J. **16** (1949), 237—246. (II)

Frank, C. R.: Parametric programming in integers. Oper. Res. Verf. **3** (1967). (I)

Fuentes-Mora, M. D., and G. Wintgen: Ausartung bei der parametrischen Optimierung. 12. Internat. wiss. Kolloqu. Techn. Hochschule Ilmenau 1967, Bd. 1, 29—34. (I)

Gaede, K.-W.: Sensivitätsanalyse für einen Markov-Process. Z. Oper. Res. **18** (1974), 197—204. (I)

Gajorsky, T.: An explicit way of finding absolute maximum of strictly concave parametric quadratic function with linear parametric dependent constraints. Acta Fac. Rerum Natur. Univ. Comenian. Math. XXVII (1972), 53—64. (I)

Gál, T.:

[1] Příspěvek k lineárnímu systémovému programování. Sledování změn prvků a_{ij} matrice soustavy podmínek simplexové úlohy lineárního programování. Ekon.-mat. obzor **3** (1967), 446—456. (I)

[2] Příspěvek k lineárnímu systémovému programování. 2: Sledování změn koeficientů a_{ij} základních strukturních priměnných simplexové úlohy lineárního programování. Ekon.-mat. obzor **4** (1968), 76—92. (I)

[3] Příspěvek k lineárnímu systémovému programování. 3: Sledování současne změny všech koefizientů a_{ij} základních strukturních proměnných simplexové úlohy lineárního programování. Ekon.-mat. obzor **4** (1968), 190—201. (I)

[4] A method for systematic simultaneous parametrization of vectors b and c in *LP*-problems. Ekon.-mat. obzor **6** (1970), 161—175. (I)

[5] Homogene mehrparametrische lineare Programmierung. Unternehmensforsch. **16** (1972), 115—136. (I)

[6] Betriebliche Entscheidungsprobleme, Sensitivitätsanalyse und parametrische Programmierung. Walter de Gruyter, Berlin—New York 1973. (I, III)

[7] Rim multiparametric linear programming. Management Sci. **21** (1975), 567—575. (I)

[8] A note on redundancy and linear parametric programming. Oper. Res. Quart. **26** (1975), 735—742. (I)

[9] A general method for determining the set of all efficient solutions to a linear vectormaximum problem. European J. Oper. Res. **1** (1977), 307—322. (I)

Gál, T., and J. Nedoma:

[1] Lineáris programosáz több paraméterell a jobb oldalon vagy a célfüggvény-koefficiensekben. Szigma **2** (1969), 213—237. (I)

[2] Multiparametric linear programming. Management Sci. **18** (1972), 406—422. (I)

[3] Methode zur Lösung mehrparametrischer linearer Programme. IV. Oberwolfach Tag. Oper. Res. 1971, Oper. Res. Verf. **12** (1972), 130—133. (I)

Gass, S. I.: Linear programming: methods and applications. McGraw-Hill, New York—Toronto—London 1958 (Russ. transl.: Moskva 1961). (III)

Gass, S. I., and T. L. Saaty:

[1] The computational algorithm for the parametric objective function. Naval Res. Logist. Quart. **2** (1955), 39—45. (I)

[2] The parametric objective function. II. Oper. Res. **3** (1955), 395—401. (I)

Gauthier, J. M.: Paramétrisation de la fonction économique d'un programme linéaire: un algorithme construit en utilisant le principe de décomposition de Dantzig et Wolfe. Revue Franç. Rech. Oper. **6** (1962) 22, 1—20. (I)

Gauvin, J., and J. W. Tolle: Differential stability in nonlinear programming. SIAM J. Control Optim. **15** (1977), 294—311. (II)

Gazik, R. J.: Convergence in spaces of subsets. Pacific J. Math. **43** (1972), 81—92. (II)

Gazon, J., and A. Tits: A structural approach to stability in linear systems. A sufficient condition. Internat. J. Systems Sci. **9** (1978), 681—694. (I)

Genuys, F.: Paramétrisation du second membre et de la fonction économique. *In*: Huard, P. (ed.): Mathématique des programmes économiques. Dunod, Paris 1964, 55—62. (I)

GEOFFRION, A. M.:
[1] A parametric programming solution to the vector maximum problem, with applications to decisions under uncertainity. Diss., Oper. Res. Progr., Stanford Univ. Febr. 1965. (I, II)
[2] Strictly concave parametric programming, part I: Basic theory. Management Sci., Ser. A, **13** (1966), 244—253. (I)
[3] Strictly concave parametric programming, part II: Additional theory and computational considerations. Management Sci., Ser. A, **13** (1967), 359—370. (I)
[4] Duality in nonlinear programming: a simplified applications oriented development. SIAM Rev. **13** (1974), 1—37. (II)

GEOFFRIAN, A. M., and R. NAUSS: Parametric and postoptimality analysis in integer linear programming. Management Sci. **23** (1977), 453—466. (I)

GIANESSI, F.: On the variation of the coefficients in a problem of linear programming. Collana ing. C. Olivetti, S.p.A. 1964. (I)

GODFREY, H. T., A. W. SPIVEY, and G. B. STILLWAGON: Production and market planning with parametric programming. Indust. Manag. Rev. **10** (1968), 61—75. (I)

GODUNOV, B. A. (Годунов, Б. А.): Некоторые параметрические задачи линейного программирования. *В*: Тр. 2 Зим. школы по мат. программир. и смежн. вопр. 1969. Вып. 2. Москва 1969, 445—465. (I)

GOLLAN, B.:
[1] Störungstheorie für abstrakte Optimierungsprobleme mit Anwendungen auf die Theorie optimaler Steuerungen. Diss., Julius-Maximilian-Univ. Würzburg 1979. (I)
[2] Perturbation theory for abstract optimization problems. Preprint No. 47, Math. Inst. d. Julius Maximilian Univ. Würzburg 1979. (I)

GOLLMER, R.:
[1] Lineare parametrische Optimierung mit mehreren Parametern in der Restriktionsmatrix. Manuskript, Humboldt Univ. Berlin, Sektion Mathematik 1978. (I)
[2] Eine Klasse linearer Optimierungsprobleme mit mehreren Parametern in der Restriktionsmatrix. Diss. A Humboldt Univ. Berlin 1980. (I)

GOL'ŠTEĬN, E. G. (Гольштейн, Е. Г.):
[1] О возможности расширения применимости частных методов линейного программирования. *В сб.*: Планир. и экон.-мат. методы. Наука, Москва 1964, 409—423. (I)
[2] Выпуклое программирование. Элементы теории. Наука, Москва 1970. (III)
[3] Теория двойственности в математическом программировании и её приложения. Наука, Москва 1971. (III)

GOL'ŠTEĬN, E. G., and D. B. JUDIN (Гольштейн, Е. Г., и Д. Б. Юдин): Новые направления в линейном программировании. Советское Радио, Москва 1966. (III)

GOL'ŠTEĬN, E. G., and S. M. MOVŠOVIČ (Гольштейн, Е. Г., и С. М. Мовшович):
[1] О некоторых современных направлениях в математическом программировании. Экон. и мат. методы **3** (1967), 766—778. (II)
[2] Непрерывная зависимость от параметра множества решений минимаксной задачи. Экон. и мат. методы **4** (1968), 920—930. (I)

GOL'ŠTEĬN, E. G., and N. V. TRET'YAKOV (Гольштейн, Е. Г., и Н. В. Третьяков):
[1] Модифицированные функции Лагранжа. Экон. и мат. методы **10** (1974), 568—591. (II)
[2] Modified Lagrangians in convex programming and their generalizations. Math. Programming Stud. **10** (1979), 86—97. (II)

GOMORY, R. E.: On the relation between integer and noninteger solutions to linear programs. Proc. Nat. Acad. Sci. USA **53** (1965) 1, 260—265. (II)

GÖPFERT, A.: Mathematische Optimierung in allgemeinen Vektorräumen. BSB B. G. Teubner Verlagsgesellschaft, Leipzig 1973. (III)

GOULD, F. J.: Extensions of Lagrange multipliers in nonlinear programming. SIAM J. Appl. Math. **17** (1969), 1280—1297. (II)

GRABOWSKI, W.:
[1] Pewne włanośći parametrycznego program liniowego. Przeglad. Statyst. **14** (1967), 267—278. (I)
[2] Metoda rozwiązywania parametrycznego programu liniowego. Przeglad. Statyst. **16** (1969), 159—173. (I)

GRAVES, R. L.: Parametric linear programming. *In*: GRAVES, R. L., and P. WOLFE (eds.): Recent adcvanes in mathematical programming. McGraw-Hill, New York—Toronto—London 1960, 201—210. (I)

GREENBERG, H. J.: The generalized penalty-function/surrogate model. Oper. Res. **21** (1973), 162—178. (II)

GREENBERG, H. J., and W. P. PIERSKALLA:

[1] A review of quasi-convex functions. Oper. Res. **19** (1971), 1553—1570. (II)

[2] Extensions of the Evans-Gould stability theorems for mathematical programs. Oper. Res. **20** (1972), 143—153. (I)

[3] Stability theorems for infinitely constrained mathematical programs. J. Optim. Theory Appl. **16** (1975), 409—428. (I)

GRIGORIADIS, M. D., and K. RITTER: A parametric method for semidefinite quadratic programs. SIAM J. Control **7** (1969), 559—577. (I)

GROSSMANN, CH., and G. SCHÖNIGER: Sensitivität und Anwendbarkeit von Straf-Barriere-Methoden. Z. Angew. Math. Mech. **57** (1977), 255—264. (II)

GRÜNBAUM, B.: Convex polytopes. John Wiley, New York 1967. (III)

GRYGAROVÁ (SOKOLOVÁ), L.:

[1] Qualitative Untersuchung des I. Optimierungsproblems in mehrparametrischer Programmierung. Apl. Mat. **15** (1970), 276—295. (I)

[2] Die Lösbarkeit eines linearen Optimierungsproblems unter Zufügung einer weiteren Restriktionsbedingung. Apl. Mat. **17** (1972), 352—387. (I)

[3] Lösungsbereich von Optimierungsproblemen mit Parametern in den Koeffizienten der Matrix der linearen Restriktionsbedingungen. Apl. Mat. **17** (1972), 388—400. (I)

[4] Über Stabilitätsbereiche bei parametrischen linearen Optimierungsproblemen. Apl. Mat. **20** (1975), 311—335. (I)

GRYGAROVÁ, L., and F. NOŽIČKA: Sphärische Abbildungen von konvexen Mengen und konvexe Optimierung. Jahrestagung Math. Optim., Vitte (Hiddensee) 1976, 37—41. (II)

GUDDAT, J.:

[1] Stabilitätsuntersuchungen in der quadratischen parametrischen Optimierung. Diss. B, Humboldt Univ. Berlin 1974. (I)

[2] On qualitative stability in the nonlinear parametric programming and some applications. *In*: PREKOPÁ, A. (ed.): Mathematical methods of operations research 1, studies on mathematical programming. Proc. 3rd Conf. Math. Progr., Matrafüred 1975. Akademiai Kiado, Budapest 1980, 69—81. (I, II)

[3] Stability in convex quadratic programming. Math. Operationsforsch. Statist. **7** (1976), 223—245. (I)

[4] Über die Nožička-Konzeption der parametrischen Optimierung und ihre Weiterentwicklung. Seminarbericht Nr. 15, Humboldt Univ. Berlin, Sektion Mathematik 1978, 38—51. (I, III)

[5] Parametrische Optimierung und Vektoroptimierung. *In*: LOMMATZSCH, K. (ed.): Anwendungen der linearen parametrischen Optimierung. Akademie-Verlag, Berlin 1979, 54—75. (II)

GUDDAT, J., and P. HOCHMUTH: Konvexitätsbereiche und Lösbarkeitsmengen in der quadratischen Optimierung. Konference o matematickych metodach v ekonomii, Zadov, Ekon.-Mat. Laboratoř 1973, 133—150. (I)

GUDDAT, J., and D. KLATTE: Qualitative stability in nonlinear optimization. *In*: PREKOPÁ, A. (ed.): Survey of mathematical programming. Proc. 9th Internat. Math. Progr. Sympos., Budapest 1976. Publishing House of the Hungarian Academy of Science, Budapest/North-Holland, Amsterdam—London 1979, 217—229. (I)

GUDDAT, J., and M. LIPS: On the theoretical basis for methods for parametric optimization problems. *In*: Proc. 9th IFIP Conf. Optim. Techn., Warsaw 1979. Lecture Notes in Control and Information Sciences, vol. 23. Springer-Verlag, Berlin—Heidelberg—New York 1980. (I)

GUDDAT, J., and K. TAMMER: Eine Modifikation der Methode von Theil und van de Panne zur Lösung einparametrischer quadratischer Optimierungsprobleme. Math. Operationsforsch. Statist. **1** (1970), 199—206. (I)

GUERRA VAZQUEZ, F.: Aplication de la optimizacion parametrica el problema de optimizacion vectorial cuadratic. Diss., Univ. Havanna 1980. (I)

GUPTA, R. P.: A simple class of parametric linear fractional functionals programming problem. Cahiers Centre Études Rech. Oper. **15** (1973), 185—196. (I)

GUSEVA, O. V. (Гусева, О. В.) Об устойчивости задачи выпуклого программирования. Мат. методы реш. экон. задач **3** (1972), 73—75. (I)

HABETLER, G. J., and M. M. KOSTREVA: On a direct algorithm for nonlinear complementarity problems. SIAM J. Control Optim. **16** (1978) 7, 504—514. (II)

HACKL, J.: Solution of optimization problems with nonlinear restrictions via continuation methods. Inst. f. Math., Johannes Kepler Univ. Linz. (I)

HAGER, W. W.: Lipschitz continuity for constraint processes. SIAM J. Control Optim. **17** (1979), 321—338. (I, II)

HAHN, H.: Reelle Funktionen, 1. Band: Punktfunktionen. Akad. Verlagsges. mbH, Leipzig 1932. (II, III)

HANSEL, R.:

[1] Niveaumengen konvexer quadratischer Optimierungsprobleme. Jahrestagung Math. Optim., Vitte (Hiddensee) 1979, 67—70. (I)

[2] Stabilität gemischt-ganzzahliger Optimierungsprobleme. Diss. A, Humboldt Univ. Berlin 1980. (I)

HARTFIEL, D. J., and G. L. CURRY: Concerning the solution vector for parametric programming problems. SIAM J. Appl. Math. **26** (1974), 294—296. (I)

HARTLEY, H. O., and L. D. LOFTSGARD: Linear programming with variable restraints. Iowa State Coll. J. Sci. **2** (1958), 161—172. (I)

HARTUNG, J.: Dualität und Sattelpunkte. Oper. Res. Verf. **12** (1972), 194—200. (II)

HAUSDORFF, F.: Mengenlehre. 3. Aufl., W. de Gruyter, Berlin 1934. (III)

HAZEWINKEL, M.: Some examples concerning linear continuity of solutions to programming problems. Oper. Res. **25** (1977), 715—719. (I)

HEADY, E. O., and W. CANDLER: Linear programming methods. Iowa State Coll. Press, Ames 1958. (I, III)

HEMPEL, D.: Abschätzung des optimalen Wertes der Zielfunktion. Z. Angew. Math. Mech. **46** (1966), 98—100. (I)

HENDZLICK, H.: Zusammenhänge zwischen Ersatzauswahlfunktionen zu nichtlinearen parametrischen Optimierungsproblemen. Dipl.-Arbeit, Humboldt Univ. Berlin, Sektion Mathematik, 1980. (I)

HENSCHKE, I.: Lineare parametrische Optimierung mit einem Parameter in der Koeffizientenmatrix. Diss. A, Humboldt Univ. Berlin 1975. (I)

HESS, H.-D.: Anwendung einer parametrischen linearen Optimierung in einem chemischen Betrieb. Industr. Organisation **35** (1966), 76—77. (I)

HESTENESS, M. R.: Optimization theory — the finite dimensional case. John Wiley, New York 1975. (III)

HILDENBRAND, W.: Core and equilibria of large economy. Princeton Univ. Press, Princeton 1974. (II)

HOANG TUY:

[1] On the convex approximation of nonlinear inequalities. Math. Operationsforsch. Statist. **5** (1974), 451—466. (II)

[2] Stability property of a system of inequalities. Math. Operationsforsch. Statist., Ser. Optim., **8** (1977), 27—39. (II)

HOCKING, R. R., and R. L. SHEPARD: Parametric solution of a class of nonconvex programs. Oper. Res. **19** (1971), 1742—1747. (I, II)

HOFFMAN, A. J.: On approximate solutions of systems of linear inequalities. J. Res. Nat. Bur. Standards **49** (1952), 263—265. (II)

HOGAN, W. W.:

[1] Directional derivatives for extremal-value-functions with applications to the completely convex case. Oper. Res. **21** (1973), 188—209. (I)

[2] The continuity of the perturbation function of a convex program. Oper. Res. **21** (1973), 351—352. (I)

[3] Point-to-set maps in mathematical programming. SIAM Rev. **15** (1973), 591—603. (I)

HOLLATZ, H.:

[1] Lineare Optimierungsprobleme mit optimalen Vektoren von gleicher Struktur. Operationsforsch. math. Statistik **1** (Schriftenr. Inst. Math. Dtsch. Akad. Wiss. Berlin, B, No. 8) (1968), 35—45. (I)

[2] Zur algorithmischen Behandlung der Stabilität linearer Optimierungsprobleme. Wiss. Z. Hochsch. Architektur Bauwesen Weimar **16** (1969), 243—245. (I)

[3] Die Konstruktion lösbarer Optimierungsprobleme. Math. Operationsforsch. Statist. **1** (1970), 255—263. (I)

[4] Parametrische Optimierung in linearen Räumen. Math. Operationsforsch. Statist. **4** (1973), 107—125. (I)

Hollatz, H., and H. Weinert: Ein Algorithmus zur Lösung des doppelt-einparametrischen linearen Optimierungsproblems. Math. Operationsforsch. Statist. **2** (1971), 181 bis 197. (I)

Holm, S., and D. Klein:

[1] Parametric analysis for integer programming problems. Paper presented at the 49th Joint National Meeting of ORSA and TIMS, Philadelphia 1976. (I)

[2] Discrete right hand side parametrization for linear integer programs. European J. Oper. Res. **2** (1978), 50—53. (I)

House, W. C.:

[1] Sensitivity analysis — a case study of the pipeline industry. Engin. Economist **12** (1966), 155—166. (I)

[2] Use of sensitivity analysis in capital budgeting. Management Services **4** (1967) 5, 37—40. (I)

Houthakker, S.: The capacity method of quadratic programming. Econometrica **28** (1960), 62—87. (II)

Howe, S.: A penalty function procedure for sensitivity analysis of concave programs. Management Sci. **21** (1974), 341—346. (I)

Hu, T. C.: Integer programming and network flows. Addison-Wesley, London 1969. (III)

Hu, T. C., and S. M. Robinson: Mathematical programming. Academic Press, New York 1973. (II, III)

Huard, P.:

[1] Resolution of mathematical programming problems with nonlinear constraints by the method of centres. *In*: Abadie, J. (ed.): Nonlinear programming. North-Holland, Amsterdam 1967, 206—219. (II)

[2] Programmation mathématique convexe. Revue Franç. Inform. Rech. opérat. **2** (1968), 43—59. (II)

[3] Optimization algorithms and point-to-set maps. Math. Programming **8** (1975), 308—331. (I, II)

Ibragimov, I. A., G. A. Kaplan, and B. S. Korš (Ибрагимов, И. А., Г. А. Каплан, и Б. С. Корш): Применение метода параметрического программирования к решению одного класса задач оптимизации. Изв. высш. учебн. заведений, Нефть и газ **5** (1968), 92—96. (I)

Ioffe, A. D., and V. M. Tichomirov (Иоффе, А. Д., и В. М. Тихомиров): Теория экстремальных задач. Наука, Москва 1974. (German transl.: Theorie der Extremalaufgaben. VEB Deutscher Verlag der Wissenschaften, Berlin 1979) (III)

Itami, H.: Expected objective value of a stochastic linear program and the degree of uncertainty of parameters. Management Sci. **21** (1974), 291—301. (I)

Ivănescu (Hammer), P.: Parametrische Transportprobleme. In: Bader, H., u. a. (eds.): Mathematik und Wirtschaft 5. Verlag Die Wirtschaft, Berlin 1968, 135—145. (I)

Jacobs, M. Q.: Some existence theorems for linear optimal control problems. SIAM J. Control **5** (1967), 418—437. (II)

Jadykin, A. B. (Ядыкин, А. Б.):

[1] Параметрический метод в задачах квадратичного программирования с вырожденной квадратичной формой. Ж. вычисл. мат. и мат. физ. **15** (1975), 1436—1446. (I)

[2] О параметризации в вырожденных задачах квадратичного программирования. Ж. вычисл. мат. и мат. физ. **17** (1977), 634—648. (I)

Jagannathan, R.:

[1] A simplex-type algorithm for linear and quadratic programming — a parametric procedure. Econometrica **34** (1966), 460—471. (II)

[2] On some properties of programming problems in parametric form pertaining to fractional programming. Management Sci., Ser A, **12** (1966), 609—615. (II)

JAGEL', A. (Ягель, А.):

[1] Основные свойства функции максимума на одном классе задач параметрического линейного программирования. ENSV Teaduste Akad. toimetised. Füüs.-mat. ja tehn. seer. **13** (1964) 4, 382—402. (I)

[2] Характеристика области параметров допустимости при одном классе задач параметрического линейного программирования. ENSV Teaduste Akad. toimetised. Füüs.-mat. ja tehn. seer. **14** (1965) 2, 223—232. (I)

JANIN, R.:

[1] Sensitivity for nonconvex optimization problems. In: AUSLENDER, A. (ed.): Convex analysis and its applications. Springer-Verlag, Berlin—Heidelberg—New York 1977, 115—119. (I)

[2] On sensitivity in an optimal control problem. J. Math. Anal. Appl. **60** (1977) 3, 631—657. (I)

JASILIONIS, R. JU. (Ясилионис, Р. Ю.): О свойствах равновесных решений одного класса сложных задач. Liet. mat. rinkingys **8** (1968), 175—184. (I), (II)

JENSEN, R. E.: Sensitivity analysis and integer linear programming. Accounting Rev. **43** (1968), 425—446. (I)

JEROSLOW, R. G.:

[1] Linear programs dependent on a single parameter. Discrete Math. **6** (1973), 119 to 140. (I)

[2] Cutting planes for complementarity constraints. SIAM J. Control Optim. **16** (1978) 1, 56—62. (II)

JOLY, J. L., and P. J. LAURENT: Stability and duality in convex minimization problems. Rev. Franç. Inform. Rech. opérat. **R2** (1971), 3—42. (I)

JONES, C. H.: Parametric production planning. Managament Sci. **13** (1967), 843—866. (II)

JUDIN, D. B., and E. G. GOLSTEIN: Lineare Optimierung I. Akademie-Verlag, Berlin 1968. (III)

KAKUTANI, S.: A generalization of Brouwer's fixed point theorem. Duke Math. J. **8** (1941), 457—459. (I, II)

KALL, P.:

[1] Stochastic linear programming. Springer-Verlag, Berlin—Heidelberg—New York 1976. (III)

[2] Stochastische Optimierung — Einige neuere Ergebnisse. Seminarbericht Nr. 15, Humboldt Univ. Berlin, Sektion Mathematik 1978. (II)

KALL, P., and W. OETTLI: Measurability theorem for stochastic extremals. SIAM J. Control **13** (1975), 994—998. (II)

KANEKO, I.:

[1] A parametric linear complementary problem involving derivatives. Math. Programming **15** (1978), 146—154. (I)

[2] A maximization problem related to parametric linear complementarity. SIAM J. Control Optim. **16** (1978) 1, 41—55. (I)

KAPLAN, A. A. (Каплан, А. А.): К вопросу о реализации методов штрафов. Новосибирск 1976. (III)

KARABEGOV, V. K. I.: A parametric problem in linear programming. U.S.S.R. Comput. Math. Math. Phys. **3** (1963), 725—741 (1965), (transl. from: Ж. вычисл. мат. и мат. физ. **3** (1963), 547—558.). (I)

KARLIN, S.: Mathematical methods and theory in games, programming and economics. Addison-Wesley, London 1959. (III)

KARMANOV, V. G. (Карманов, В. Г.): Математическое программирование. Наука, Москва 1975. (III)

KAŠKA, J.:

[1] O jedné úloze parametrického programováni. Conf. E.S./T.I.M.S., Warszawa 1966, Ekon.-mat. obzor **3** (1967), 298—307. (I)

[2] Přispěvek k parametrickému programováni. Ekon.-mat. obzor **10** (1974), 33—49. (I)

KAŠKA, J., and M. PIŠEK: Parametrické kvadratické programováni. Ekon.-mat. obzor **1** (1965), 383—390. (I)

Kassay, F.: An approach to sensitivity analysis of transportation problems. Ekon.-mat. obzor **13** (1977), 213—223. (I)

Kausmann, U.: Mehrparametrische lineare Optimierung. Diss. A, Humboldt Univ. Berlin 1971. (I)

Kausmann, U., K. Lommatzsch, and F. Nožička: Lineare parametrische Optimierung. Akademie-Verlag, Berlin 1976. (I, III)

Kazarjan, N. G., and S. A. Sukiasjan (Казарян, Н. Г., и С. А. Сукиасян): Алгоритм решения простейшей параметрической задачи линейного программирования. Тр. вычисл. центра АН АрмССР и Ереванск. унив. **2** (1964), 17—21. (I)

Kelley Jr., J. E.: Parametric programming and the primal-dual algorithm. Oper. Res. **7** (1959), 327—334. (I)

Keng Chieh Wang, P.: Stabilisabilité d'une classe de systèmes stochastiques linéaire. Comptes Rendus de l'Académie des Sciences, Paris, **263** (1966) 14, A467—A469. (I)

Kim, C.: Parametrizing an activity vector in linear programming. Oper. Res. **19** (1971), 1632—1646. (I)

Kirsch, A., Zur Störung von Optimierungsaufgaben unter besonderer Berücksichtigung von optimalen Steuerungsproblemen. Diss., Georg August Univ. Göttingen 1978. (II)

Klatte, D.:

[1] Untersuchungen zur lokalen Stabilität konvexer parametrischer Optimierungsprobleme. Diss. A, Humboldt Univ. Berlin 1977. (I)

[2] Lineare Optimierungsprobleme mit Parametern in allen Koeffizienten der Zielfunktion und der Restriktionen. Wiss. Z. Humboldt Univ. Berlin, Math.-Naturw. Reihe, **5** (1977), 555—560. (I)

[3] On the lower semicontinuity of optimal sets in convex parametric optimization. Math. Programming Stud. **10** (1979), 104—109. (I)

[4] Lineare Optimierungsprobleme mit Parametern in der Koeffizientenmatrix der Restriktionen. *In*: Lommatzsch, K. (ed.): Anwendungen der linearen parametrischen Optimierung. Akademie-Verlag, Berlin 1979, 23—53. (I)

[5] Zum Beweis von Stabilitätseigenschaften linearer parametrischer Optimierungsaufgaben mit variabler Koeffizientenmatrix. Jahrestagung Math. Optim., Vitte/Hiddensee 1979, 215—218. (I)

Klee, V. L.: Convex sets in linear spaces, I. Duke Math. J. **18** (1951), 443—466. (II)

Kleinmann, P.: Quantitative Sensitivitätsanalyse bei parametrischen Optimierungsaufgaben. Seminarbericht Nr. 9, Humboldt Univ. Berlin 1978. (I)

Kojima, M.:

[1] A unification of the existence theorem of the nonlinear complementarity problem. Math. Programming **9** (1975), 257—277. (II)

[2] A complementary pivoting approach to parametric nonlinear programming. Math. Oper. Res. **4** (1979) 4, 464—477. (II)

Kojima, M., H. Nishino, and T. Sekine: An extension of Lemke's method to the piecewise linear complementarity problem. SIAM J. Appl. Math. **31** (1976), 600—613. (II)

Kokotovič, P. V., J. B. Cruz, J. E. Heller, and P. Sannuti: Synthesis of optimally sensitive systems. Proc. IEEE **56** (1968), 1318—1324. (I, II)

Kokotovič, P. V., J. E. Heller, and P. Sannuti: Sensitivity comparison of optimal control. Internat. J. Control **9** (1969), 111—115. (I, II)

Korbut, A. A., and J. J. Finkelstein: Diskrete Optimierung. Akademie-Verlag, Berlin 1971 (transl. from Russ.). (III)

Kornbluth, J. S. H.: Duality, indifference and sensitivity analysis in multiobjective linear programming. Oper. Res. Quart. **25** (1975), 599—614. (I, II)

Kosmol, P.: Optimierung konvexer Funktionen mit Stabilitätsbetrachtungen. Habilitationsschrift, Univ. Kiel 1973 (see also: Dissertationes Mathematicae, No. CXL, Rozprawy matematyczne, Warszawa 1976). (I, II)

Krabs, W.:

[1] Zur stetigen Abhängigkeit des Extremalwertes eines konvexen Optimierungsproblems von einer stetigen Änderung des Problems. Z. Angew. Math. Mech. **52** (1972), 359—368. (I)

[2] Stetigkeitsfragen bei der Diskretisierung konvexer Optimierungsprobleme. *In*: Collatz, L., and W. Wetterling (eds.): Numerische Methoden bei Optimierungsaufgaben. Birkhäuser-Verlag, Basel—Stuttgart 1973, 73—89. (I, II)

[3] Stabilität und Stetigkeit bei nichtlinearer Optimierung. *In*: Henn, R., H. P. Künzi, and H. Schubert (eds.): Operations-Research-Verfahren XVII, Teil 2. Verlag Anton Hain, Meisenheim/Glan 1973, 207—228. (I)

[4] Discretization of approximation problems in the view of optimization. *In*: Lorentz, G. G. (ed.): Approximation Theory. Academic Press, New York—London 1973, 395—401. (I, II)

[5] Optimierung und Approximation. Teubner-Verlag, Stuttgart 1975. (III)

[6] Stetige Abänderung der Daten bei nichtlinearer Optimierung und ihre Konsequenzen. *In*: Operations-Research-Verfahren XXV, Teil 1. Verlag Anton Hain, Meisenheim/Glan 1976, 99—113. (I, II)

Kreutzberger, O., and L. Krigler: Bestimmung von Stabilitätsbereichen in der stochastischen linearen Optimierung mit Hilfe parametrischer Optimierung und Regressionsrechnung. Wiss. Z. Martin Luther Univ. Halle-Wittenberg 5 (1977); 71. (II)

Krasner, N. Ja. (Краснер, Н. Я.): О сведении некоторых задач нелинейного программирования к линейным параметрическим. *В сб.*: Вопросы оптимального программирования в производственных задачах. Воронеж 1968, 42—46. (I, II)

Krasner, N. J., and T. I. Pisareva (Краснер, Н. Я., и Т. И. Писарева): Отыскание относительно-оптимальных планов и решение параметрических задач линейного программирования. *В сб.*: Вопросы оптимального программирования в производственных задач 3. Воронеж 1969, 88-98. (I)

Kuhn, M. W., and A. W. Tucker: Nonlinear programming. In: Proc. of the 2nd Berkeley Symposium on Math. Statistics and Probability, 1950. Univ. of California Press, Berkeley—Los Angeles 1951, 481—492. (III)

Kummer, B.:

[1] Global stability of optimization problems. Math. Operationsforsch. Stat. Ser. Optim. 8 (1977), 367—383. (I, III)

[2] Globale Stabilitätsuntersuchungen für parametrische Optimierungsprobleme. Diss. B, Humboldt Univ. Berlin, 1977. (I)

[3] Stabilität quadratischer Optimierungsprobleme. Wiss. Z. Humboldt Univ. Berlin, Math.-Naturw. Reihe, 5 (1977), 365—569. (I)

[4] Stetigkeitsverhalten von Punkt-Menge-Abbildungen in der mathematischen Optimierung. Seminarbericht Nr. 15, Humboldt Univ. Berlin 1978, 76—87. (I, III)

[5] Stability and weak duality in convex programming without regularity. Preprint, Humboldt Univ. Berlin 1978. (I, II)

[6] A note on the continuity of the solution set of special dual optimization problems. Math. Programming Stud. 10 (1979), 110—114. (I)

Künzi, H. P., and W. Krelle: Nichtlineare Programmierung. Springer-Verlag, Berlin—Göttingen-Heidelberg 1962. (III)

Kurata, R.: Notes on parametric quadratic programming. J. Operations Res. Japan 8 (1966), 150—153. (I)

Kurata, S.: Primal-dual method of parametric programming and Iri's theory on network flow problems. J. Operations Res. Japan 2 (1965), 104—144. (I)

Kuratowski, K. (Куратовский, К.): Топология, том I. Мир, Москва 1966. (transl. from French) (III)

Laperche, Ch.: Multiparametric linear programming algorithmic aspects. Mémoire présenté de l'obtenu du grade d'Ingénieur Civil en mathématiques appliquées, Univ. Catholique de Louvain 1970. (II)

Larčenko, N. E. (Ларченко, Н. Е.): Об одной задаче параметрического программирования. Экон. и мат. методы 2 (1966), 448—451. (I)

Lasota, A., and C. Olech: On Cesari's semicontinuity condition for set valued mappings. Bull. Acad. Pol. Sci. 16 (1968), 711—717. (II)

Lasry, J. M., and R. Robert: Analyse nonlinéaire multivoque. Cahiers math. décision No. II Univ. of Paris-Dauphine 1976. (II)

Laurent, P.-J. (Лоран, П.-Ж.): Аппроксимизация и оптимизация. Мир. Москва 1975 (transl. from French).

Leitmann, G.: Guaranteed asymptotic stability for some linear systems with bounded uncertainties. Trans. ASME Ser. G.J. Dynamic systems measurement control 101 (1979), 895—897. (I, II)

LEMKE, C. E.:
[1] Bimatrix equilibrium points and mathematical programming. Management Sci. **11** (1965), 681—689. (II)
[2] Recent results on complementarity problems. In: ROSEN, J. B., O. L. MANGASARIAN, and K. RITTER (eds.): Nonlinear programming. Proc. Sympos., Univ. of Wisconsin, Madison (Wisc.) 1970. Academic Press, New York 1970, 349—384. (II)

LEMPIO, F., and H. MAURER: Differential stability in infinite-dimensional nonlinear programming. Appl. Math. Optim. **6** (1980), 139—152. (II)

LEVIĬ, V. D., and A. A. VOLODIN (Левий, В. Д., и А. А. Володин): Об одной задаче целочисленного программирования с переменными коэффициентами. Экон. и мат. методы **10** (1974), 1172—1177. (I)

LEVITIN, E. S. (Левитин, Е. С.):
[1] О корректности ограничений и устойчивости в экстремальных задачах. II Вестн. Московск унив. **2** (1968), 8—22. (I, II)
[2] О дифференциальных свойствах оптимального значения параметрических задач математического программирования. Докл. Акад. Наук СССР **215** (1974), 4. (II)
[3] О коррекции решений задач нелинейного программирования с непольной информации. Всесоюзная летняя школа о методах оптимизации и их применении. Акад. Наук СССР, сиб. отдел., Иркутск 1974. (I, II)
[4] Об условиях локального минимума в экстремальных задачах с ограничесниями. Докл. Акад. Наук СССР **221** (1975), 1031—1034. (II)
[5] Об оценках устойчивости по решению в задаче безусловной оптимизации. Ж. вычисл. мат. и мат. физ. **16** (1976), 585—596. (I, II)
[6] О дифференцируемости по параметру оптимального значения параметрических задач математического программирования. Кибернетика **1** (1976), 44—59. (I)

LEVITIN, E. S., A. A. MILJUTIN, and N. P. OSMOLOVSKIĬ (Левитин, Е. С., А. А. Милютин и Н. П. Осмоловский): Условия высших порядков локального минимума в задачах с ограничениями. Усп. мат. наук **33** (1978) 6, 85—148. (III)

LEVITIN, E. S., and B. T. POLJAK (Левитин, Е. С., и Б. Т. Поляк): Методы минимизации при наличии ограничений. Ж. вычисл. мат. и мат. физ. **6** (1966), 787 до 823. (II)

LIPS, M.: Näherungsweise Bestimmung von Auswahlfunktionen für nichtlineare parametrische Optimierungsprobleme. Dipl.-Arbeit, Humboldt Univ. Berlin, Sektion Mathematik 1979. (I, II)

LJOM, K. S.: Parametrische quadratische Optimierung. Suchak ka Mulli **11** (1967), 2—7 (Korean). (I)

LOMMATZSCH, K.:
[1] (1) Ein notwendiges und hinreichendes Optimalitätskriterium für allgemeine quadratische Optimierungsprobleme.
(2) Über die Lage lokaler Minima quadratischer Optimierungsaufgaben.
(3) Lösungsalgorithmen für quadratische Optimierungsaufgaben mit nicht notwendig konvexer Zielfunktion. Apl. Mat. **19** (1974), 193—209. (I)
[2] (ed.): Anwendungen der linearen parametrischen Optimierung. Akademie-Verlag, Berlin 1979. (I, III)

LORENZEN, G.: Parametrische Optimierung und einige Anwendungen. R. Oldenbourg Verlag, München—Wien 1974. (I, III)

LUENBERGER, D. G.:
[1] Convergence rate of a penalty-function scheme. J. Optim. Theory Appl. 7 (1971), 39—51. (II)
[2] Introduction to linear and nonlinear programming. Addison-Wesley, London 1973. (III)

LÜTHI, H.-J.: Komplementaritäts- und Fixpunktalgorithmen in der mathematischen Programmierung, Spieltheorie und Ökonomie. Lecture Notes in Economics and Mathematical Systems, vol. 129. Springer-Verlag, Berlin—New York 1976. (II)

MACKENROTH, U.: Dualität und Approximation bei konvexen Optimierungsproblemen. Z. Angew. Math. Mech. **58** (1978), 459—460. (II)

Maghout, K.: Une méthode pour la résolution des programmes linéaires. Programmes paramétriques. Comptes Rendus de l'Académie des Sciences, Paris, **250** (1960), 2837 to 2839. (I)

Makarov, V. L., and A. M. Rubinov (Макаров, В. Л., и А. М. Рубинов): Математическая теория экономической динамики и равновесия. Наука, Москва 1973. (III)

Mancino, O. G., and G. Stampacchia: Convex programming and variational inequalities. J. Optim. Theory Appl. **9** (1972), 3—23. (II)

Mangasarian, O. L.: Nonlinear programming. McGraw-Hill, New York 1969. (III)

Manjurov, D. M. (Манюров, Д. М.):
[1] Об одном классе задач параметрической линейной программировании. Изв. Акад. наук АзерССР, Сер. физ.-техн. и мат. наук, **4** (1966), 62—66. (I)
[2] Об одной задаче квадратичного параметрического программирования. Изв. Акад. наук АзерССР, Сер. физ.-техн. и мат. наук, **6** (1967), 66—68. (I)
[3] Исследование задачи дробно-линейного программирования при варьировании некоторых параметров. Изв. Акад. наук АзерССР, Сер. физ.-техн. и матем. наук, **4** (1968), 89—95. (I)
[4] Параметрическое блочное программирование. В: Вопр. вычисл. мат., АН АзерССР, Баку 1968, 157—167. (I)

Manne, A. S.: Note on parametric linear programming. RAND-Corp. Rev. P-468 (1953). (I)

Manteuffel, K., and E. Seiffart: Über den praktischen Einsatz von Methoden der parametrischen Optimierung in einem metallurgischen Betrieb. Fertigungstechnik und Betrieb **6** (1966), 322—327. (II)

Marsten, R. E., and T. L. Morin: Parametric integer programming: The right-hand-side case. *In*: Hammer, P. L., E. L. Johnson, B. H. Korte, and G. L. Nemhauser (eds.): Studies in integer programming. Annals of Discrete Mathematics 1. North-Holland, Amsterdam 1977, 375—390. (I)

Marti, J.: Konvexe Analysis. Birkhäuser-Verlag, Basel-Stuttgart 1977. (III)

Marti, K.: Über stochastische Optimierungsprobleme bei Unsicherheit über die Verteilung des zufälligen Parameters. Z. Angew. Math. Mech. **57** (1977), T 335—T 337. (II)

Martin, D. H.: On the continuity of the max mum in parametric linear programming. J. Optim. Theory Appl. **17** (1975), 205—210. (I)

Martinet, B.: Perturbation des méthodes d'optimisation. Applications. Rev. Française d'Automatique, d'Informatique et de Recherche Opérationelle, Analyse Numérique **12** (1978), 153—171. (I)

Marunová, E.: Problém nejednoznačného řešení úloh lineárního programování s lineární parametrizazí vektoru pozădavků. Ekon.-mat. obzor **6** (1970), 409—420. (I)

Maurin, H.: Paramétrisation générale d'un programme linéaire. Rev. Franç. Rech. Opér. **8** (1964) 32, 277—292. (I)

Megiddo, N.:
[1] On monotonicity in parametric linear complementarity problems. Math. Programming **12** (1977), 60—66. (I)
[2] On the parametric nonlinear complementarity problem. Math. Programming Stud. 7 (1978), 142—150. (I)

Melikadze, N. L. (Меликадзе, Н. Л.): Оптимизация одной многопараметровой функции. Сакартвелос ССР Мецниерсбата академиис моамбе, Сообщ. АН ГрузССР **54** (1969) 1, 41—44. (I)

Meyer, G. G. L.:
[1] A systematic approach to the synthesis of algorithms. Num. Math. **24** (1975), 277 to 289. (II)
[2] Convergence conditions for a type of algorithm model. SIAM J. Control Optim. **15** (1977), 779—784. (II)
[3] Asymptotic properties of sequences iteratively generated by point-to-point maps. Math. Programming Stud. **10** (1979), 115—127. (II)

Meyer, G. G. L., and E. Polak: Abstract models for the synthesis of optimization algorithms. SIAM J. Control Optim. **9** (1971), 547—560. (II)

MEYER, R. R.:
[1] The validity of a family of optimization methods. SIAM J. Control 8 (1970), 41—54. (I, II)
[2] Integer and mixed-integer programming models: General properties. J. Optim. Theory Appl. **16** (1975) 3/4. (II)
[3] Mixed-integer models for piecewise-linear functions of a single variable. Discrete Math. **16** (1976), 163—171. (II)
[4] A comparision of the forcing function and point-to-set mapping approaches to convergence analysis. SIAM J. Control Optim. **15** (1977), 699—715. (II)

MICHAEL, E.:
[1] Topologies on spaces of subsets. Trans. Amer. Math. Soc. **71** (1951), 152—182. (II, III)
[2] Continuous selections I. Ann. of Math. **63** (1956), 361—382. (II, III)
[3] Continuous selections II. Ann. of Math. **64** (1956), 562—580. (II, III)
[4] Continuous selections III. Ann. of Math. **65** (1957), 375—390. (II, III)

MIHOC, GH., and I. NĂDEJDE: Programarea matematică. Programarea parametrică si neliniară. Ed. ştiint., Bucureşti 1966. (III)

MILLS, H. D.: Marginal values of matrix games and linear programs. *In*: KUHN, H. W., and A. W. TUCKER (eds.): Linear inequalities and related systems. Princeton Univ. Press, Princeton (N.J.) 1956, 183—193. (II)

MOHAN, S. R.: Existence of solution rays for linear complementarity problems with Z-matrices. Math. Programming Stud. 7 (1978), 108—119. (II)

MOISEEV, N. N. (ed.) (Моисеев, Н. Н.): Современное состояние исследования операции. Наука, Москва 1979. (III)

MOORE, R. L.:
[1] Concerning upper semicontinuous collections of continua. Proc. Nat. Acad. Sci. U.S.A. **10** (1924), 356—360. (II)
[2] Concerning upper semicontinuous collections of continua. Trans. Amer. Math. Soc. **27** (1925), 416. (II)

MOSCO, U.: On the continuity of the Young-Fenchel transform. J. Math. Anal. Appl. **35** (1971), 518—535. (II)

MÜLLER-MERBACH, H.:
[1] A parametric linear programming approach for the project cost curves. Working Paper, Operations Res. Center, Univ. of California, Berkeley 1964. (I)
[2] Lineare Planungsrechnung mit parametrisch veränderten Koeffizienten der Bedingungsmatrix. Ablaufs- und Planungsforsch. 8 (1967), 341—354. (I)
[3] Sensitivitätsanalyse von Transportproblemen der linearen Planungsrechnung (mit ALGOL-Programm). Elektron. Datenverarbeit. **10** (1968), 184—188. (I)

NACCHACHE, P. H.: Stability in multicriteria optimization. J. Math. Anal. Appl. **68** (1979), 441—453. (II)

NAUSS, R. M.: Parametric integer programming. Ph. D. Diss. Working Paper No. 226, Western Management Sci. Inst., UCLA, Jan. 1976. (I)

NEMHAUSER, G. L., and R. S. GARFINKEL: Integer programming. John Wiley, New York 1972. (III)

NEUSTADT, L. W.: Optimization. Princeton Univ. Press, Princeton (N.J.) 1974. (III)

NOLTEMEIER, H.:
[1] Bemerkungen zur Parametrisierung und zum Verteilungsproblem bei speziellen graphentheoretischen Algorithmen. Oper. Res. Verf. **6** (1969), 181—185. (I, II)
[2] Sensitivitätsanalyse bei diskreten linearen Entscheidungsmodellen. Oper. Res. Verf. **8** (1970), 202—210. (I)
[3] Sensitivitätsanalyse bei diskreten linearen Optimierungsproblemen. Lecture Notes in Operations Research and Mathematical Systems Economics, Computer Science, Information and Control, vol. 30. Springer-Verlag, Berlin—Heidelberg—New York 1970. (I)

NOŽIČKA, F.:
[1] Ein geometrischer Beweis des Dualitätssatzes der linearen Optimierung. Math. Operationsforsch. Statist. **2** (1971) 4, 3—17. (II)

[2] Über eine Klasse von linearen einparametrischen Optimierungsproblemen. Math. Operationsforsch. Statist. **3** (1972) 3, 159—194. (I)
[3] Lineare parametrische Optimierung — ein Problem der Stabilität der optimalen Lösung. *In*: Konference o matematickych metodach v ekonomii, Praha 1972, 45—82. (I)

Nožička, F., J. Guddat, and H. Hollatz: Theorie der linearen Optimierung. Akademie-Verlag, Berlin 1972. (III)

Nožička, F., J. Guddat, H. Hollatz, and B. Bank: Theorie der linearen parametrischen Optimierung. Akademie-Verlag, Berlin 1974. (I, III)

Nykowski, I. (Ныковский, И.):
[1] Dwuparametryczny dualny problem liniowy. I, II. Przeglad. Statyst. **12** (1965), 203—217, 311—323. (I)
[2] Линейная задача с двупараметрической целевой функцией. Экон. и мат. методы **2** (1966), 548—557. (I)

Oelschlägel, D.:
[1] Parameterabhängige konvexe Optimierungsprobleme. Wiss. Z. Techn. Hochsch. Carl Schorlemmer Leuna-Merseburg **10** (1968), 28—34. (I)
[2] Ein Lösungsverfahren für parameterabhängige, nichtlineare, konvexe Optimierungsprobleme. Wiss. Z. Techn. Hochsch. Carl Schorlemmer Leuna-Merseburg **10** (1968), 325—330. (I)
[3] Zur Existenz von Lösungen parameterabhängiger konvexer Optimierungsprobleme. Wiss. Z. Techn. Hochsch. Carl Schorlemmer Leuna-Merseburg **13** (1971), 194—199. (I)

Orkisz, T.: The analysis of stability of the optimal solution in linear programs. Przeglad. Statyst. **11** (1964), 49—62. (I)

Ortega, J. M., and Rheinboldt, W. C.: Iterative solution of nonlinear equations in several variables. Academic Press, New York 1970. (III)

Ospanov, Ś. S. (Оспанов, С. С.): Каноническая форма задачи линейного программирования с параметром. *В сб.*: Материалы первой научно-теоретической конференции молодых математиков и механиков города Алма-Аты. Алма-Ата 1969, 177—179. (I)

Panwalkar, S. S.: Parametric analysis of linear programs with upper bounded variables. Naval Res. Logist. Quart. **20** (1973), 83—93. (I)

Perron, O.: Irrationalzahlen. W. de Gruyter, Berlin 1947. (III)

Peschel, M., and C. Riedel: Polyoptimierung, eine Entscheidungshilfe für ingenieurtechnische Kompromißlösungen. Verlag Technik, Berlin 1976. (III)

Peterson, E. L.: The duality between suboptimization and parameter deletion, with application to parametric programming and decomposition theory in geometric programming. Math. Oper. Res. **2** (1977), 311—319. (II)

Pirzl, J.:
[1] Optimierung unter Nebenbedingungen. Struktur einer Klasse von Algorithmen I. Computing **8** (1971), 121—142. (II)
[2] Optimierung unter Nebenbedingungen. Struktur einer Klasse von Algorithmen II. Computing **8** (1971), 272—283. (II)

Podkaminer, L.: The influence of the change of the matrix elements in a linear programming problem upon the optimal value of the objective function. Ekon.-mat. obzor **6** (1970), 310—314. (I)

Podrebarac, M. L., and S. S. Sengupta: Parametric linear programming: some extensions. INFOR **9** (1971), 305—319. (I)

Polak, E.: Computational methods in optimization. A unified approach. Academic Press, New York 1971. (III)

Poljak, B. T. (Поляк, Б. Т.): Градиентные методы минимизации функционалов. Ж. вычисл. мат. и мат. физ. **3** (1963), 643—654. (II)

Powell, M. J. D.: A method for nonlinear constraints in minimization problems. In: Fletcher, R. (ed.): Optimization. Academic Press, New York 1969, 283—298. (II)

PŠENIČNYĬ, B. N. (Пшеничный, Б. Н.):
[1] Notwendige Optimalitätsbedingungen. BSB B. G. Teubner Verlagsgesellschaft, Leipzig 1972 (transl. from Russ.). (III)
[2] Выпуклый анализ и экстремальные задачи. Наука, Москва 1980. (III)

RADKE, M. A.: Sensitivity analysis in discrete optimization. Ph. D. Diss. Working Paper No. 240, Western Management Sci. Inst., Univ. of California, Los Angeles, Sept. 1975. (I)

RAPOPORT, R. I., L. A. BUJANOVSKIĬ, I. JA. VAHUTINSKIĬ, S. A. KAC, and V. M. SLOBODKIN (Рапопорт, Р. И., Л. А. Буяновский, И. Я. Вахутинский, С. А. Кац и В. М. Слободкин): Анализ устойчивости экстремальных задач с ограничениями. *В сб.*: Методологическое моделлирование процессов функц. болш. сис., Москва 1968. (I)

RASTRIGIN, L. A., and K. K. RIPA (Растригин, Л. А., и К. К. Рипа): Некоторые статистические свойства дискретного накопления при многопараметрической оптимизации в обстановке больших помех. *В сб.*: Автоматика и вычисл. техн. **13**, Рига, 1966, 93—101. (I)

REDETZKI, R.-R.: Stabilitätsanalyse der stochastischen linearen Optimierung I/II. Wiss. Z. Techn. Hochsch. Ilmenau **4** (1976), 31; **6** (1977), 111. (II)

REINDLMEIER, J.: Zur stetigen Parameterabhängigkeit der Menge der Minimalpunkte bei Minimax-Aufgaben. Diss., Georg August Univ. Göttingen 1975. (I)

REINSCHKE, K.: Ein Lösungsverfahren der parametrischen linearen Optimierung. Wiss. Z. Hochsch. Architektur Bauwesen Weimar **16** (1969), 284—286. (I)

RICHTER, C.: Über die numerische Behandlung nichtlinearer Optimierungsaufgaben mit Hilfe von verallgemeinerten Variationsungleichungen und von Nichtoptimalitätsmaßen. Diss. B, Techn. Univ. Dresden 1980. (II)

RITTER, K.:
[1] Ein Verfahren zur Lösung parameter-abhängiger, nichtlinearer Maximumprobleme. Unternehmensforsch. **6** (1962), 149—166. (I)
[2] Über Probleme parameterabhängiger Planungsrechnung. (DVL-Bericht Nr. 23) Vereinigte Universitäts- und Fachbuchhandlungen: O. Müller-Trewendt—H. Sach GmbH., Köln 1963. (I)
[3] A method for solving maximum-problems with a nonconcave quadratic objective funktion. Z. Wahrsch. Verw. Gebiete **4** (1965), 340—351. (II)
[4] A method for solving nonlinear maximum-problems depending on parameters. Naval Res. Logist. Quart. **14** (1967), 147—162. (II)
[5] A parametric method for solving certain nonconcave maximum problems. J. Computer System Sci. **1** (1967), 44—54. (II)

ROBINSON, S. M.:
[1] A quadratically-convergent algorithm for general nonlinear programming problems. Math. Programming **3** (1972), 145—156. (II)
[2] Computable error bounds for nonlinear programming. Math. Programming **5** (1973), 235—242. (II)
[3] Perturbed Kuhn-Tucker points and rates of convergence for a class of nonlinear programming algorithms. Math. Programming **7** (1974), 1—16. (I, II)
[4] Stability theory for systems of inequalities, Part I: Linear systems. SIAM J. Numer. Anal. **12** (1975), 754—769. (I, II)
[5] Stability theory for systems of inequalities, Part II: Differentiable nonlinear systems. SIAM J. Numer. Anal. **13** (1976), 497—513. (I, II)
[6] An implicit-function theorem for generalized variational inequalities. MRC Techn. Summary Rep. No. 1672, Math. Res. Cent. Univ. of Wisconsin, Madison 1976. (II)
[7] Regularity and stability for convex multivalued functions. Math. Oper. Res. **1** (1976), 130—143. (II)
[8] A characterization of stability in linear programming. Oper. Res. **25** (1977), 435 to 447. (I)
[9] Strongly regular generalized equations. MRC Techn. Summary Rep. No. 1877, Math. Res. Cent. Univ. of Wisconsin, Madison 1978; Math. Oper. Res. **5** (1980), 43—62. (II)
[10] Generalized equations and their solutions, Part I: Basic theory. Math. Programming Stud. **10** (1979), 128—141. (II)
[11] Some continuity properties of polyhedral multifunctions. MRC Techn. Summary Rep. No. 2014, Math. Res. Cent. Univ. of Wisconsin, Madison 1979. (II)

[12] Generalized equations and their solutions, Part II: Applications to nonlinear programming. MRC Techn. Summary Rep. No. 2048, Math. Res. Cent. Univ. of Wisconsin, Madison 1980. (II)

ROBINSON, S. M., and R. H. DAY: A sufficient condition for continuity of optimal sets in mathematical programming. J. Math. Anal. Appl. **45** (1974), 506—511. (I)

ROBINSON, S. M., and R. R. MEYER: Lower semicontinuity of multivalued linearization mappings. SIAM J. Control **11** (1973), 525—533. (I, II)

ROCKAFELLAR, R. T.:

[1] Duality and stability in extremum problems involving convex functions. Pacific J. Math. **21** (1967), 167—187. (I, II)

[2] Convex analysis, Princeton Univ. Press, Princeton (N.J.) 1970. (III)

[3] Some convex programs whose duals are linearly constrained. *In*: ROSEN, J. B., O. L. MANGASARIAN, and K. RITTER (eds.): Nonlinear programming. Academic Press, New York 1970, 293—322. (II)

[4] Ordinary convex programs without a duality gap. J. Optim. Theory Appl. **7** (1971), 143—148. (II)

[5] Ordinary convex programs with a duality gap. J. Optim. Theory Appl. **12** (1973), 555—562. (II)

[6] Penalty methods and augmented Lagrangians in nonlinear programming. *In*: RUBERTI, A., and R. CONTI (eds.): Fifth Conf. on Optim. Techniques. Part 1. Lecture Notes in Computer Science, vol. 3. Springer-Verlag, Berlin—Heidelberg—New York 1973, 418—425. (II)

[7] Augmented Lagrange multipliers functions and duality in nonconvex programming. SIAM J. Control **19** (1974), 268—285. (II)

[8] Monotone operators and the proximal point algorithm. SIAM J. Control Optim. **14** (1976), 877—898. (II)

[9] Proximal subgradients, marginal values and augmented Lagrangians in nonconvex optimization. Manuskript, Univ. of Washington, Seattle 1980. (II)

[10] Lagrange multipliers and subderivatives of optimal value functions in nonlinear programming. Manuskript, Laxenburg 1980. (II)

ROLEWICZ, S.: Funktionalanalysis und Steuerungstheorie. Springer-Verlag, Berlin—Heidelberg—New York 1976. (III)

ROMANOVSKIĬ, I. V. (Романовский, И. В.): Алгоритмы решения экстремальных задач. Наука, Москва 1977. (III)

ROODE, J. D.: Generalized Lagrangian functions and mathematical programming. *In*: FLETCHER, R. (ed.): Optimization. Sympos. of the Inst. of Math. and its Appl. Univ. of Keele. Academic Press, London 1969. (II)

RÖSSLER, M.:

[1] Algorithmen zur Lösung des parametrischen nichtlinearen Komplementaritätsproblems. *In*: Proceedings in Operations Research 5. Physica-Verlag, Würzburg—Wien, 22—27. (I)

[2] Linear programming with nonlinear parametrization. Stud.-Berichte Eidgenöss. Techn. Hochsch Zürich, IFOR 5, 1976 (I)

ROST, D., and S. SCHMUNTZSCH: Ermittlung optimaler Futterrationen für Mastschweine unter Berücksichtigung variierender Kosten einiger Futtermittel (Anwendung von Modellen der linearen parametrischen Optimierung). Kühn-Archiv **84** (1970), 311—327. (II)

RUBINŠTEĬN, G. S. (Рубинштейн, Г. С.): Несколько примеров двойственных экстренальных задач. *В сб.*: Математическое программирование. Наука, Москва 1966. (III)

SAATY, T. L.: Coefficient perturbation of a constrained extremum. Oper. Res. **7** (1959), 294—302. (I)

SAATY, T. L., and S. GASS: Parametric objective function I. Oper. Res. **2** (1954), 316—319. (I)

SAIGAL, R.: A note on a special linear complementarity problem. Opsearch **7** (1970), 179—183. (II)

SAINT-PIERRE, J.: Borel convex-valued multifunctions. *In*: AUSLENDER, A. (ed.): Convex analysis and its applications. Springer-Verlag, Berlin—Heidelberg—New York 1977. (II)

Samelson, H., R. M. Thrall, and O. Wesler: A partition theorem for Euclidean n-space. Proc. Amer. Math. Soc. 9 (1958), 805—807. (II)

Sarkisjan, S. D. (Саркисян, С. Д.): Об одной задаче параметрического линейного программирования, когда зависимость функционала от параметра нелинейна. Тр. вычисл. центра Акад. наук АрмССР и Ереванск. унив., Ереван 1964, 10—16. (I)

Scharf, V.: Optimuminvariante Bereiche der Zielfunktionskoeffizienten linearer Programme. Z. Angew. Math. Mech. 54 (1974), 58—62. (I)

Schäuble, M., and K.-H. Seidler: Lagrange-Theorie, Lagrange-Methoden und Verallgemeinerungen. Wiss. Z. Techn. Hochsch. Ilmenau 18 (1972) 3, 69—109. (II)

Schmidt, B.: Stochastische mehrparametrische lineare Optimierung. Dipl.-Arbeit, Humboldt Univ. Berlin, Sektion Mathematik 1975. (II)

Schmuntzsch, S.: Untersuchungen zur parametrischen Optimierung (dargestellt an einem Beispiel zur Berechnung ertragsabhängiger Futterrationen). Agrarökonomik 9 (1966), 478—489. (II)

Schmuntzsch, S., and D. Rost: Ermittlung optimaler lebendgewichtsabhängiger Futterrationen für Mastschweine mit Hilfe der Regressionsrechnung und der linearen parametrischen Optimierung. Wiss. Z. Martin Luther Univ. Halle-Wittenberg 16 (1968), 161—189. (II)

Schreiter, D.: Die parametrische Linearprogrammierung — Anwendungsmöglichkeiten und Lösungsalgorithmus. Wirtschaftswiss. 12 (1964), 1300—1324. (I)

Schubert, H.: Topologie. Eine Einführung. Teubner-Verlag, Stuttgart 1964. (III)

Schwetlick, H : Numerische Lösung nichtlinearer Gleichungen. VEB Deutscher Verlag der Wissensch·aften, Berlin 1979. (III)

Seiffart, E.: Ein Lösungsalgorithmus eines parametrischen Verteilungsproblems. Ekon.-mat. obzor 2 (1966), 263—270. (I)

Sengupta, J. K.:

[1] The stability of truncated solutions of stochastic linear programming. Econometrica 34 (1966) 1, 77—104. (II)

[2] Recursive constraints and stochastic linear programming. Metrika 10 (1966), 6—12. (II)

Sengupta, J. K., and T. K. Kumar: An application of sensitivity analysis to a linear programming problem. Unternehmensforsch. 9 (1965), 18—36. (II)

Sengupta, J. K., C. Millham, and G. Tintner: On the stability of solutions under error in stochastic linear programming. Metrika 9 (1965), 47—60. (II)

Sengupta, J. K., and B. C. Semyal: Sensitivity analysis methods for a crog-mix-problem in linear programming. Unternehmensforsch. 14 (1970), 2—26. (II)

Sengupta, J. K., and G. Tintner: On the stability of solutions under recursive programming. Unternehmensforsch. 10 (1966), 1—14. (II)

Sertel, M. R.: The fundamental continuity theory of optimization on a compact space (1). J. Optim. Theory Appl. 16 (1975), 549—558. (I)

Shapiro, J. F.: Sensitivity analysis in integer programming. *In*: Hammer, P. L., E. L. Johnson, B. H. Korte, and G. L. Nemhauser (eds.): Studies in integer programming. Annals of discrete mathematics 1. North-Holland, Amsterdam 1977, 467—478. (I)

Shetty, C. M.:

[1] Sensitivity analysis in linear programming. J. industr. Engin. 10 (1959), 379—386. (I)

[2] Solving linear programming problems with variable parameters. J. industr. Engin. 10 (1959), 433—438. (I)

[3] On analysis of the solution to a linear programming problem. Oper. Res. Quart. 12 (1961), 89—104. (I)

Siassi, J.: Eine parametrische Formulierung des Verfahrens von Hildreth und d'Esposo. Ekon.-mat. obzor 3 (1972), 277—283. (II)

Simons, E.: A note on parametric linear programming. Management Sci. 8 (1962), 355 to 358. (I)

Skarin, V. D. (Скарин, В. Д.): К регуляризации минимаксных задач, возникающих в выпуклом программировании. Ж. вычисл. мат. и мат. физ. 17 (1977), 1408—1420. (II)

Sokolová (Grygarová), L.: Problém viceparametrického lineárného programování. Ekon.-mat. obzor 4 (1968), 44—68. (I)

SOMMERMEIER, K.:

[1] Die parametrische Optimierung bei linearer Abhängigkeit der Koeffizienten einer Reihe von einem Parameter. Wiss. Z. Hochsch. f. Ökonomie Berlin (1968), 171 bis 181. (I)

[2] Koeffizientenänderung in der linearen Optimierung. Math. Operationsforsch. Statist. 2 (1971), 271–281. (I)

ŠOR, N. Z., and M. B. ŠČEPAKIN (Шор, Н. З., и М. Б. Щепакин): Задачи многоэтапного стохастического программирования в параметрической форме. *В сб.*: Теория оптимальных решений. Труды семинара 3. Киев 1969, 64—73. (II)

SRINIVASAN, V.: An operator theory of parametric programming for the transportation problem II. Naval Res. Logist. Quart. 19 (1972), 227–252. (I, II)

SRINIVASAN, V., and G. L. THOMPSON: An operator theory of parametric programming for the transportation problem I. Naval Res. Logist. Quart. 19 (1972), 205–225. (I, II)

STAHL, W.: Sensitivitätsanalyse mehrstufiger Standortprobleme. Mathematical Systems in Economics 30. Verlag Anton Hain, Meisenheim/Glan 1976. (II)

STAROSEL'SKIĬ, V. A. (Старосельский, В. А.). Параметрическое линейное программирование. Двойственная задача линейного программирования. СЕМИ Акад. Наук СССР, Москва 1965, 48—50. (I)

STERN, M. H., and D. M. TOPKIS: Rates of stability in nonlinear programming. Oper. Res. 24 (1974), 462–476 (I, III)

STOER, J., and C. WITZGALL: Convexity and optimization in finite dimensions I. Springer-Verlag, Berlin —Heidelberg—New York 1970. (III)

SWARUP, K.: On varying all the parameters in a linear fractional functionals programming problem. Metrika 13 (1968), 196–205 (I)

SWERSEY, R I.: Simultaneous parametric programming. Oper. Res. 12 (1964), 781–783. (I)

SZWARC, W.: The stability of the transportation problem. Mathematica, Cluj, 4 (1962), 397—400. (I)

SZWARC, W., and A CALCYNSKI: A remark on basic solutions of the linear parametric problem. Mathematica, Cluj, 6 (1964), 129–130. (I)

TAHA, H. A.: Operations research IV. Parametric programming. Memo 601, The Inst. of Nat. Planning, Cairo 1965. (I)

TAMMER, E.-CH.:

[1] Ein Problem der parametrischen hyperbolischen Optimierung *In*: Konference o matematických metodach v ekonomii, Zadov, Ekon. mat. Laboratoř 1973, 285 bis 305. (I)

[2] Dualität und Stabilität in der hyperbolischen Optimierung. Diss. A, Humboldt Univ. Berlin, 1974. (I)

[3] Parametrische hyperbolische Optimierungsprobleme mit Parametern in der Zielfunktion. Math. Operationsforsch. Statist., Ser. Optim., 8 (1977), 207—226. (I)

TAMMER, K.:

[1] Einige Bemerkungen zur M-Methode der linearen Optimierung. Ekon.-mat. obzor 7 (1971), 419—425 (I, II)

[2] Die Abhängigkeit eines quadratischen Optimierungsproblems von einem Parameter in der Zielfunktion. Math. Operationsforsch. Statist. 5 (1974), 573—590. (I)

[3] Ein Verfahren zur Lösung einer Klasse von indefiniten quadratischen Optimierungsproblemen. VII. IKM der Hochsch. Architektur Bauwesen Weimar — Berichte, Berlin 1976, 217—221. (I, II)

[4] On the solution of the distribution problem of stochastic programming. *In*: PREKOPÁ, A. (ed.): Progress in Operations Research. Coll. Math. Soc. Janos Bolyai 12. Janos Bolyai Math. Soc., Budapest/North-Holland, Amsterdam—London 1976, 907—920. (I, II)

[5] Möglichkeiten zur Anwendung der Erkenntnisse der parametrischen Optimierung für die Lösung indefiniter quadratischer Optimierungsprobleme, Math. Operationsforsch. Statist. 7 (1976), 209—222. (I, II)

[6] Über die Menge der Kuhn-Tucker-Punkte zu einem konvexen quadratischen Optimierungsproblem. Z. Angew. Math. Mech. 58 (1978), 211—213. (II)

[7] Relations between stochastic and parametric programming for decision problems with a random objective function. Math. Operationsforsch. Statist , Ser. Optim., 9 (1978), 523—535. (I, II)

[8] Über den Zusammenhang von parametrischer Optimierung und Entscheidungsproblemen der stochastischen Optimierung. *In*: LOMMATZSCH, K (ed.): Anwendungen der linearen parametrischen Optimierung. Akademie-Verlag, Berlin 1979, 76 bis 91. (I, II)

[9] Beiträge zur Theorie der parametrischen Optimierung, zu den mathematischen Grundlagen ihrer Anwendung und zu den Lösungsverfahren. Diss. B, Humboldt Univ. Berlin 1979. (I, II)

[10] Theorie und Verfahren für parameterabhängige lineare Komplementaritätsprobleme. Jahrestagung Math. Optim., Vitte (Hiddensee) 1979, 205—213. (I)

[11] Behandlung stochastischer Optimierungsprobleme nach den Gesichtspunkten der Vektoroptimierung. Wiss. Z. Techn. Hochsch. Leipzig.

THAEMELT, V. (Темельт, В.):

[1] О пространстве параметров общей задачи линейного программирования. Сибир. мат. ж. **8** (1967), 680—686. (I)

[2] Двойственность бесконечномерных задач линейного программирования и некоторые смежные вопросы. Кибернетика **4** (1969), 80—87. (II)

TIHONOV, A. N. (Тихонов, А. Н.):

[1] О некоторых задачах оптимального планирования и устойчивых методах их решения. Докл. Акад. Наук СССР **164** (1965), 507—510. (II)

[2] О некорректных задачах оптимального программирования. Ж. вычисл. мат. и мат. физ. **6** (1966), 81—89. (II)

[3] Об устойчивости задач оптимизации функционалов. Ж. вычисл. мат. и мат. физ. **6** (1966), 631—634. (II)

TIHONOV, A. N., and V. JA. ARSENIN (Тихонов, А. Н., и В. Я. Арсенин): Методы решения некорректных задач. Наука, Москва 1974. (III)

TIHONOV, A. N., V. G. KARMANOV, and T. I. RUDNEVA (Тихонов, А. Н., В. Г. Карманов и Т. Л. Руднева): Об устойчивости задач линейного программирования. В сб.: Вычислительные методы программирования. 12. Московск. унив. 1969, 3—9. (I, II)

TISHYADHIGAMA, S., E. POLAK, and R. KLESSIG: A comparative study of several general convergence conditions for algorithms modeled by point-to-set maps. Math. Programming Stud. **10** (1979), 172—190. (II)

TODD, M. J.: The computation of fixed points and applications. Springer-Verlag, Berlin—Heidelberg—New York 1976. (III)

TONE, K.: Postoptimization of quadratic programming. KEID Engin. Reports **29** (1976), 1—6. (I)

TOWNSLEY, R. J., and W. CANDLER: Quadratic as parametric linear programming. Naval Res. Logist. Quart. **19** (1972), 183—189. (I)

TURNOVEC, F.: Global and local sensitivity analysis. Internat. Wiss. Koll. Techn. Hochsch. Ilmenau, Vortragsreihe Mathematische Optimierungsmethoden und ihre Anwendungen. 1976 (I)

VAN MOESEKE, P., and G. TINTNER: Base duality theorem for stochastic and parametric programming. Unternehmensforsch. 8 (1964), 75—79. (II)

VAN DE PANNE, C.:

[1] An alternative interpretation of the primal-dual method and some related parametric methods. Internat. Econ. Rev. **9** (1968), 87—99. (II)

[2] A complementary variant of Lemke's method for the linear complementary problem. Math. Programming **7** (1974), 283—310. (II)

[3] Methods for linear and quadratic programming. North-Holland, Amsterdam—Oxford 1975. (I, III)

VAN DE PANNE, C., and P. BOSJE: Sensitivity analysis of the cost coefficients estimates: the case of linear decision rules for employment and production. Management Sci. **9** (1963), 82—107. (I)

VARDANJAN, V. B. (Варданян, В. Б.): О решении одной параметрической задачи линейного программирования. *В*: Тр. Вычисл. центра Акад. Наук АрмССР и Ереванск. унив., Ереван 1964, 5—9. (I)

VARMA, G. K.:
[1] On parametric linear fractional functionals programming. Trabajos Estadist. Investigacion Oper. **23** (1972) 1/2, 149—157. (I)
[2] General parametric linear fractional programming. Metrica **19** (1972), 11—17. (I)
VEINOTT Jr., A. F.: Production planning with convex costs: a parametric study. Management Sci. **10** (1964), 441—446. (II)
VOGEL, W.: Vektoroptimierung in Produkträumen. Working Paper No. P 76/193, Univ. Bonn 1975. (II)

WALKUP, D. W., and R. J.-B. WETS:
[1] Continuity of some convex-cone-valued mappings. Proc. Amer. Math. Soc. **18** (1967), 229—235. (I)
[2] A Lipschitzian characterization of convex polyhedra. Proc. Amer. Math. Soc. **20** (1969), 167—173. (I)
[3] Lifting-projections of convex polyhedra. Pacific J. Math. **28** (1969), 465—475. (I)
WATSON, L. T.: Some perturbation theorems for Q-Matrices. SIAM J. Appl. Math. **31** (1976), 379—384. (I)
WEBB, K. W.:
[1] The mathematical theory of sensitivity. 18th Nat. Meeting Oper. Res. Soc. America, Detroit (Mich.) 1960 (mimeo); Abstract: Oper. Res. **8** (1960), B 120—B 121. (II)
[2] Some aspects of the Saaty linear programming sensitivity equation. Oper. Res. **10** (1962), 266—267. (II)
WEICKENMEIER, E.: Zur Lösung parametrischer linearer Programme mit polynomischer Parameterfunktion. Z. Oper. Res. **22** (1978), 131—149. (I)
WEINERT, H.:
[1] Probleme der zweifach-mehrparametrischen linearen Optimierung. Wiss. Z. Hochsch. Architektur Bauwesen Weimar **16** (1969), 307—309. (I)
[2] Probleme der linearen Optimierung mit nichtlinear einparametrischen Koeffizienten in der Zielfunktion. Math. Operationsforsch. Statist. **1** (1970), 21—43. (I)
[3] Doppelt-einparametrische lineare Optimierung I: Unabhängige Parameter. Math. Operationsforsch. Statist. **1** (1970), 173—197. (I)
[4] Doppelt-einparametrische lineare Optimierung II: Abhängige Parameter. Math. Operationsforsch. Statist. **2** (1971), 19—39. (I)
WENDLER, K. (I):
Die ε-Störung linearer und quadratischer Optimierungsprobleme und ihre Anwendung auf das Hildreth-Verfahren. Unternehmensforsch. **15** (1971), 1—14. (I)
WENDLER, K. (II):
Die rechentechnische Realisierung von Verfahren zur Lösung von linearen Optimierungsproblemen mit Parametern als Koeffizienten der Zielfunktion oder als rechte Seiten der Restriktionen. Wiss. Z. Humboldt Univ. Berlin, Math.-Naturw. Reihe, **5** (1977) 603—607. (I)
WETS, R. J.-B.:
[1] Programming under uncertainty: The equivalent convex program. SIAM J. Appl. Math. **14** (1966), 89—105. (II)
[2] On inf-compact mathematical programs. *In*: RUBERTI, A., and R. CONTI (eds.): 5th Conf. on Optim. Techn. Part 1. Springer-Verlag, Berlin—Heidelberg—New York 1973, 426—436. (II)
[3] On the convergence of random convex sets. *In*: AUSLENDER, A. (ed.): Convex analysis and its applications. Springer-Verlag, Berlin—Heidelberg—New York 1977, 191—206. (II)
WIERZBICKI, A.:
[1] Unified approach to the sensitivity of optimal control systems. Proc. of the IFAC Fourth World Congress Warsaw 1969. (II)
[2] Differences in structure and sensitivity of optimal control systems. Arch. Autom. Telemech. **2** (1970), 111—136. (II)
[3] A penalty function shifting method in constrained static optimization and convergence properties. Avtomat. i Telemeh. **16** (1971), 395—416. (II)
[4] Coordination and sensitivity analysis of large scale system with performance interaction. Proc. of the IFAC Fifth World Congress Paris 1972, Part 3, Internat. Fed. Automat. Control, Düsseldorf 1971. (II)

[5] Modele i wrazliwość układów sterowania. Wydawnistwa naukow-techniczne, Warszawa 1977. (III)

Wierzbicki, A., and A. Dontchev: Basic relations in performance sensitivity analysis of optimal control systems. Control Cybernet. 4 (1974). (II)

Wierzbicki, A., and S. Kurcyusz: Projection on a cone, generalized penalty functionals and duality theory. SIAM J. Control 1 (1977), 25—56. (II)

Williams, A. C.: Marginal values in linear programming. J. Soc. Indust. Appl. Math. **11** (1963), 82—94. (I)

Willner, L. B.: On parametric linear programming. SIAM J. Appl. Math. **15** (1967), 1253—1257. (I)

Wintgen, G.:

[1] Methoden zur einfachen Berücksichtigung von Veränderungen in der Zielfunktionen oder in den Einschränkungen bei Problemen der Linearprogrammierung. Wiss. Z. Hochsch. Architektur Bauwesen Weimar **8** (1961), 486—491. (I)

[2] Indifferente Optimierungsprobleme. Oper. Res. Verf. **6** (1969), 233—236. (II)

Wolfe, P.: The simplex-method for quadratic programming.Econometrica **27** (1959), 382—398. (II)

Wolkowicz, H.:

[1] Calculating the cone of directions of constancy. J. Optim. Theory Appl. **25** (1978), 451—457. (II)

[2] Geometry of optimality conditions and constraint qualifications: the convex case. Math. Programming **19** (1980) 1, 32—60. (II)

Wollmer, R. D.:

[1] Sensitivity analysis in networks. Oper. Res. Center, Univ. of California, Berkeley 1965, 65—68. (I)

[2] Stochastic sensitivity analysis of maximum flow and shortest route networks. Management Sci. **14** (1968), 551—564. (II)

Yu, P. L.: Cone convexity, cone extremum points and nondominated solutions of decision problems with multiobjectives. J. Optim. Theory Appl. **14** (1974), 319—379. (II)

Yu, P. L., and M. Zeleny: Linear multiparametric programming by multicriteria simplex method. Management Sci. **23** (1976), 159—170. (II)

Žak, S. V., and E. L. Litver (Жак, С. В., и Е. Л. Литвер): О некоторых вопросах параметрического линейного программирования. *В сб.*: Вопросы вычислительной математики и вычислительной техники. Ростовск. унив., Ростов-на-Дону 1965, 8—22. (I)

Zang, I., and M. Avriel: On functions whose local minima are global. J. Optim. Theory Appl. **16** (1975) 3/4, 183—189. (I, II)

Zang, I., E. U. Choo, and M. Avriel:

[1] A note on functions whose local minima are global. J. Optim. Theory Appl. **18** (1976), 555—559. (I, II)

[2] On functions whose stationary points are global minima. J. Optim. Theory Appl. **22** (1977), 195—208. (I, II)

Zangwill, W. J.: Nonlinear programming: a unified approach. Prentice Hall, Englewood Cliffs (N.J.) 1969. (III)

Zeidler, E.: Vorlesungen über nichtlineare Funktionalanalysis Teil I, II, III. BSB B. G. Teubner Verlagsgesellschaft, Leipzig 1976/1977. (III)

Zeleny, M.: Linear multiobjective programming. Springer-Verlag, Berlin—Heidelberg—New York 1974. (III)

Zlobec, S.: Feasible direction methods in the absence of Slater's condition. Math. Operationsforsch. Statist., Ser. Optim., **9** (1978), 189—200. (II)

Zoutendijk, G.: Methods of feasible directions. Elsevier, Amsterdam 1960. (III)

Zowe, J., and S. Kurcyusz: Regularity and stability for the mathematical programming problem in Banach spaces. Appl. Math. Optim. **5** (1979), 49—62. (II)

Zsigmond, I.: Parametrization of all the coefficients of a linear programming problem. Computing Center for Universities, ESZK 4, Budapest 1976. (I)

Further references after finishing the manuscript

BANK, B., and R. HANSEL: Stability of mixed-integer quadratic programming problems. *In*: FIACCO, A. V. (ed.): Mathematical programming with data perturbations. Math. Programming Stud. (*to appear*). (I)

BENVISTE, R.: One way to solve the parametric quadratic problem. Math. Programming **21** (1981), 224–228. (I)

MCBRIDE, R. D., and J. S. YORMAK: Finding all solutions for a class of parametric integer programming problems. Management Sci. **26** (1980), 784–795. (I)

BROSOWSKI, B.:

[1] Parametric semi-infinite optimization. Verlag Peter D. Lang, Frankfurt/M.—Bern 1981. (I, III)

[2] Parametric semi-infinite linear programming. Math. Programming Stud. (*to appear*). (I)

DORMANY, M.: A parametric method for the solution of indefinite quadratic programming problems. Szigma **13** (1980), 285–303. (II)

ERMOL'JEV, JU. M., and L. G. ERMOL'JEVA (Ермольев, Ю. М., и Л. Г. Ермольева): Метод параметрической декомпозиции. Кибернетика **2** (1973), 66—69. (II)

FIACCO, A. V., and W. P. HUTZLER: Basic results in the development of sensitivity and stability analysis in nonlinear programming. Computers and Oper. Res. **9** (1982), 9–28. (I, II)

FRIEDRICH, B.: Untersuchungen zur Stabilität einiger Ersatzprobleme der stochastischen Optimierung in Bezug auf Änderungen der zugrundegelegten Zufallsvariablen. Diss. A, Humboldt Univ. Berlin 1982. (II)

FRIEDRICH, B., and K. TAMMER: Untersuchungen zur Stabilität einiger Ersatzprobleme der stochastischen Optimierung in Bezug auf Änderungen des zugrundegelegten Zufallsvektors. Wiss. Z. Humboldt Univ. Berlin, Math.-Naturw. Reihe **5** (1981), 373–376. (II)

GÁL, T.: Postoptimal analysis, parametric programming, and related topics. McGraw-Hill, New York 1979. (I, III)

GARCIA, C. B., and F. J. GOULD: A theorem on homotopy paths. Math. of Oper. Res. **3** (1978), 282–289. (II)

ISTOMIN, L. A. (Истомин, Л. А.): О согласновании решений прямой и двойственной задач линейного параметрического программирования. Тр. инст. мат. и мех. Уральск. науч. центр АН СССР (1979) 27, 15—19. (I)

JONGEN, H. T.: On nonconvex optimization. Diss., Twente Univ. of Technology, The Netherlands, 1977. (I)

JONGEN, H. T., and F. TWILT: On decomposition and structural stability in non-convex optimization. *In*: COLLATZ, H. L., and G. M. W. WETTERLING (eds.): Numerische Methoden bei graphentheoretischen und kombinatorischen Problemen. Birkhäuser-Verlag, Basel 1979, 162–183. (I)

JONGEN, H. T., P. JONKER, and F. TWILT: On deformation in optimization. Dept. of Appl. Math., Twente Univ. of Technology, The Netherlands, 1979. (I)

KIRSCH, A.: Continuous perturbations of indefinite optimization problems. J. Optim. Theory Appl. **32** (1980), 171–182. (I)

KLATTE, D.:

[1] Über lokale Stabilitätsmengen in der parametrischen quadratischen Optimierung. Math. Operationsforsch. Statist., Ser. Optim., **10** (1979), 511–521. (I)

[2] A sufficient condition for lower semicontinuity of solution sets of systems of convex inequalities. Math. Programming Stud. (*to appear*). (I)

KOJIMA, M.:

[1] On the homotopic approach to systems of equations with separable mappings. Math. Programming Stud. 7 (1978), 170–184. (II)

[2] Studies on piecewise-linear approximations of piecewise-C mapping in fixed points and complementarity theory. Math. of Oper. Res. **3** (1978), 17–36. (II)

[3] Strongly stable stationary solutions in nonlinear programs. *In*: ROBINSON, S. M. (ed.): Analysis and computation of fixed points. Academic Press, New York 1980, 93–138. (I)

[4] An introduction to variable dimension algorithms for solving systems of equations. *In*: Allgower, E. L., K. Glashoff, and H. O. Peitgen (eds.): Numerical solution of nonlinear equations. Lecture Notes in Mathematics, vol. 878. Springer-Verlag, Berlin—Heidelberg—New York 1981, 200—237. (II)

Kojima, M., and R. Hirakayashi:

[1] Some results on the strong stability in nonlinear programs. Techn. Rep. 4, Tokyo Inst. of Technology, Tokyo, March 1980, (I)

[2] Continuous deformation of nonlinear programs. Res. Rep. B-101, Tokyo Inst. of Technology, Tokyo, Sept. 1981. (II)

Kojima, M., and S. Mizumo: Computation of all solutions to a system of polynomial equations. Res. Rep. B-105, Tokyo Inst. of Technology, Tokyo 1981. (II)

Kojima, M., H. Nishino, and N. Arima: A PL homotopy for finding all the roots of a polynomial. Math. Programming **16** (1979), 37—62. (II)

Kozerackaja, L. N. (Козерацкая, Л. Н.): О решении одного типа задач линейного параметрического программирования. Кибернетика (1981) 1, 75—81. (I)

Kummer, B.: Generalized equations: Solvability and regularity. Math. Programming Stud. **30** (*to appear*). (II)

Levin, G. M., and V. S. Tanajev (Левин, Г. М., и В. С. Танаев): О параметрической декомпозиции экстремальных задач. Кибернетика (1977) 3, 123—128. (II)

Mandel, R.:

[1] Über die Existenz von Lösungen ganzzahliger Optimierungsaufgaben. Math. Operationsforsch. Statist., Ser. Optim., **12** (1981), 33—39. (I, II)

[2] Über die Beschränktheit nach unten von Funktionen über der Menge der (gemischt) ganzzahligen Punkte eines konvexen Polyeders. Math. Operationsforsch. Statist., Ser. Optim. **12** (1981), 193—202. (I, II)

Pham canh Duong and Hoang Tuy: Stability, surjectivity, and local invertibility of non-differentiable mappings. Acta Math. Vietnamica **3** (1978) 1, 89—105. (I)

Šironin, V. M. (Широнин, В. М.): О расположении целочисленных точек в неограниченных выпуклых множествах евклидого пространства. Дисс., МГУ Москва 1980. (II)

Spingarn, J., and R. T. Rockafellar: The generic nature of optimality conditions in nonlinear programming. Math. of Oper. Res. **4** (1979), 464—477. (II)

Tamino, T., and Y. Sawaragi: Stability of nondominated solutions in multicriteria decision-making. J. Optim. Theory Appl. **32** (1980), 171—182. (I)

Tammer, K.:

[1] Behandlung stochastischer Optimierungsprobleme unter dem Gesichtspunkt der Strategie der Vektoroptimierung. Wiss. Z. Techn. Hochsch. Leipzig **4** (1980), 295 bis 302. (II)

[2] Möglichkeiten zur Lösung nichtkonvexer Optimierungsprobleme mit Zweiebenenstruktur mittels lokal-dualer Methoden. Wiss. Ber. Techn. Hochsch. Leipzig **23** (1981), 4—9. (II)

Verina, L. F. (Верина, Л. Ф.): О решении задачи параметрического квадратичного программирования. Весцi АН БССР, Сер. фiз.-мат. н., **2** (1980), 17—22. (I)

Wolkowicz, H.: A strengthened test for optimality. Preprint, Dep. of Math., Univ. of Alberta, Edmonton, Alberta, Nov. 1979. J. Optim. Theory Appl. (*to appear*). (II)

Zlobec, S.:

[1] Regions of stability for ill-posed convex programs. Apl. Mat. **27** (1981). (I)

[2] Estimates of the optimal value changes for ill-posed convex programs. Preprint, Dept. of Math., McGill Univ., Montreal, Quebec, April 1981. (I)

Zlobec, S., and A. Ben-Israel: Perturbed convex programs: continuity of optimal solutions and optimal values. Oper. Res. Verf. **31** (1979), 737—749. (I)

Zlobec, S., R. Gardner, and A. Ben-Israel: Regions of stability for arbitrarily perturbed convex programs. *In*: Fiacco, A. V. (ed.): Mathematical programming with data perturbations. M. Dekker, New York 1981. (I)

Main Statements in Relation to Stability of Special Classes of Parametric Optimization Problems

1. $\inf \{\underline{c}^{\mathsf{T}}x \mid Ax \leqq \underline{b}\}$ [1]
   cf. Theorem 5.2.5; Corollaries 5.5.2.1, 5.5.2.2
2. $\inf \{\underline{c}^{\mathsf{T}}x \mid Ax \leqq \underline{b},\ x \in E_n,\ x_1, \dots, x_s \text{ integer}\}$
   cf. Theorems 5.6.1, ..., 5.6.5; Lemma 5.6.2
3. $\inf \{x^{\mathsf{T}}\underline{C}x + \underline{p}^{\mathsf{T}}x \mid Ax \leqq \underline{b}\}$, C positive semidefinite, symmetric
   cf. Theorems 5.3.1, ..., 5.3.4, 5.2.6
4. $\inf \{x^{\mathsf{T}}Cx + \underline{p}^{\mathsf{T}}x \mid Ax \leqq \underline{b},\ x \in E_n,\ x_1, \dots, x_s \text{ integer}\}, \quad 0 \leqq s \leqq n$
   cf. Theorems 4.5.3, 4.5.5; Note 5.5.3
5. $\inf \{x^{\mathsf{T}}Cx + \underline{p}^{\mathsf{T}}x \mid Ax \leqq \underline{b}\}$, C positive semidefinite, symmetric
   cf. Theorems 5.5.1, 5.5.2; Corollary 5.5.2.3
6. *Linear Complementarity*
   $-Kx + y = \underline{b}\,, \quad x \geqq 0\,, \quad y \geqq 0\,, \quad x^{\mathsf{T}}y = 0$
   cf. Theorems 5.4.1, ..., 5.4.4
7. $\inf \{f(x) \mid g_i(x) \leqq \underline{\lambda}_i,\ i = 1, \dots, m,\ x \in E_n\}$, f, g_i convex
   cf. Theorem 4.3.5
8. $\inf \left\{f(x) + \sum_{i=1}^{m} \underline{\lambda}_i g_i(x) \,\middle|\, x \in M\right\}$, f, g_i convex, $M \subset E_n$
   cf. Theorems 4.3.8, 5.2.1, ..., 5.2.5
9. $\inf \{f(x) \mid x \in M(\underline{\lambda})\}$, $M\colon \Lambda \to 2^X$, X normed space, f convex
   cf. Theorems 4.3.1, 4.3.2
10. $\inf \{f(x, \underline{\lambda}) \mid g_i(x, \underline{\lambda}) \leqq 0,\ i = 1, \dots, m,\ x \in E_n\}, \quad f(\cdot, \lambda), g_i(\cdot, \lambda)$ convex
    cf. Theorems 4.3.6, 4.3.7; Lemma 4.3.1
11. $\inf \{f(x, \underline{\lambda}) \mid g_i(x, \underline{\lambda}) \leqq 0,\ i \in I,\ x \in E_n\}, \quad f(\cdot, \lambda), g_i(\cdot, \lambda)$ quasiconvex
    cf. Theorems 4.3.3, 4.3.4
12. $\inf \{f(x, \underline{\lambda}) \mid x \in M(\underline{\lambda})\}$, $M\colon \Lambda \to 2^X$, X metric space
    cf. Theorems 4.2.1, ..., 4.2.5; 4.3.3 ($X = E_n$)

[1]) The underlined symbols are considered to be parameters.

Index

Bei Fragen zur Produktsicherheit wenden Sie sich bitte an:
If you have any questions regarding product safety,
please contact:

Walter de Gruyter GmbH
Genthiner Straße 13
10785 Berlin
productsafety@degruyterbrill.com